Engaging Algebra

Introductory Algebra

College Edition

Scott Storla

Introduction

Algebra is fun when it's going well and frustrating when it's not. One way to increase the fun is to have lots of ways of thinking about mathematics. The more options you have, the more often you'll succeed. To increase your options, I'll ask you to think in ways you may find uncomfortable at first. Being uncomfortable is a good thing, it means you're increasing your options. So, when you become uncomfortable, embrace the topic, and remind yourself that mathematical control increases as you develop different points of view.

Why do you need algebra?

There's no one answer to this question since every student has different needs. For some students, algebra is used as a screening tool. Many majors, like business, and many technical programs, like heating and air conditioning (HVAC), have found that students who are successful at algebra more often complete their business or HVAC program. For other students, algebra is a prerequisite for required classes. Nursing students for instance often need some level of algebra before taking chemistry. Some of you won't realize you need algebra until later in your undergraduate or graduate work. I've seen students enter a program like psychology because it, "doesn't have math" only to fall behind as functions become increasingly important. Of course, students involved with a science like physics or economics need algebra to speak the language of science. Finally, for those heading towards a major in mathematics, being automatic at algebra is necessary for higher mathematics.

How to be successful

Regardless of why you need algebra, it's important this course be a successful course. Here are some habits successful students have.

1) There's no such thing as conscious multitasking. Turn off all electric distractions. The best way to practice algebra is with a textbook and a homework notebook.
2) Always pre-read the next day's material before attending class. You won't understand everything you read, but you'll have a better feel for what's happening in class.
3) Attend every class session. During your pre-reading you developed ideas about the material. When you attend class, and see the material for the second time, you'll realize there are misunderstandings between how you thought things were and how they actually are. Make sure you take thorough notes on these misconceptions.
4) After pre-reading a section, and attending class where the section was discussed, return to the book and practice the material in that section. Since this is the third time you've seen the material you will often be comfortable with the topic. Always spend 15 to 20 minutes reviewing old material before you begin your after-class homework. Some best practices for review are at http://www.eamath.com under the student resources tab.
5) In your homework notebook highlight misconceptions so you can get help the next day.

Even if you've had algebra in the past it wasn't like this. Trust the materials, engage in the homework and you will be successful.

Chapter 1

Expressions

1.1 An Introduction to the Order of Operations

"Don't put the cart before the horse." is an old saying that makes the point you want to get things in the right order. In mathematics, at first, this isn't easy to do. For example, to discuss operators I'll need to use a mathematical expression and to discuss a mathematical expression, I'll need to refer to operators. In today's lesson I'll be introducing three important ideas, whole numbers, operators and properties. I'd recommend re-reading this section occasionally as you work on chapter 1 to help appreciate how these ideas fit together.

1.1.1 The Whole Numbers

In mathematics, a **set** is a well-defined and distinct collection of objects. We use braces, { },to indicate a set and a comma to separate the objects. The set, {1,2,3,...}, where the ellipsis, (the three dots ...) imply the numbers continue in the same pattern forever (in this case 4, 5, 6 etc.) is known as the set of **natural numbers**. Expanding the elements in our set to also include 0, {0,1,2,3,...}, gives us the set of **whole numbers**.

Numbers are often visualized using the "number line", 0 1 2 3 4 5 6 ,

where the distance on the line from 0 to 1 is thought of as one "unit", the distance from 0 to 2 as two lengths of one unit and so on. The arrow to the right on the number line implies that 7, 8, 9 etc. will follow. So far, the line is just the "points" at 0, at 1, at 2 etc. That's why the distances between the points are greyed out. By the end of the course we'll "fill in" the rest of the line.

Two different numbers are considered "ordered" with the number to the right being "greater than" and the number on the left being "less than". For instance, 2 is less than 3 since 2 is to the left of 3.

Although the whole numbers themselves are mathematically fascinating, we'll spend most of our time in this course combining numbers and operators.

1.1.2 Operations , Operators and Expressions

Operations transform numbers, and **operators** tells us which operation to perform. For instance, the addition operator, +, tells us to perform the operation of addition. From past experience you know that 5 + 15 can be replaced with the single number 20. Today you'll only operate on numbers, but later in the course you'll also operate on variables (letters).

Numbers, letters and operators are used to build expressions. In mathematics, an **expression** is a meaningful collection of numbers, letters, operations and the idea of grouping. (We'll look at the idea of grouping in a few minutes.) I'll use the word **simplify** when I want you to perform all the allowable operations in an expression. Before we begin simplifying expressions though, it's important that you and I share a common vocabulary for discussing operations.

1.1.3 Some Vocabulary for Operations

We add **terms** to get a **sum**. In the expression 2 + 3, 2 is a term, 3 is a term and 5 is the sum. 2 + 3 is also a sum but the word "sum" usually refers to the single number.

We multiply **factors** to get a **product**. Three common ways to represent multiplication are $\times$ or $\bullet$ or ( ), so 2×5 or $2 \bullet 5$ or 2(5) all result in the product 10.

When we subtract a **subtrahend** from a **minuend** (minuend $-$ subtrahend) we have a **difference**. For example, with $5-3$, 5 is the minuend, 3 is the subtrahend and the difference would be 2. We use a "dash", $-$, to show the operation of subtraction.

We divide a **dividend** by a **divisor** to get a **quotient** and a **remainder**. To show the operation of division we might use the $\div$ symbol (dividend $\div$ divisor), a bar ($\frac{\text{dividend}}{\text{divisor}}$) or a slash, $\text{Dividend}/\text{Divisor}$. Usually we don't write remainders that happen to be 0.

Mathematics is especially dependent on working memory. To use your working memory well, you have to become confident with your vocabulary. A good way to memorize vocabulary is to start a set of note cards with the word on the front and the definition and an example on the back. Words that are highlighted in **bold** are good candidates for note cards. Every day, quiz yourself using the cards, until you're automatic with the vocabulary of mathematics.

Think like an expert

For over 80 years psychologists have studied the differences between experts and novices. Some of this research has included how people think about arithmetic and algebra. When experts use their working memory to process arithmetic and algebra they use words like term, factor, sum and product correctly. Begin becoming an expert. Memorize the vocabulary of mathematics.

One last point, if an expression can be described using two different words, the order you choose to list those words, isn't important. For instance, if we were working with $\frac{4-2}{6+1}$, and the question asked, "$4-2$ is a _____ and also the _____", my answer might be difference and dividend and your answer might be dividend and difference. Either answer is fine.

Practice 1.1.3 Some Vocabulary for Operations

Fill in the blanks using the words term, sum, factor, product, minuend, subtrahend, difference, dividend, divisor or quotient.

---

a) Before simplifying $\frac{4-2}{6+1}$;

4 is the ____, 2 is the ____, and $4-2$ is a ____ and also the ____, 6 is a ____, 1 is a ____, and $6+1$ is a ____ and also the ____.

Minuend, subtrahend, difference, dividend, term, term, sum, divisor	$\Rightarrow$	Since 4 is to the left of the subtraction symbol, it's the minuend. Since 2 is to the right of the subtraction symbol, it's the subtrahend. $4-2$ is both a difference and, since it's above the division bar, it's also the dividend. 6 is being added so it's a term. 1 is being added, so it's a term. $6+1$ is a sum and, since it's below the division bar, it's also the divisor.

Homework 1.1 Fill in the blanks using the words term, sum, factor, product, minuend, subtrahend, difference, dividend, divisor or quotient.

1) Before simplifying 2×3;

   2 is a ____, 3 is a ____, and 2×3 is a ____.

2) Before simplifying $20\div 2$;

   20 is the ____, 2 is the ____ and $20\div 2$ is a ____.

3) Before simplifying $18-12$;

   18 is the ____, 12 is the ____, and $18-12$ is a ____.

4) Before simplifying $\frac{7\times 2}{4-3}$;

   7 is a ____, 2 is a ____ and 7×2 is a ____ and also the ____, 4 is the ____, 3 is the ____ and $4-3$ is both a ____ and also the ____.

1.1.4 Ordering The Four Basic Operations

If I ask someone, who's good at algebra, the best name for $\frac{12+4(3)}{12}-\frac{4(3)+12}{12}$ they'll immediately answer, "It's a difference." That's because they've (usually unconsciously), ordered the seven operators and realized the last operation they'll do is the subtraction. Shifting the majority of your attention from numbers and letters to operations is one mark of an expert.

Think like an expert

When novices first view an expression, they tend to focus on the numbers and letters. When experts first view an expression, they notice the numbers and letters but put the majority of their attention on the operators and grouping symbols. Think like an expert, consciously analyze operations and their order.

To practice ordering operations, I'm going to ask you to count the number of operators and then name the operations using the correct order. Before we start though, I want to make a comment about PEMDAS (Parentheses, Exponents, Multiply, Divide, Add, Subtract) or other similar memory aids which many students use to order operations.

Procedures base on memory aids like PEMDAS are too limited to handle all but the simplest problems in arithmetic or algebra. So if you're using something like PEMDAS it's time to expand your tools. It will take time to understand all the subtleties when ordering operations, so we'll begin slowly. For the next few weeks you should have a printed copy of the full order of operations next to you and consciously consider each step as you simplify expressions.

Procedure **– Order of Operations**
Begin with the innermost grouping idea and work out; Explicit grouping ( ), [ ], { } Implicit grouping Operations; in the dividend or divisor. inside absolute value bars. in radicands or exponents. 1. Start to the left and work right simplifying each operation, different from the basic four, as you come to them. 2. Start again to the left and work right simplifying each multiplication or division as you come to them. 3. Start again to the left and work right simplifying each addition or subtraction as you come to them.

Before we start with the four basic operations, addition, subtraction, multiplication and division, I want to mention a big difference in how experts and novices use worked examples. Experts become automatic with the worked examples and then use the homework to help generalize what they've learned from the worked examples. Novices start with the homework and only look back at worked examples when they get stuck. They hope to find a pattern in one of the worked examples that will help them finish their specific problem. Trying to use worked examples as a type of pattern matching is a very weak strategy for "learning" algebra.

To get the most out of a worked example, copy the problem and, without looking at the textbook, write your own process and answer. Then look at the book and make sure both the answer <u>and the process</u> (the explanation to the right of $\Rightarrow$) match. If there wasn't a match, figure out what went wrong, (it's easy to do with a step by step worked example) copy the problem to a blank piece of paper, and try again. Only move to the next worked example when you're able to correctly work your current worked example without hesitation.

Think like an expert
Novices consider a problem "done" when they get an answer. Experts consider a problem done when they understand the <u>process</u> that leads to a <u>correct</u> answer. Only consider an answer correct when you're comfortable with each step in the process that lead to your correct answer.

One last point, don't use a calculator today when you practice operating on whole numbers. You need to personally become automatic with the order of operations so the transition to operating on variables will be easier.

Practice 1.1.4 Ordering the Four Basic Operations

Count the number of operators, discuss the correct order and then simplify the expression.

a) $1+2\times3+4$

There are three operators. I'll multiply first and then add left to right.

$1+6+4 \Rightarrow$ Started with line 2 and multiplied. Line 2 is finished.

$7+4 \Rightarrow$ Moved to line 3 and began adding left to right.

$11 \Rightarrow$ Added.

b) $20\div2-\frac{6}{2}$

There are three operators. I'll divide first left to right and then subtract.

$10-3 \Rightarrow$ Simplified $20\div2$ and then $\frac{6}{2}$. Line 2 is finished.

$7 \Rightarrow$ Moved to line 3 and subtracted.

c) $14-10+8\div4\times2$

There are four operators. First, I'll multiply and divide left to right. Next, I'll add and subtract left to right.

$14-10+2\times2$
$14-10+4$ $\Rightarrow$ Moving left to right, divided first and then multiplied. Line 2 is finished.

$4+4$
8 $\Rightarrow$ Moved to line 3. Subtracted first and then added.

Homework 1.1 *Count the number of operators, discuss the correct order and then simplify the expression.*

5) $15-5\times3$	6) $\frac{6}{3}-1+1$	7) $4\times5+3\times2-2\times4$	8) $8\times6-8+6\div3$
9) $24-12\div6\times2$	10) $\frac{6}{3}\times\frac{14}{2}$	11) $18-{}^{14}/_{2}-7$	12) $80\div10\times4-2+2\times2$

1.1.5 Explicit Grouping

In the last topic, you saw that 1 + 2 x 3 simplifies to 7 (and not 9) because we multiply first and add second. If I wanted you to add first and multiply second, (so the result would be 9) I'd use an explicit grouping symbol like parentheses, $(1+2)\times3$ to force the addition to be performed first. With explicit grouping, you can see the grouping symbols. Parentheses ( ), brackets [ ], and braces { } are all examples of explicit grouping symbols.

Sometimes grouping symbols are **nested** within other grouping symbols. We often use different symbols for the inner and outer grouping so we can see what belongs together. To discuss $3[30-7(3+1)]$ we'd say, "The parentheses are nested inside the brackets."

One last point about grouping symbols. As we discussed earlier, a number written next to a grouping symbol implies multiplication. For instance, 7(4), $7[4]$, and $7\{4\}$ all imply 7×4.

Practice 1.1.5 Explicit Grouping

Count the number of operators, discuss the correct order and then simplify the expression.

a) $(6+3)-(6-3)$

There are three operators. First, I'll move left to right completing the operations inside the parentheses. Last, I'll subtract.

$9-(6-3)$	$\Rightarrow$	Added inside the left-most parentheses first. The first set of parentheses are no longer necessary.
$9-3$	$\Rightarrow$	Subtracted inside the parentheses.
6	$\Rightarrow$	Subtracted.

b) $12-\left(\dfrac{16}{4}+2\right)$

There are three operators. Starting inside the parentheses I'll divide first and add second. The subtraction will be done last.

$12-(4+2)$ $12-6$	$\Rightarrow$	Divided and then added inside the parentheses.
6	$\Rightarrow$	Subtracted.

c) $3[30-7(3+1)]$

There are four operators. First, I'll add inside the parentheses. Next, I'll multiply and then subtract inside the brackets. Last, I'll multiply by 3.

$3[30-7(4)]$	$\Rightarrow$	Since the parentheses are nested inside the brackets the addition is simplified first.
$3[30-28]$ $3[2]$	$\Rightarrow$	Inside the brackets multiplication is first followed by subtraction.
6	$\Rightarrow$	Multiplied. Remembered, $3[2]$ implies 3×2.

Homework: 1.1 Count the number of operators, discuss the correct order and then simplify the expression.

13) $8+4-(8+4)$	14) $(15-8)(10-6)$	15) $4\{3(7-3)\}$
16) $36 \div 3(2+2)$	17) $8-2(5-3)$	18) $(1+3)[3(1+3)]$

1.1.6 Implicit Grouping

With implicit grouping we don't see a grouping symbols but the idea of grouping still applies. One common use for implicit grouping involves quotients. To simplify a quotient, perform all the operations in the dividend (the top) and the divisor (the bottom) before you find the quotient and remainder. For example, to simplify the quotient $\frac{18+6}{18-6}$, I'd add in the dividend $\frac{24}{18-6}$, subtract in the divisor $\frac{24}{12}$, and only then divide to get a quotient of 2.

Before simplifying this next set of problems, please take this advice to heart.

Think like an expert
Everyone does math, "In their head". Experts keep a written record of their thought process, novice often don't. Only by keeping a step by step record of what you thought was true, can you review your work and learn from your mistakes. Do math like an expert and keep a well-organized written record of what's "in your head".

Practice 1.1.6 Implicit Grouping

Count the number of operators, discuss the correct order and then simplify the expression.

a) $\frac{14-2}{2(4-2)}$

There are four operators. First, I'll subtract in the dividend and then subtract inside the parentheses in the divisor. Next, I'll multiply in the divisor. Last, I'll find the quotient.

$\frac{12}{2(4-2)}$ $\frac{12}{2(2)}$	$\Rightarrow$	Subtracted in the dividend and then subtracted inside the parentheses.
$\frac{12}{4}$	$\Rightarrow$	Multiplied in the divisor.
3	$\Rightarrow$	Found the quotient.

Homework 1.1 Count the number of operators, discuss the correct order and then simplify the expression.

19) $\frac{6-1}{2+3}$ 20) $\frac{(26-12)(2)}{26-12(2)}$ 21) $\frac{2+2\times 3}{8\div 4\times 2}$ 22) $\frac{8}{5(2)-10}$ 23) $\frac{12+4(3)}{12}-\frac{4(3)+12}{12}$

1.1.7 The Idea of a Property

Our last big idea today involves properties. A **property** is a general idea that we'll often use in specific situations. For instance, a property of fire is that it needs oxygen to burn. We're using this general property when we blow on a struggling campfire or extinguish a frying pan fire with a cover. In our course, properties will be the set of mathematical ideas that we take for granted to be true.

Using properties to help make decisions is one of the biggest differences (if not the biggest) between people who are in control of their algebra and those who feel algebra is kind of magical where things unpredictably appear, disappear or change appearance. Properties help reduce this feeling of randomness by explaining the "why" that underlies the "how".

I'm bringing up properties because you may have noticed that with addition and multiplication the number to the left and to the right of the operator both had the same name, (with addition both were terms and with multiplication both were factors) while with subtraction and division they had different names. This idea, that the order of the numbers doesn't affect the outcome for addition and multiplication, is an example of the commutative property.

The Commutative Property for Addition
The **order** of the terms doesn't affect the sum.
Example: $a+b=b+a$
Note: Subtraction is not commutative.

The Commutative Property for Multiplication
The order of the factors doesn't affect the product.
Example: $a \times b = b \times a$
Note: Division is not commutative.

Try to keep this property in mind as you work through this last set of vocabulary questions.

1.1.8 More Vocabulary for Operations

Now that you've practiced the order of operations, you're able to discuss expressions in more detail. For example, should $13(2)+17$ be called a product, because of the multiplication, or a sum, because of the addition? The answer is that $13(2)+17$ is a sum because, if you follow the order of operations, the last operation you'd perform would be the addition.

Practice 1.1.8 More Vocabulary for Operations

Fill in the blanks using the words term, sum, factor, product, minuend, subtrahend, difference, dividend, divisor or quotient.

a) Before simplifying $(3+4)(11-8)$;

3 is a ____, 4 is a ____, $3+4$ is a ____ and $(3+4)$ is a ____. 11 is the ____, 8 is the ____, $11-8$ is a ____ and $(11-8)$ is a ____. Last, $(3+4)(11-8)$ is a ____.

term, term, sum factor, minuend, subtrahend, difference, factor, product	⇒	Since 3 and 4 are being added both are terms and $3+4$ is a sum. Enclosing the sum in parentheses, $(3+4)$, results in a factor. Since 11 is to the left of the dash it's the minuend, since 8 is to the right it's the subtrahend which makes $11-8$, a difference. Enclosing the difference in parentheses, $(11-8)$ results in a factor. Since both $(3+4)$ and $(11-8)$ are factors $(3+4)(11-8)$ is a product.

Homework 1.1 Fill in the blanks using the words term, sum, factor, product, minuend, subtrahend, difference, dividend, divisor or quotient.

24) Before simplifying $\frac{7+4}{5-2}$;

7 is a ____, 4 is a____ and $7+4$ is a ____ and also the ____. 5 is the____, 2 is the ____, $5-2$ is a ____ and also the ____ and $\frac{7+4}{5-2}$ is a ____.

25) Before simplifying $3(5)+2(5)$;

3 is a____, 5 is a____ , $3(5)$ is a____ and also a____. 2 is a____, 5 is a ____, $2(5)$ is a ____ and also a ____ and $3(5)+2(5)$ is a ____.

26) Before simplifying $\left(\frac{6}{1}\right)\left(\frac{4}{3}\right)$;

6 is the____, 1 is the____, $\frac{6}{1}$ is a ____ and $\left(\frac{6}{1}\right)$ is a ____. 4 is the ____, 3 is the ____, $\frac{4}{3}$ is a____ and $\left(\frac{4}{3}\right)$ is a____. Finally, $\left(\frac{6}{1}\right)\left(\frac{4}{3}\right)$ is a ____.

27) Before simplifying $\frac{2(1+3)}{5+4}$;

1 is a ___, 3 is a ___, $1+3$ is a ___, $(1+3)$ is a ___, 2 is a____ and $2(1+3)$ is a ___ and also the___. 5 is a ___, 4 is a ___, $5+4$ is a ___ and also the ___ and $\frac{2(1+3)}{5+3}$ is a ___.

28) Before simplifying $2(15-1)-3(6+4)$;

15 is the ____, 1 is the ____, $15-1$ is a ____, $(15-1)$ is a ____, 2 is a ____, and $2(15-1)$ is both a ____ and the ____. 6 is a ____, 4 is a ____, $6+4$ is a ____, $(6+4)$ is a ____, 3 is a ____ and $3(6+4)$ is both a ____ and the ____. Finally, $2(15-1)-3(6+4)$ is a ____.

29) Before simplifying $25-4(1)(2)$;

4 is a ____, 1 is a ____, 2 is a ____ and $4(1)(2)$ is both a ____ and the ____, 25 is the ____ and $25-4(2)(2)$ is a ____.

1.1.9 The Distributive Property

It might surprise you to learn that most of what we do in this text relies on only nine properties. Four properties explain what's happening with sums, four explain what's happening with products and the **distributive property** allows us, under certain conditions, to change our point of view from a sum to a product, or from a product to a sum.

Property – The Distributive Property of Multiplication over Addition
A sum of terms, each with a common factor, can be replaced by the product of the sum of the remaining factors and the common factor. Example: $2(3)+5(3)=(2+5)(3)$

Notice that in this context we're using the equality symbol, $=$, to imply that one expression can be replaced by the other.

We can use the distributive property to explain why $2+3(5)$ simplifies to 17 and not 25. In the expression $2+3(5)$ the first term counts two quantities of 1 and the second term counts three quantities of 5, that is, our sum is really $2(1)+3(5)$. Notice that I've included a factor of 1 in the first term that we don't "see" in the original expression. It's common in algebra to not explicitly write factors of 1 unless it's important for the discussion.

Currently $2(1)+3(5)$ can't be added using the distributive property since the first term counts quantities of 1 and the second term counts quantities of 5. If we multiply the second term though, we go from counting 3 quantities of 5 to counting 15 quantities of 1 and our sum becomes $2(1)+15(1)$. Thinking of $2(1)+15(1)$ as the "left" side of the distributive property I'll replace the expression with $(2+15)(1)$ which is the "right" side of the distributive property. People often explain this replacement using the words, "factoring out". That is, they'll say something like, "I factored out the common factor of one." Once 1 is factored out I see the sum $(2+15)$ counts 17 quantities of 1 which we'll write as 17. $2(1)+15(1)=(2+15)(1)=17(1)=17$. Here's some practice finding sums using the distributive property.

Practice 1.1.9 The Distributive Property

Use the distributive property to simplify the following sums.

a) $20+30$

1. Using quantities of 2 2. Using quantities of 5 3. Using quantities of 10
4. Using quantities of 1

1. $10(2)+15(2)$
$(10+15)(2)$
$(25)(2)$
50

$\Rightarrow$ In each case, I'll write the original terms as a product where the second factor is the quantity. Since the second factor in both terms will be common, I'll be able to use the distributive property to simplify.

2.	3.	4.
$4(5)+6(5)$	$2(10)+3(10)$	$20(1)+30(1)$
$(4+6)(5)$	$(2+3)(10)$	$(20+30)(1)$
$(10)(5)$	$(5)(10)$	$(50)(1)$
50	50	50

Homework 1.1 Use the distributive property to simplify the following sums.

30) $12+24$	31) $60+75$	32) $21+42+63$
a) Using quantities of 2	a) Using quantities of 1	a) Using quantities of 3
b) Using quantities of 3	b) Using quantities of 3	b) Using quantities of 7
c) Using quantities of 6	c) Using quantities of 5	c) Using quantities of 21
d) Using quantities of 12	d) Using quantities of 15	d) Using quantities of 1
e) Using quantities of 1		

Homework 1.1 Answers

1) factor, factor, product 2) dividend, divisor, quotient 3) minuend, subtrahend, difference

4) factor, factor, product, dividend, minuend, subtrahend, difference, divisor

5) There are two operations. Multiplication is first followed by subtraction. The answer is 0.

6) There are three operations. Division is first followed by subtraction and then addition. The answer is 2.

7) There are five operations. The products are simplified first left to right, addition is next followed by subtraction. The answer is 18.

8) There are four operations. Multiplication is first, division is next and subtraction and addition left to right. The answer is 42.

9) There are three operations. The division and multiplication are simplified left to right followed by the subtraction. The answer is 20.

10) Currently we think of this as three operations. The correct order of operations is to divide the 6 by the 3, multiply the result by 14 and finally divide by 2. You'll often see people divide left to right first and multiply last. As we'll see later, the results will be the same. The answer is 14.

11) There are three operations. Division is first followed by subtracting left to right. The answer is 4.

12) There are five operations. The division is first followed by the multiplications left to right. The subtraction is next and the addition last. The answer is 34.

13) There are three operations. First add inside the parentheses, then add and subtract left to right. The answer is 0.

14) There are three operations. Simplify inside each set of parentheses left to right and finally multiply. The answer is 28.

15) There are three operations. The subtraction is first followed by multiplication by 3 and finally multiplication by 4. The answer is 48.

16) There are three operations. The addition is first, the division is next and the multiplication is last. The answer is 48.

17) There are three operations. The subtraction inside the parentheses is first, the multiplication by 2 is next and the subtraction is last. The answer is 4.

18) There are four operations. The additions inside parentheses are done first left to right. Then the multiplication by 3 inside the brackets is followed by the last multiplication. The answer is 48.

19) There are three operations. First, subtract in the dividend and add in the divisor, then divide. The answer is 1.

20) There are five operations. In the dividend subtract and multiply, in the divisor multiply and subtract. Last, divide the results. The answer is 14.

21) There are five operations. In the dividend multiply first and then add. In the divisor divide and multiply left to right. Finally divide. The answer is 2.

22) There are three operations. In the divisor multiply and subtract, then divide 8 by the result. After simplifying, the divisor becomes 0. The expression is undefined.

23) There are seven operations. For the first quotient multiply and add in the dividend, then divide. For the second quotient, multiply and add in the dividend, then divide. Last, subtract the quotients. The answer is 0.

24) term, term, sum, dividend, minuend, subtrahend, difference, divisor, quotient

25) factor, factor, product, term, factor, factor, product, term, sum

26) dividend, divisor, quotient, factor, dividend, divisor, quotient, factor, product

27) term, term, sum, factor, factor, product, dividend, term, term, sum, divisor, quotient

28) minuend, subtrahend, difference, factor, factor, product, minuend, term, term, sum, factor, factor, product, subtrahend, difference

29) factor, factor, factor, product subtrahend, minuend, difference

30)

a) $\begin{gathered}6(2)+12(2)\\(6+12)(2)\\(18)(2)\\36\end{gathered}$ b) $\begin{gathered}4(3)+8(3)\\(4+8)(3)\\(12)(3)\\36\end{gathered}$ c) $\begin{gathered}2(6)+4(6)\\(2+4)(6)\\(6)(6)\\36\end{gathered}$ d) $\begin{gathered}1(12)+2(12)\\(1+2)(12)\\(3)(12)\\36\end{gathered}$ e) $\begin{gathered}12(1)+24(1)\\(12+24)(1)\\(36)(1)\\36\end{gathered}$

31)

a) $\begin{gathered}60(1)+75(1)\\(60+75)(1)\\(135)(1)\\135\end{gathered}$ b) $\begin{gathered}20(3)+25(3)\\(20+25)(3)\\(45)(3)\\135\end{gathered}$ c) $\begin{gathered}12(5)+15(5)\\(12+15)(5)\\(27)(5)\\135\end{gathered}$ d) $\begin{gathered}4(15)+5(15)\\(4+5)(15)\\(9)(15)\\135\end{gathered}$

32)

a) $\begin{gathered}7(3)+14(3)+21(3)\\(7+14+21)(3)\\(42)(3)\\126\end{gathered}$ b) $\begin{gathered}3(7)+6(7)+9(7)\\(3+6+9)(7)\\(18)(7)\\126\end{gathered}$ c) $\begin{gathered}1(21)+2(21)+3(21)\\(1+2+3)(21)\\(6)(21)\\126\end{gathered}$

d) $\begin{gathered}21(1)+42(1)+63(1)\\(21+42+63)(1)\\(126)(1)\\126\end{gathered}$

1.2 Multiplying and Dividing Integers

Over time, for a number of reasons, people found a need to move beyond the set of whole numbers. Merchants, for example, realized the natural numbers easily counted the six sheep they had loaded on a ship, but, if the ship sank, the natural numbers couldn't count the loss of six sheep. To describe this loss, some merchants began to include a dash symbol, $-$, along with the number. A natural number without the dash meant something "positive" (I have 6 sheep to sell), while the same number with a dash meant something "negative" (I've lost 6 sheep that I could have sold). In this section, we'll begin working with the integers which allows us to count both positive and negative quantities.

1.2.1 The Integers

To build the set of **integers** we begin with the natural numbers which we rename the, "positive integers". Next, we include the "negative integers" which is each natural number preceded by a dash. Last, we include 0. Zero is considered neither positive nor negative. When you're considering issues of positive or negative you're considering the **sign** of the number.

***Definition* – Integers**
The set of positive integers, negative integers and 0. Example $\{\ldots -3, -2, -1, 0, 1, 2, 3\ldots\}$

Visually, the integers run left to right from "negative infinity" (since there are an infinite number of negative integers) to "positive infinity" (since there are an infinite number of positive integers) with the negative integers to the left of zero and the positive integers to the right of zero. Notice our points, whether to the left or right of zero, are still one unit apart.

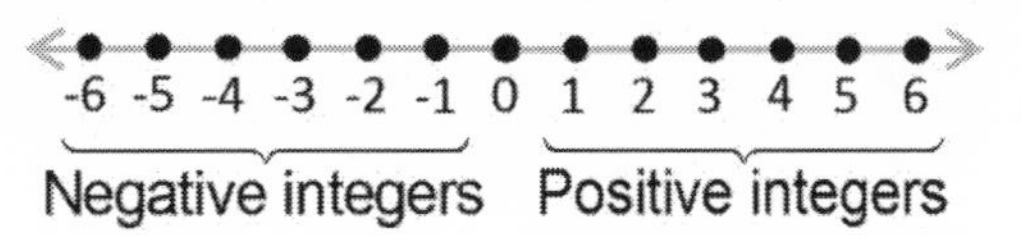

Since the natural numbers and the positive integers are the same set, we can think of each positive integer as a product that counts a quantity of 1. A logical extension for negative integers would be that they count quantities of -1. For instance, we can think of -5 as the product $5(-1)$, which counts five quantities of negative one.

1.2.2 A Note About the Dash Symbol

When you operate on integers you'll need to wrestle with the different ways we might interpret a dash symbol. In the previous section, we used the dash symbol to imply the operation of subtraction. Now we're using the same symbol to imply a negative number. As you'll see in a few minutes, a couple other common uses for the dash symbol is an implicit factor of -1 and to represent the idea of opposite (negation). In hindsight, using the same symbol for an operation and also for the sign of a number might not have been the best choice, but it's what we did.

1.2.3 A Quick Look at Absolute Value

When I interview people who consistently process signs correctly, they have an unconscious (automatic) process based on memorized sign rules, and a second conscious process, based on the idea of separating issues of size (magnitude) from issues of sign (positive or negative). If I ask them to find the product $(-2)(-3)$, they'll use their memorized sign rules to automatically answer, "Six.". If I ask them to describe how they know the product is six most will say something like, "Well two times three is six and, since both numbers are negative, my answer's positive six." Notice they usually treat both numbers as positive to find the size of the product and then they find the sign.

Learning to correctly separate the issue of sign, which for us will have to do with factors of 1 or -1, from the issue of size, which has to do with the idea of an "absolute" value, is an important skill as you move forward with algebra. For now, we'll think of a number's **absolute value** as it's distance from 0. This means both positive five and negative five would have an absolute value of 5 since they're both five units from 0. Notice the absolute value of a non-zero number will always be positive.

We use two vertical bars, $|\ \ |$, as the operator for the operation of absolute value. Both $|-5|$ and $|5|$ would simplify to 5. I don't want to spend much time discussing absolute value right now for a couple reasons. First, people usually don't actually write and then simplify absolute value as part of the procedure for operating on signed numbers. Instead, they begin by assuming that both numbers are positive or zero. Second, working with the mathematical definition of absolute value takes time and it's a topic I feel is better left for intermediate algebra.

1.2.4 A Procedure for Multiplying Integers

In practice, your goal is to become automatic at processing signs when you multiply any two integers. Here's the procedure we'll use.

Procedure **– Multiplying Two Non-Zero Integers**
1. Treat both integers as positive and find their product. 2. If both integers had the same sign, the product is positive. If they originally had different signs, the product is negative.

Notice our procedure doesn't address factors of 0. For that situation, there's a separate rule;

The Zero-Product Rule
The product of an integer and 0 can be replaced by 0. Example: $0 \times 6 = 0$

As you work on the homework, please keep this next idea in mind.

Think like an expert

Experts are automatic at operations with signs, novices often rely on a calculator. To be successful later, you need to become automatic with integer operations now. Automaticity only comes from repeated practice using your brain, not a calculator.

Practice 1.2.4 A Procedure for Multiplying Integers
Simplify.

a) $(-3)(5)$

$(3)(5)=15$	$\Rightarrow$	Thought of both integers as positive and found their product.
-15	$\Rightarrow$	The final product is negative since the original integers had different signs.

b) $(-7)(2)(-2)$

$(-14)(-2)$	$\Rightarrow$	Began multiplying left to right. 7×2 is 14 and since the signs are different the product is -14.
28	$\Rightarrow$	Continued to multiply. 14×2 is 28 and the product is positive since the signs were the same.

Homework 1.2 Simplify.

1) 7×-2	2) $-12(-10)$	3) $-2(-4)(-3)$	4) $(-3)(3)\times-2$
5) $-1\times-8\times0\times2$	6) $-(5\times-4)$	7) $-[(-2)(7)(-3)]$	8) $-[-1(-5)]$

1.2.5 Dividing Integers

When we divide two signed numbers we can reuse the ideas from multiplication to find the sign of the quotient. Here's the procedure for dividing integers.

***Procedure* – Dividing Two Non-Zero Integers**

1. Treat both integers as positive and find their quotient.
2. If originally both integers had the same sign, the quotient is positive, if they had different signs, the quotient is negative.

To simplify $\frac{-18}{9}$ using the procedure, I'd divide 18 by 9 to get 2, and then, since the signs are different, the final quotient would be -2.

If the dividend is 0, and the divisor isn't 0, the quotient is 0. For example, $0/8=0$. If the divisor is 0, then regardless of the dividend, the quotient is undefined. So, $8/0$ is undefined.

Practice 1.2.5 Dividing Integers

Simplify.

a) $\dfrac{-18}{3}$

-6	$\Rightarrow$	18 divided by 3 is 6 and the quotient is negative since the signs were different.

b) $-\dfrac{8}{2}$

$-1\left(\dfrac{8}{2}\right)$	$\Rightarrow$	Whenever I'm unsure of how to interpret the dash symbol, I view the dash symbol as a factor of -1.
$-1(4)$	$\Rightarrow$	8 divided by 2 is 4 and the quotient stays positive since the signs are the same.
-4	$\Rightarrow$	Simplified using the procedure for multiplication.

Homework 1.2 Simplify.

9) $\dfrac{25}{-5}$ 10) $-\dfrac{14}{7}$ 11) $\dfrac{-18}{-9}$ 12) $-\dfrac{-15}{3}$

13) $-\dfrac{-12}{-4}$ 14) $-\dfrac{0}{-5}$ 15) $\dfrac{-(-30)}{-3}$ 16) $-\dfrac{-21}{-(-7)}$

1.2.6 Combining Operations

Let's end this section with some problems that combine multiplication and division.

Practice 1.2.6 Combining Operations

Simplify.

a) $\dfrac{(-2)(-10)}{-4}$

$\dfrac{20}{-4}$	$\Rightarrow$	Implicit grouping implies the dividend must be simplified first. Two times ten is twenty and the signs are the same so the product is positive twenty.
-5	$\Rightarrow$	Twenty divided by four is five and the signs are different so the quotient is negative five.

b) $-\left(\dfrac{-30}{6}\right)\left(\dfrac{15}{-3}\right)$

$-(-5)\left(\dfrac{15}{-3}\right)$	$\Rightarrow$	Simplified the first quotient. Thirty divided by six is five and the signs were different so the quotient is negative five.
$-(-5)(-5)$	$\Rightarrow$	Simplified the second quotient. Fifteen divided by three is five and the signs were different so the quotient is negative five.
$(5)(-5)$	$\Rightarrow$	Saw the dash outside the parentheses as a factor of negative one, $-1(-5)(-5)$, and multiplied the first two factors to get positive five.
-25	$\Rightarrow$	Continued multiplying to find my final product.

c) $-\dfrac{16-12}{12-10}$

$-1\left(\dfrac{4}{2}\right)$	$\Rightarrow$	Thought of the dash in front of the division as a factor of -1, then implicit grouping tells me to simplify the dividend and divisor first.
$-1(2)$	$\Rightarrow$	The quotient inside the parentheses simplifies to 2.
-2	$\Rightarrow$	Wrote $-1(2)$ as -2.

Homework 1.2 Simplify.

17) $\dfrac{-(12\times-12)}{24}$ 18) $-\left(\dfrac{-14}{2}\right)\left(\dfrac{14}{-2}\right)$ 19) $\dfrac{(6)(-4)}{6-4}$ 20) $-\dfrac{-2(-6)}{-4}$

21) $-\dfrac{28-14}{-14}$ 22) $\dfrac{-2\times-8+8}{4\times-2}$ 23) $\left(\dfrac{-9}{3}\right)\left(-\dfrac{9}{3}\right)\left(\dfrac{9}{-3}\right)$ 24) $-6\times-4\div-2$

25) $-[-1(5)]+(-1)(-5)$ 26) $-\dfrac{5+6}{-(18-7)}$

Homework 1.2 Answers

1) –14	2) 120	3) –24	4) 18	5) 0	6) 20	7) –42
8) –5	9) –5	10) –2	11) 2	12) 5	13) –3	14) 0
15) –10	16) 3	17) 6	18) –49	19) –12	20) 3	21) 1
22) –3	23) –27	24) –12	25) 10	26) 1		

1.3 Adding Integers

Earlier we took the point of view that the positive integers (natural numbers) count quantities of positive one and the negative integers count quantities of negative one. We'll continue this point of view as we begin adding integers.

1.3.1 Adding Integers Automatically

With adults, I've found that money is the most intuitive way to discuss adding two integers. If we think of $3+5$ as counting dollars we have, $3(1)+5(1)$, then it's reasonable we'll end with eight dollars, $3(1)+5(1)=8(1)=8$. On the other hand, if we think of $-3+-5$ as counting dollars we owe, $3(-1)+5(-1)$, then it's reasonable that we'll end owing eight dollars, $3(-1)+5(-1)=8(-1)=-8$. Notice that unlike multiplication, where the product of two negative numbers was positive, the sum of two negative numbers remains negative.

When the two terms have different signs the sign of the sum will be the same as the term with the largest absolute value. This make sense since, if we begin owing more than we have, our sum should be negative. On the other hand, if we have more than we owe, the sum should be positive. That's why the sum $3+-5$ is -2 while the sum $-3+5$ is 2. Here's a procedure that puts these ideas together.

***Procedure* – Adding Two Non-Zero Integers**
1. If the signs are the same, treat both integers as positive, find their sum and include the common sign. 2. If the signs are different, treat both integers as positive, subtract the smaller value from the larger and include the sign of whichever original integer had the larger absolute value.

Practice 1.3.1 Adding Integers Automatically

Simplify.

a) $-3+5$		
$5-3=2$	$\Rightarrow$	The original signs were different so I thought of both integers as positive and subtracted the smaller value from the larger.
2	$\Rightarrow$	The sum is positive since 5 had the larger absolute value.
b) $-7+2+-3$		
$-7+2+-3$ $-5+-3$	$\Rightarrow$	Adding left to right, the first two terms have different signs so I thought of both as positive and subtracted the smaller from the larger, $7-2=5$. The sign of the sum is negative since -7 had the larger absolute value.
-8	$\Rightarrow$	Continued to add. Since the signs are the same thought of both terms as positive, found their sum, $5+3=8$, and included the common sign.

Homework 1.3 Simplify.

1) $12+-19$	2) $-45+82$	3) $-12+-8$	4) $-7+3+4$
5) $30+-18+-7$	6) $2+-9+7$	7) $-13+9+-2$	8) $-22+-11+-3$
9) $-8+10+-8+5$	10) $-12+-4+6+4$	11) $14+6+-22+-2$	

1.3.2 Combining Operations

Now let's do some problems that also include products and quotients.

Practice 1.3.1 Adding Integers Automatically

Simplify.

a) $-3(5)+(-2)(-6)$

$-15+12$	$\Rightarrow$	Multiplication, left to right, is first.
-3	$\Rightarrow$	Added. Since the signs are different I thought of both integers as positive, found their difference, $15-12=3$, and made the sum negative since -15 had the larger absolute value.

b) $\dfrac{-3(-4)}{3+-4}$

$\dfrac{12}{-1}$	$\Rightarrow$	Multiplied in the dividend and added in the divisor.
-12	$\Rightarrow$	Divided

Homework 1.3 Simplify.

12) $-(-7+3)$ 13) $12+\dfrac{-12}{3}$ 14) $-\left(\dfrac{-14}{2}\right)+\left(\dfrac{14}{-2}\right)$ 15) $\dfrac{-7+-3}{-7+5}$

16) $-6\times-2+-3\times5$ 17) $-8+2(-3)$ 18) $-6\times-4+-2$ 19) $-(7+-4)+-7+4$

20) $(-4)(15+-20)+8(2)$ 21) $\dfrac{-15+-4+12}{15+4+-12}$ 22) $-[-1(-5)]+(1)(-5)$

23) $\dfrac{-15}{15}+-\dfrac{4}{4}+\dfrac{12}{-12}$ 24) $\dfrac{5(-5)+(-3)(-3)}{(-5+3)(5+-3)}$

Homework 1.3 Answers

1) –7	2) 37	3) –20	4) 0	5) 5	6) 0
7) –6	8) –36	9) –1	10) –6	11) –4	12) 4
13) 8	14) 0	15) 5	16) –3	17) –14	18) 22
19) –6	20) 36	21) –1	22) –10	23) –3	24) 4

1.4 Subtracting Integers

Earlier I mentioned that merchants started using negative numbers to represent the idea of a loss. For mathematicians, negative numbers helped extend the idea of a difference. Unlike addition, where the sum of two whole numbers is always another whole number, the difference of two whole numbers may or may not be a whole number. For instance, the difference $3-5$ doesn't simplify to a whole number.

Before we begin subtracting integers, I want to again raise the issue of the different uses for the dash symbol;

Meanings for the – symbol
Depending on the situation you may find it easier to think of the dash symbol, –, as the subtraction operator, as a (negative) sign, as an opposite or as a factor of -1. Sometimes it's helpful to change your point of view as you move from one step to another within the same problem. It's important not to look for a single way or a "right" way to process the dash symbol. Instead, with practice, you'll develop ways of using the context to decide which meaning is the most helpful for you.

1.4.1 Viewing Subtracting as Adding the Opposite

Recall that subtraction involves finding the difference between a minuend and a subtrahend. When students first encounter a problem where the minuend is greater, for instance $5-3$, they're often given an explanation like, "If you start with five apples, and take three apples away, you're left with two apples." That is, $5-3=2$. Unfortunately, this reasoning doesn't extend intuitively to a problem like $3-5$ where the minuend is less. Students are often uncomfortable if they're asked to, "Start with three apples and take away five apples." To simplify subtraction, we'll use a definition that remains consistent whether the minuend is greater than, less than, or equal to, the subtrahend.

Definition **– Subtraction as Adding the Opposite**
The difference $a-b$ is the sum of *a* and the opposite of *b*. Example: $3-5$ is $3+-5$.

Here's a useful procedure that's based on this definition.

Procedure **– Subtracting Two Integers**
1. Identify the minuend and the subtrahend. 2. Change the subtraction to addition. 3. Change the subtrahend to its opposite. 4. Follow the procedure for adding two integers.

Practice 1.4.1 Viewing Subtraction as Adding the Opposite

Find the difference by changing the subtraction to addition and the subtrahend to its opposite.

a) $9-17$

$9-17$	$\Rightarrow$	The minuend is 9 and the subtrahend is 17.
$9+-17$	$\Rightarrow$	Changed the subtraction to addition and 17 to its opposite.
-8	$\Rightarrow$	Followed the procedure for addition.

b) $-12-4$

$-12-4$	$\Rightarrow$	The minuend is -12 and the subtrahend is 4.
$-12+-4$	$\Rightarrow$	Changed the subtraction to addition and 4 to its opposite.
-16	$\Rightarrow$	Followed the procedure for addition.

c) $-7--3$

$-7--3$	$\Rightarrow$	The minuend is -7 and the subtrahend is -3.
$-7+3$	$\Rightarrow$	Changed the subtraction to addition and -3 to its opposite.
-4	$\Rightarrow$	Followed the procedure for addition.

Homework 1.4 *Find the difference by changing the subtraction to addition and the subtrahend to its opposite.*

1) $-8-10$	2) $5--2$	3) $12-12$	4) $-4--6$	5) $-9--2$
6) $-10-4$	7) $1--6$	8) $-12--14$	9) $13--15$	10) $16-12$

1.4.2 Combining Operations

Most students find it takes a lot of practice to become comfortable simplifying signs when subtraction is involved. For instance, when the class begins simplifying $5-4(2+1)$, it's common for a number of students to ask, "Is that a subtract four or a negative four?" The answer is, "It depends on your point of view."

If you take the point of view that the dash is a subtraction then, following the order of operations, you'd add first, multiply second and subtract last.

$$5-4(2+1)$$
$$5-4(3)$$
$$5-12$$
$$-7$$

On the other hand, if you decide to use the definition and change the subtraction to adding the opposite, then you're taking the point of view that it's a negative four. Notice that after following the order of operations your answer will again be -7.

$$5-4(2+1)$$
$$5+-4(2+1)$$
$$5+-4(3)$$
$$5+-12$$
$$-7$$

With arithmetic, one of these approaches isn't "better" than the other. When it comes to algebra though, it's the second approach (adding the opposite) that's used most often so that's the procedure you should use for the homework. After finishing each of the following problems, I'd like you to check your answer using your calculator. Here's a couple of important points about using a calculator to simplify signed expressions.

Careful!	The dash and your calculator
	On most calculators, the subtraction key – performs subtraction while the change sign key, usually (–) or ±, changes the sign of a number. Trying to use the subtraction key to change the sign of a number will usually result in an error. Also, on some calculators you need to enter the number first and then the dash while on other calculators you enter the dash first and then the number.

Practice 1.4.3 Combining Operations

Working left to right identify whether each dash implies negative or subtraction. Next, change all subtractions to adding the opposite. Last, simplify the expression.

a) $-9-(-7)$

$-9-(-7)$	$\Rightarrow$	Looking left to right the first dash implies negative, the second implies subtraction and the third implies negative.
$-9+7$	$\Rightarrow$	Changed the subtraction to adding the opposite.
-2	$\Rightarrow$	Simplified the addition.

b) $(-4)(-3)-3(5)$

$(-4)(-3)-3(5)$	$\Rightarrow$	Left to right the first and second dash imply negative and the third dash implies subtraction.
$(-4)(-3)+-3(5)$	$\Rightarrow$	Changed the subtraction to adding the opposite.
$(-4)(-3)+-3(5)$ $12+-15$ -3	$\Rightarrow$	Simplified using the order of operations.

c) $8-(4-11)$

$8-(4-11)$	$\Rightarrow$	Both the first and second dash imply subtraction.
$8-1(4-11)$	$\Rightarrow$	Made the factor of 1 after the subtraction explicit. Including this "unseen" factor of 1 at this point is common in algebra.
$8+-1(4+-11)$	$\Rightarrow$	Changed all subtractions to addition of the opposite.
$8+-1(-7)$	$\Rightarrow$	Added inside the parentheses.
$8+7$	$\Rightarrow$	By the order of operations multiplication comes before addition.
15	$\Rightarrow$	Added.

Homework 1.4 *Working left to right identify whether each dash implies negative or subtraction. Next, change all subtractions to adding the opposite. Last, simplify the expression.*

11) $-4-10-6$ 12) $(-2)-2(-7+3)$ 13) $-3(1-5)\div 6$ 14) $(-7)(2)-(-4)(3)$

15) $(-2)(-2)-2(3-7)$ 16) $\dfrac{-16}{4(1-(-3))}$ (Start with the dividend.)

17) $-(3-11)\div -2(1-5)$ 18) $7-[2-2(6-9)]$ 19) $-2-[-2-(2-6)]$

20) $7-5(2-(-8))+(-5)(3)$ 21) $4-4[8-8(3-1)-4]$

22) $\dfrac{2-14}{-4(5-8)}-\dfrac{36}{-4}$ (Start with the dividend.)

Homework 1.4 Answers

1) The minuend is -8 and the subtrahend is 10. $-8+-10=-18$.

2) The minuend is 5 and the subtrahend is -2. $5+2=7$.

3) The minuend is 12 and the subtrahend is also 12. $12+-12=0$.

4) The minuend is -4 and the subtrahend is -6. $-4+6=2$

5) The minuend is -9 and the subtrahend is -2. $-9+2=-7$.

6) The minuend is -10 and the subtrahend is 4. $-10+-4=-14$.

7) The minuend is 1 and the subtrahend is -6. $1+6=7$.

8) The minuend is -12 and the subtrahend is -14. $-12+14=2$.

9) The minuend is 13 and the subtrahend is -15. $13+15=28$.

10) The minuend is 16 and the subtrahend is 12. $16+-12=4$.

11) The first dash implies negative, the second and third imply subtraction. $-4+-10+-6=-20$

12) The first dash implies negative, the second implies subtraction and third implies negative.
$(-2)+-2(-7+3)=6$

13) The first dash implies negative and the second implies subtraction. $-3(1+-5)\div6=2$

14) The first dash implies negative, the second implies subtraction and the third implies negative. The number following the subtraction is 1 (which, as usual, isn't written).
$(-7)(2)+-1(-4)(3)=-2$.

15) The first and second dash both imply negative, the third and fourth dash imply subtraction. The number following the first subtraction is 4 and the number following the second subtraction is 7. $(-2)(-2)+-2(3+-7)=12$.

16) The dash in the dividend implies negative. In the divisor, the first dash implies subtraction and the second implies negative. The number following the subtraction is -3.
$$\frac{-16}{4(1+3)}=-1.$$

17) The first dash implies negative (a factor of -1), the second implies subtraction the third implies negative and the fourth implies subtraction. The number following the first subtraction is 11 and the number following the second subtraction is 5.
$-(3+-11)\div-2(1+-5)=16$. (If you thought at first that the answer was 1, I did too when I was checking my answers. 16 is the correct answer though.)

18) The three dashes all imply subtraction. The number following the first subtraction is 1, the number following the second subtraction is 2 and the number following the third subtraction is 9. $7+-1[2+-2(6+-9)]=-1$.

19) The first dash implies negative, the second implies subtraction the third implies negative and the fourth and fifth imply subtraction. The number following the first subtraction is 1 (which, as usual, isn't written). In the same way, the number following the second subtraction is also 1. The number following the third subtraction is a 6.
$-2+-1[-2+-1(2+-6)]=-4$.

20) The first and second dash imply subtraction, the third implies negative. The number following the first subtraction is 5. The number following the second subtraction is -8.
$7+-5(2+8)+(-5)(3)=-58$.

21) All four dashes imply subtraction. The number following the first subtraction is 4. The number following the second subtraction is 8. The number following the third subtraction is 1 and the number following the fourth subtraction is 4. $4+-4[8+-8(3+-1)+-4]=52$.

22) Starting in the dividend of the first term the first dash implies subtraction then, in the divisor, the next dash implies negative and the third dash implies subtraction. The dash between the rational expressions implies subtraction and the last dash in the divisor of the second term implies negative. The number following the first subtraction is 14. The number following the second subtraction is 8 and the number following the third subtraction is 1. $\frac{2+-14}{-4(5+-8)}+-1\left(\frac{36}{-4}\right)=8$.

1.5 Reducing Fractions

The integers allow us to count something “positive”, like profit and something “negative”, like debt. In both cases, though, we are counting discreet units of 1 or -1. A nice visualization of this is the number line where both the negative integers and the positive integers are counting points that are a discreet distance to the left or right of 0. What if we wanted to describe a distance that ended up “between” two points? For that, we’ll need the idea of a rational number.

1.5.1 The Idea of a Rational Number

If your thermometer only had integers you could measure a temperature of 98° or 99° but you couldn’t measure a temperature of 98.6°. To measure the temperature 98.6° we divide the distance between 98 and 99 into 10 parts of equal distance and

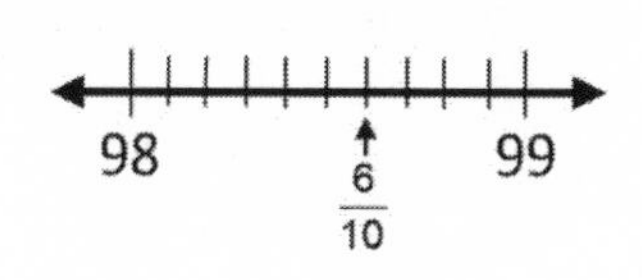

then use 6 of the 10 parts or $\frac{6}{10}$ parts. The number $\frac{6}{10}$ is an example of a rational number.

Definition **– Rational Number**
A number that can be written as the quotient of two integers. Example: $3/5$, -0.1, 15% Note: Rational numbers with a divisor of 0 are undefined.

Fractions are a common type of rational number. In this section, we’ll begin reviewing operations with fractions. With a fraction the dividend is often called the numerator and the divisor is often called the denominator. Notice the integers are a subset of the rational numbers since every integer can be written as a fraction with a denominator of 1. For instance, the integer -3 can be written as the fraction $\frac{-3}{1}$.

1.5.2 Prime Factorization

Many procedures with fractions rely on prime factorization. To discuss prime factoring, we need a little vocabulary. A **prime number** is a natural number, greater than 1, which only has factors 1 and itself. The first few prime numbers are 2, 3, 5, 7, 11, 13, and 17. A **composite number** is a natural number, greater than 1, which is not prime. The first few composite numbers are 4, 6, 8, 9, 10, 12, and 14.

You have prime factored a number when the number is written as the product of only prime factors. We consider $2\times2\times3$, the **prime factorization** of 12 since the factors 2 and 3 are prime. We do not consider 2×6 a prime factorization of 12 because 6 is composite. To help organize a prime factorization it’s common to write the prime factors left to right from smallest to largest using the commutative property of multiplication.

Practice 1.5.2 Prime Factorization

Prime factor any composite factors and use the commutative property to reorder the final prime factors from smallest to largest (left to right).

a) $3 \times 6 \times 2$

$3 \times 2 \times 3 \times 2$	$\Rightarrow$	Prime factored 6. (The factorization $3 \times 3 \times 2 \times 2$ is also fine.)
$2 \times 2 \times 3 \times 3$	$\Rightarrow$	Reorder the factors.

Homework 1.5 *Prime factor any composite factors and use the commutative property to write the final prime factors from smallest to largest (left to right).*

1) $5 \times 5 \times 15$	2) $4 \times 3 \times 6 \times 2$	3) $14 \times 9 \times 21$	4) $15 \times 6 \times 8$

1.5.3 Factor Rules

It's easy to see that the prime factorization of 6 is 2×3. It's harder to see that the prime factorization of 120 is $2 \times 2 \times 2 \times 3 \times 5$. The factor rules will assist in prime factoring larger numbers by helping you decide if 2, 3 or 5 are factors of your number. You should be aware that it's probably more common to call the factor rules the divisibility rules.

Factor Rules for the Natural Numbers 2, 3 and 5
2 is a factor if the number is even. Even numbers end in 0,2,4,6 or 8.
3 is a factor if the sum of the number's digits is a multiple of 3.
5 is a factor if the number ends in 5 or 0.

Practice 1.5.3 Factor Rules

Decide if the following numbers have 2, 3, or 5 as factors.

a) 12,345

2 is not a factor	$\Rightarrow$	Since the number does not end in 0,2,4,6 or 8 the number is not even.
3 is a factor	$\Rightarrow$	Since the sum of the digits is 15, $1+2+3+4+5=15$, and 15 has a factor of 3, 12,345 also has at least one factor of 3.
5 is a factor	$\Rightarrow$	12,345 ends in 5 so it has a factor of 5.

b) 120

2 is a factor	$\Rightarrow$	Since the number ends in 0 it's even and has 2 as a factor.
3 is a factor	$\Rightarrow$	The sum of the digits is 3, $1+2+0=3$, so 3 is a factor.
5 is a factor	$\Rightarrow$	120 has a factor of 5 since it ends in 0.

Homework 1.5 *Decide if the following numbers have 2, 3, or 5 as factors.*

5) 18	6) 540	7) 375	8) 119

1.5.4 Factor Trees

Sometimes, a prime factorization will "pop" into your head. The factor rules, along with a factor tree, can help if nothing pops.

For example, to build a factor tree for 20, I might start with the factors 5 and 4. Since 5 is prime I'll circle it and since 4 is composite I'll continue to factor and then circle the 2's since they're prime. I can stop now since all the circled factors are prime. The prime factorization of 20 is $2 \times 2 \times 5$.

Notice that starting with 2 and 10 again leads to the prime factorization $2 \times 2 \times 5$. The idea that every composite number has a unique prime factorization, is often called the Fundamental Theorem of Arithmetic.

Here's another example. When I see 120 I think of 12 times 10 so that's how I'll start. After continuing until only prime factors remain I find that the prime factorization for 120 is $2 \times 2 \times 2 \times 3 \times 5$.

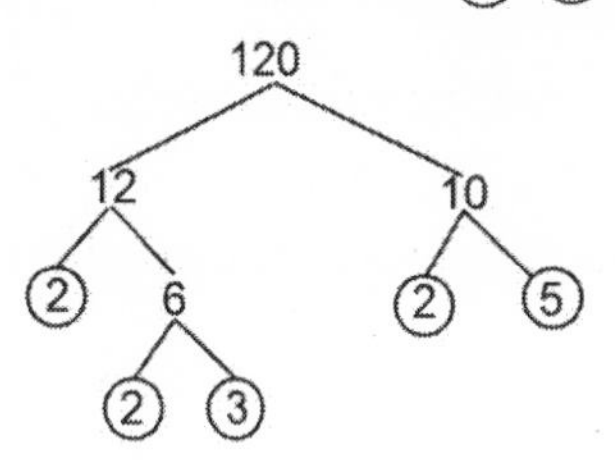

Sometimes finding factors of 2, 3 and 5 isn't enough. For instance, 119 has prime factorization 7×17. Here's a procedure that will help you find any prime factorization.

Procedure **– To Prime Factor a Natural Number**
1. Build a factor tree using the factor rules for 2, 3, and 5. 2. After step 1 begin dividing uncircled factors by the prime numbers from 7 up to the natural number part of the number's square root. 3. Write your prime factors in order from smallest to largest.

Homework 1.5 Prime factor the following numbers.

9) 24	10) 42	11) 80	12) 91	13) 102	14) 429

1.5.5 Reducing Fractions

If the numerator and denominator of a fraction have a common prime factor, we can use the fact that any non-zero number divided by itself is 1, and the fact that the product of a number and 1 can be replaced by the number itself, to "**reduce**" the fraction.

For instance, to reduce $\frac{18}{30}$, I'd first prime factor the numerator and denominator and write the prime factors from smallest to largest. Next, I'd "cross out" the common factor of 2 and a common factor of 3 since each is a factor of 1. After reducing, I'm left with three-fifths.

$$\frac{2 \times 3 \times 3}{2 \times 3 \times 5}$$

$$\frac{\cancel{2} \times \cancel{3} \times 3}{\cancel{2} \times \cancel{3} \times 5} = \frac{3}{5}$$

Notice that reducing "removes" unneeded factors of 1 from the product, it doesn't make the value of the fraction smaller. Here's a procedure to help reduce fractions.

Procedure **– Reducing Fractions**
1. Prime factor the numerator and denominator.
2. Reduce factors common to both the numerator and denominator.
3. Find the product of any remaining factors in the numerator and then find the product of any remaining factors in the denominator.

Before you start on the homework, I want to caution you about an issue that comes up in class. Often, students reduce a fraction like $\frac{12}{18}$ using a process that starts something like this, "Well two goes into twelve six times and two goes into eighteen two times and then three goes into six twice and... etc." Although this idea can be used with arithmetic, it doesn't transfer well to algebra. For instance, later you'll be reducing $\frac{x^2-10x-24}{x^2-13x+12}$, and no one does this using the process, "Well ex minus twelve goes into ex squared minus ten ex minus twenty four ex plus two times...etc." On the other hand, prime factoring and reducing common factors transfers directly to algebra.

Practice 1.5.5 Reducing Fractions

Reduce using prime factorization.

a) $\frac{-36}{30}$

$\frac{-1\times2\times2\times3\times3}{2\times3\times5}$	$\Rightarrow$	Prime factored the numerator and denominator. Made the factor of -1 explicit.
$\frac{-1\times\cancel{2}\times2\times\cancel{3}\times3}{\cancel{2}\times\cancel{3}\times5}$	$\Rightarrow$	Reduced common factors of 1.
$-\frac{6}{5}$	$\Rightarrow$	Multiplied the unreduced factors in the numerator and then in the denominator.

b) $18\div54$

$\frac{2\times3\times3}{2\times3\times3\times3}$	$\Rightarrow$	Prime factored the numerator and denominator.
$\frac{\cancel{2}\times\cancel{3}\times\cancel{3}}{\cancel{2}\times\cancel{3}\times\cancel{3}\times3}=\frac{1}{3}$	$\Rightarrow$	Reduced and multiplied. Remembered to include a factor of 1 in the numerator (common factors are reduced to 1, not 0) since the remaining factor of 3 is in the denominator.

c) $6/180$

$\frac{\cancel{2}\times\cancel{3}}{2\times\cancel{2}\times\cancel{3}\times3\times5}$	$\Rightarrow$	Prime factored the numerator and denominator, reduced common factors and found the remaining products. Remembered to include a factor of 1 in the numerator.

Homework 1.5 Reduce using prime factorization..

15) $\frac{52}{36}$ 16) $-8 \div 40$ 17) $42/36$ 18) $63 \div 84$ 19) $\frac{-56}{120}$ 20) $\frac{105}{126}$

21) $28/42$ 22) $-16 \div 20$ 23) $\frac{102}{34}$ 24) $84 \div -98$ 25) $-105/-126$

26) $\frac{45}{135}$ 27) $\frac{143}{165}$ 28) $-\frac{221}{-153}$

Homework 1.5 Answers

1) $3\times5\times5\times5$ 2) $2\times2\times2\times2\times3\times3$ 3) $2\times3\times3\times3\times7\times7$ 4) $2\times2\times2\times2\times3\times3\times5$

5) 2 and 3 are factors. 6) 2, 3, and 5 are factors. 7) 3 and 5 are factors.

8) All three are **not** factors. 9) $2\times2\times2\times3$ 10) $2\times3\times7$ 11) $2\times2\times2\times2\times5$

12) 7×13 13) $2\times3\times17$ 14) $3\times11\times13$ 15) $\frac{2}{2}\times\frac{2}{2}\times\frac{13}{9}=\frac{13}{9}$

16) $\frac{2}{2}\times\frac{2}{2}\times\frac{2}{2}\times\frac{-1}{5}=-\frac{1}{5}$ 17) $\frac{2}{2}\times\frac{3}{3}\times\frac{7}{6}=\frac{7}{6}$ 18) $\frac{3}{3}\times\frac{7}{7}\times\frac{3}{4}=\frac{3}{4}$ 19) $\frac{2}{2}\times\frac{2}{2}\times\frac{2}{2}\times\frac{-7}{15}=\frac{-7}{15}$

20) $\frac{3}{3}\times\frac{7}{7}\times\frac{5}{6}=\frac{5}{6}$ 21) $\frac{\cancel{2}\times2\times\cancel{7}}{\cancel{2}\times3\times\cancel{7}}=\frac{2}{3}$ 22) $\frac{-1\times\cancel{2}\times\cancel{2}\times2\times2}{\cancel{2}\times\cancel{2}\times5}=-\frac{4}{5}$ 23) $\frac{\cancel{2}\times3\times\cancel{17}}{\cancel{2}\times\cancel{17}}=3$

24) $\frac{\cancel{2}\times2\times3\times\cancel{7}}{-1\times\cancel{2}\times7\times\cancel{7}}=-\frac{6}{7}$ 25) $\frac{\cancel{-1}\times\cancel{3}\times5\times\cancel{7}}{\cancel{-1}\times2\times\cancel{3}\times3\times\cancel{7}}=\frac{5}{6}$ 26) $\frac{\cancel{3}\times\cancel{3}\times\cancel{5}}{\cancel{3}\times\cancel{3}\times\cancel{5}\times3}=\frac{1}{3}$

27) $\frac{\cancel{11}\times13}{3\times5\times\cancel{11}}=\frac{13}{15}$ 28) $\cancel{-1}\times\frac{13\times\cancel{17}}{\cancel{-1}\times3\times3\times\cancel{17}}=\frac{13}{9}$

1.6 Multiplying and Dividing Fractions

In this section, we'll practice multiplying and dividing fractions.

1.6.1 Multiplying Fractions

Here's a fun way to multiply fractions using prime factorizations.

Procedure **– Multiplying Fractions**
1. Multiply the prime factored form of the numerators and denominators together. 2. Reduce common factors. 3. Multiply the remaining factors in the numerator and the remaining factors in the denominator.

Practice 1.6.1 Multiplying Fractions

Multiply using prime factorization.

a) $\frac{6}{15}\times\frac{25}{28}$

$$\frac{2\times3\times5\times5}{3\times5\times2\times2\times7}$$
$\Rightarrow$ Multiplied together the prime factored form of the numerator and denominator.

$$\frac{\cancel{2}\times\cancel{3}\times\cancel{5}\times5}{\cancel{3}\times\cancel{5}\times\cancel{2}\times2\times7}=\frac{5}{14}$$
$\Rightarrow$ Reduced common factors and multiplied remaining factors.

b) $\frac{5}{9}\times\frac{-21}{15}\times3$

$$\frac{-1\times5\times3\times7\times3}{3\times3\times3\times5}$$
$\Rightarrow$ Prime factored the numerator and denominator. Thought of -21 as $-1\times3\times7$ and 3 as $3/1$.

$$\frac{-1\times\cancel{5}\times\cancel{3}\times7\times\cancel{3}}{\cancel{3}\times\cancel{3}\times3\times\cancel{5}}=-\frac{7}{3}$$
$\Rightarrow$ Reduced common factors and multiplied.

c) $2\frac{1}{3}\times-1\frac{2}{7}$

$$\frac{7}{3}\times-\frac{9}{7}$$
$\Rightarrow$ Wrote the mixed number as an improper fraction.

$$\frac{\cancel{7}\times-1\times\cancel{3}\times3}{\cancel{3}\times\cancel{7}}=-3$$
$\Rightarrow$ Multiplied, reduced common factors and multiplied the remaining factors.

Homework 1.6 Multiply using prime factorization.

1) $-\frac{3}{4}\times\frac{12}{15}$	2) $\frac{24}{25}\times\frac{35}{30}$	3) $\frac{-5}{24}\times18$	4) $7\times\frac{10}{21}\times-6$
5) $2\frac{4}{7}\times1\frac{1}{27}$	6) $\frac{-4}{9}\times\frac{6}{5}\times\frac{-15}{18}$	7) $-3\frac{3}{4}\times\frac{1}{5}\times-1\frac{4}{9}$	8) $2\frac{1}{7}\times\frac{14}{45}\times1\frac{1}{2}$
9) $-\frac{21}{4}\times\frac{2}{-9}\times\frac{3}{5}\times\frac{-6}{15}$	10) $1\frac{5}{8}\times\frac{25}{49}\times4\times\frac{21}{39}$		

1.6.2 The Reciprocal

Before we begin dividing fractions there's a couple skills you'll need. First, you'll need to find the **reciprocal** of a fraction. The reciprocal of a fraction is a second fraction which when multiplied to the first, gives a product of 1. For example, the reciprocal of $\frac{2}{3}$ is $\frac{3}{2}$ because $\frac{2}{3}\times\frac{3}{2}=\frac{6}{6}=1$. The reciprocal of -7 would be $-\frac{1}{7}$ since $-\frac{7}{1}\times-\frac{1}{7}=1$. Students often say that the reciprocal, "Is the original number flipped upside down." which is true, (as long as the original number isn't 0), but not very precise. Here's some practice finding a reciprocal.

Practice 1.6.2 The Reciprocal

Find the reciprocal of the given number.

a) $\frac{14}{15}$

$\frac{15}{14}$ $\Rightarrow$ Since $\frac{14}{15}\times\frac{15}{14}$ reduces to 1.

b) $-2\frac{3}{8}$

$-\frac{19}{8}$ $\Rightarrow$ Made sure to think of $-2\frac{3}{8}$ as $-1\times2\frac{3}{8}$ before carrying out the procedure for rewriting as an improper fraction.

$-\frac{8}{19}$ $\Rightarrow$ The reciprocal is also negative since. $\left(-\frac{19}{8}\right)\left(-\frac{8}{19}\right)=1$.

c) 0

$\frac{1}{0}$ $\Rightarrow$ This is undefined, 0 has no reciprocal.

Homework 1.6 Find the reciprocal of the given number.

11) $\frac{40}{7}$ 12) $-\frac{5}{9}$ 13) $1\frac{3}{5}$ 14) -1 15) $-\frac{15}{14}$ 16) 8

1.6.3 Identifying the Numerator and Denominator of a Complex Fraction

The second skill you'll need to divide fractions is being able to recognize the numerator and denominator of the complex fraction.

A fraction, where the numerator, the denominator, or both are also fractions, is often called a **complex fraction**. An important skill when you're dividing complex fractions is being able to separate the numerator and denominator of the major fraction. For example, if we start with the complex fraction $\frac{\frac{2}{3}}{\frac{5}{6}}$ the numerator of the major fraction would be the minor fraction $\frac{2}{3}$

and the denominator would be the minor fraction $\frac{5}{6}$. Notice how the fraction bar separating the numerator and denominator of the major fraction is longer than the fraction bar for either of the minor fractions.

Sometimes, when students begin simplifying a complex fraction like $\dfrac{-8}{\frac{40}{7}}$, they have a difficult time figuring out the numerator of the major fraction. If you put your attention on the larger fraction bar though, I hope you can see that the numerator is -8 and the denominator is $\frac{40}{7}$. It's common for people to rewrite the major fraction as $\dfrac{\frac{-8}{1}}{\frac{40}{7}}$ before beginning the procedure for dividing.

1.6.4 Dividing Fractions

Here's the procedure for dividing fractions.

Procedure **– Dividing Fractions**
To divide two fractions, multiply the minor fraction in the numerator by the reciprocal of the minor fraction in the denominator.

For instance, to simplify $\dfrac{\frac{2}{3}}{\frac{5}{6}}$, I'd first recognize the numerator of the major fraction is $\frac{2}{3}$ and the denominator is $\frac{5}{6}$. Next, I'd multiply the numerator of the major fraction by the reciprocal of the denominator, $\frac{2}{3}\times\frac{6}{5}$. Last, I'd find the reduced product, $\frac{2}{3}\times\frac{6}{5}=\frac{2\times2\times\cancel{3}}{\cancel{3}\times5}=\frac{4}{5}$.

Practice 1.6.4 Simplifying a Complex Fraction Using the Multiplicative Identity

First, identify the numerator and the denominator of the major fraction.
Then, simplify using the procedure for dividing fractions.

a) $\dfrac{\frac{2}{5}}{\frac{4}{15}}$

$\frac{2}{5}\times\frac{15}{4}$	$\Rightarrow$	Multiplied the numerator of the major fraction by the reciprocal of the denominator.
$\frac{\cancel{2}\times3\times\cancel{5}}{\cancel{5}\times\cancel{2}\times2}=\frac{3}{2}$	$\Rightarrow$	Reduced common factors and multiplied in the numerator.

b) $\frac{36}{5} \div \frac{4}{15}$

The numerator is $\frac{36}{5}$ and the denominator is $\frac{4}{15}$. $\Rightarrow$ The numerator is to the left of $\div$ and the denominator is to the right.

$\frac{36}{5} \times \frac{15}{4}$ $\Rightarrow$ Multiplied the numerator by the reciprocal of the denominator.

$\frac{\cancel{2} \times \cancel{2} \times 3 \times 3 \times 3 \times \cancel{5}}{\cancel{5} \times \cancel{2} \times \cancel{2}} = 27$ $\Rightarrow$ Reduced common factors and multiplied.

c) $\frac{-4}{2\frac{2}{3}}$

$\frac{-\frac{4}{1}}{\frac{8}{3}}$ $\Rightarrow$ Thought of the 4 in the numerator as $\frac{4}{1}$ and wrote the denominator as an improper fraction.

$-\frac{4}{1} \times \frac{3}{8} = -\frac{3}{2}$ $\Rightarrow$ Multiplied by the reciprocal of the denominator and reduced.

Homework 1.6 First, identify the numerator and the denominator of the major fraction. Then, simplify using the procedure for dividing fractions.

17) $\frac{\frac{8}{9}}{\frac{2}{3}}$ 18) $\frac{4}{21} \div \frac{8}{7}$ 19) $\frac{-8}{\frac{40}{7}}$ 20) $\frac{-1\frac{6}{7}}{3\frac{5}{7}}$ 21) $-\frac{\frac{9}{2}}{63}$

22) $\frac{4}{21} \div \frac{8}{7}$ 23) $\frac{14/15}{24/25}$ 24) $\frac{-4\frac{4}{5}}{-2\frac{2}{5}}$ 25) $-\frac{14}{3} \div \frac{7}{6}$ 26) $\frac{-\frac{15}{14}}{\frac{10}{21}}$

We'll finish with some problems where we combine multiplication and division.

Homework 1.6 Simplify.

27) $\frac{12}{20} \times \frac{7}{9} \div 14$ 28) $\frac{\frac{5}{12} \times \frac{9}{15}}{\frac{5}{9} \times \frac{6}{14}}$ 29) $\frac{3}{4} \div 1\frac{5}{9} \div \frac{-27}{4}$ 30) $\frac{8}{-13} \div -\frac{3}{26} \times \frac{-3}{5}$

Homework 1.6 Answers

1) $\dfrac{-1\times\cancel{3}\times\cancel{2}\times\cancel{2}\times3}{\cancel{2}\times\cancel{2}\times\cancel{3}\times5}=-\dfrac{3}{5}$

2) $\dfrac{\cancel{2}\times2\times2\times\cancel{3}\times\cancel{5}\times7}{\cancel{5}\times5\times\cancel{2}\times\cancel{3}\times5}=\dfrac{28}{25}$

3) $\dfrac{-1\times5\times\cancel{2}\times\cancel{3}\times3}{\cancel{2}\times2\times2\times\cancel{3}}=-\dfrac{15}{4}$

4) $\dfrac{\cancel{7}\times2\times5\times-1\times2\times\cancel{3}}{\cancel{7}\times\cancel{3}}=-20$

5) $\dfrac{18}{7}\times\dfrac{28}{27}=\dfrac{\cancel{3}\times\cancel{3}\times2\times2\times2\times\cancel{7}}{\cancel{7}\times\cancel{3}\times\cancel{3}\times3}=\dfrac{8}{3}$

6) $\dfrac{\cancel{-1}\times\cancel{2}\times2\times2\times\cancel{3}\times\cancel{-1}\times\cancel{3}\times\cancel{5}}{\cancel{3}\times\cancel{3}\times\cancel{5}\times\cancel{2}\times3\times3}=\dfrac{4}{9}$

7) $-\dfrac{15}{4}\times\dfrac{1}{5}\times-\dfrac{13}{9}=\dfrac{\cancel{-1}\times\cancel{3}\times\cancel{5}\times\cancel{-1}\times13}{2\times2\times\cancel{5}\times\cancel{3}\times3}=\dfrac{13}{12}=1\dfrac{1}{12}$

8) $\dfrac{15}{7}\times\dfrac{14}{45}\times\dfrac{3}{2}=\dfrac{\cancel{3}\times\cancel{5}\times\cancel{2}\times\cancel{7}\times\cancel{3}}{\cancel{7}\times\cancel{3}\times\cancel{3}\times\cancel{5}\times\cancel{2}}=1$

9) $\dfrac{\cancel{-1}\times\cancel{3}\times7\times\cancel{2}\times\cancel{3}\times-1\times\cancel{2}\times\cancel{3}}{\cancel{2}\times\cancel{2}\times\cancel{-1}\times\cancel{3}\times\cancel{3}\times5\times\cancel{3}\times5}=\dfrac{-7}{25}$

10) $\dfrac{13}{8}\times\dfrac{25}{49}\times\dfrac{4}{1}\times\dfrac{21}{39}=\dfrac{\cancel{13}\times5\times5\times\cancel{2}\times\cancel{2}\times\cancel{3}\times\cancel{7}}{\cancel{2}\times\cancel{2}\times2\times\cancel{7}\times7\times\cancel{3}\times\cancel{13}}=\dfrac{25}{14}=1\dfrac{11}{14}$

11) $\dfrac{7}{40}$ 12) $-\dfrac{9}{5}$ 13) $\dfrac{5}{8}$ 14) -1 15) $-\dfrac{14}{15}$ 16) $\dfrac{1}{8}$

17) The numerator is $\dfrac{8}{9}$ and the denominator is $\dfrac{2}{3}$. The quotient is $\dfrac{4}{3}$

18) The numerator is $\dfrac{4}{21}$ and the denominator is $\dfrac{8}{7}$. The quotient is $\dfrac{1}{6}$

19) The numerator is -8 and the denominator is $\dfrac{40}{7}$. The quotient is $\dfrac{-7}{5}$

20) The numerator is $-\dfrac{13}{7}$ and the denominator is $\dfrac{26}{7}$. The quotient is $-\dfrac{1}{2}$

21) The numerator is $\dfrac{9}{2}$ and the denominator is 63. The quotient is $\dfrac{-1}{14}$

22) The numerator is $\dfrac{4}{21}$ and the denominator is $\dfrac{8}{7}$. The quotient is $\dfrac{1}{6}$

23) The numerator is $14/15$ and the denominator is $24/25$. The quotient is $\dfrac{35}{36}$

24) The numerator is $\dfrac{-24}{5}$ and the denominator is $\dfrac{-12}{5}$. The quotient is 2

25) The numerator is $-\dfrac{14}{3}$ or $\dfrac{14}{3}$ and the denominator is $\dfrac{7}{6}$. The quotient is -4

26) The numerator is $-\dfrac{15}{14}$ and the denominator is $\dfrac{10}{21}$. The quotient is $-\dfrac{9}{4}$

27) $\dfrac{1}{30}$ 28) $\dfrac{21}{20}$ 29) $-\dfrac{1}{14}$ 30) $-\dfrac{16}{5}$

1.7 Adding and Subtracting Fractions

Adding and subtracting fractions that don't share a common denominator is one of the more difficult topics in arithmetic. In this section we'll begin with fractions that have a common denominator and then we'll work with fractions that don't share a common denominator.

1.7.1 Adding and Subtracting Fractions with Common Denominators

For me, a fraction is a product where the denominator tells me the "size" of the fraction and the numerator counts how many I have. For example, I see $\frac{4}{7}$ as the product $\left(\frac{4}{1}\right)\left(\frac{1}{7}\right)$ which I'll write as $4\left(\frac{1}{7}\right)$. The first factor, the 4 (the original numerator), tells me I have four quantities of the second factor, the $\left(\frac{1}{7}\right)$ (the original denominator). So, when I see $\frac{4}{7}$, I visualize that I've cut the distance on the number line from 0 to 1 into seven equal pieces of size $\frac{1}{7}$ and I'm interested in the distance from 0 to the end of the fourth piece.

With this idea in mind it makes sense that to find the sum $\frac{4}{7}+\frac{3}{7}$ I should add the numerators and keep the common denominator since four one-sevenths and three more one-sevenths should be seven one-sevenths, $4\left(\frac{1}{7}\right)+3\left(\frac{1}{7}\right)=(4+3)\left(\frac{1}{7}\right)=7\left(\frac{1}{7}\right)$. Notice how simplifying the expression, $7\left(\frac{1}{7}\right)=\left(\frac{7}{1}\right)\left(\frac{1}{7}\right)=\frac{7}{7}=1$ supports the idea that seven distances of one-seventh moves us one whole unit to the right of 0 on the number line. Here's a procedure to add or subtract fractions that have a common denominator.

***Procedure* – Adding or Subtracting Fractions with Common Denominators**
1. Add or subtract the numerators and keep the common denominator. 2. Reduce if possible.

Practice 1.7.1 Adding and Subtracting Fractions with Common Denominators
Simplify.

a) $\frac{4}{5}-\frac{7}{5}$

$\frac{4-7}{5}$	$\Rightarrow$	Since the fractions have a common denominator subtracted the numerator values and kept the common denominator.
$\frac{-3}{5}$	$\Rightarrow$	Subtracted in the numerator.

b) $\frac{1}{3}+\frac{4}{3}-\frac{2}{3}$	
$\frac{1+4-2}{3}$ $\Rightarrow$	All the fractions have a common denominator. Added and subtracted the numerator values and kept the common denominator.
$\frac{5-2}{3}=\frac{3}{3}=1$ $\Rightarrow$	The final answer.

c) $2\frac{1}{5}-3\frac{2}{5}$	
$\frac{11}{5}-\frac{17}{5}$ $\Rightarrow$	Wrote the mixed numbers as improper fractions. (It's also possible to do this problem using mixed numbers and borrowing.)
$\frac{11-17}{5}=\frac{-6}{5}=-1\frac{1}{5}$ $\Rightarrow$	Subtracted in the numerator and kept the common denominator. Problems that begin with mixed numbers often end with mixed numbers.

Homework 1.7 Simplify.

1) $\frac{4}{15}+\frac{11}{15}$	2) $-\frac{7}{9}-\frac{11}{9}$	3) $5\frac{2}{3}+2\frac{1}{3}$	4) $\frac{7}{4}-\frac{5}{4}+\frac{1}{4}$
5) $2\frac{4}{5}-\frac{1}{5}-1\frac{8}{5}$	6) $-\frac{1}{8}-\frac{9}{8}+\frac{5}{8}$	7) $-\frac{1}{11}-\frac{2}{11}-\frac{3}{11}+\frac{4}{11}$	8) $\frac{32}{40}-\frac{15}{40}-\frac{28}{40}$

1.7.2 Building the LCD

Using our procedure to add or subtract fractions works well, as long as our fractions share a common denominator. What should we do with a problem like $\frac{2}{3}+\frac{1}{2}$ where the denominators are different? The answer is, we'll use our previous procedure after we replace any fraction without the common denominator with an "equivalent" fraction that does have the common denominator. When building a common denominator, we like to use the least (the smallest) common denominator which is usually referred to as the **LCD**. Here's a procedure for building the LCD.

***Procedure* – Building the LCD (Least Common Denominator)**
1. Prime factor each denominator. 2. Write down the prime factorization of the first denominator. This is the start of your LCD. 3. Now go through the remaining prime factorizations one by one and only include in your LCD the factors you need, but don't already have. 4. After considering all the denominators, multiply together the factors you included from steps 2 and 3. This is your LCD.

As you build the LCD I'd like you to practice a skill that will speed up your work when dealing with factors and products. The idea is based on the associative property of multiplication.

***Property* – The Associative Property of Multiplication**
The grouping of the factors doesn't affect the product. Example: $(2\times 3)(4)=(2)(3\times 4)$

To simplify a product like $2\times 3\times 3\times 5$, I'd use the associative property to regroup the factors, "in my head" as $(2\times 5)(3\times 3)$ and quickly find the product 10×9 or 90. With a factorization like $2\times 7\times 2\times 3$ I notice my initial grouping $(2\times 7)(2\times 3)=(14)(6)$ doesn't help since the final product doesn't "pop" into my head, but if I group the factors as $(7\times 3)\times 2\times 2$, I see I'm doubling 21 to get 42, and then doubling 42 to get a final product of 84.

If this seems a bit much today, it's fine to continue finding the product using a calculator. If you do use a calculator though, it's important to return to this section a few days from now and practice with the associative property. As you'll see in chapter 5, one of the advantages of finding prime factorizations is using the commutative and associative properties to quickly change your point of view when it comes to finding products from prime factorizations.

Practice 1.7.2 Building the LCD

Find the LCD of the following fractions.

a) $\frac{1}{12}$ and $\frac{5}{18}$

$12=2\times 2\times 3$ $18=2\times 3\times 3$	$\Rightarrow$	Prime factored each denominator.
$2\times 2\times 3$	$\Rightarrow$	Wrote down the prime factorization of the first denominator. 12 is now included in the LCD.
$2\times 2\times 3\times 3$	$\Rightarrow$	Moved on to 18 which has one factor of 2 and two factors of 3. Since the LCD already has one factor of 2 and one factor of 3, I only needed to include one more factor of 3 to have 18 in the LCD.
36	$\Rightarrow$	Since I've now considered all the denominators I grouped the factors as $(2\times 2)(3\times 3)$ and multiplied $(4)(9)$ to find the LCD is thirty-six. Another grouping would be $(2\times 3)(2\times 3)=(6)(6)$ which again gives the product thirty-six.

b) $\frac{-9}{14}$ and $1\frac{5}{12}$ and $\frac{2}{21}$

$14=2\times 7$ $12=2\times 2\times 3$ $21=7\times 3$	$\Rightarrow$	Prime factored each denominator.
2×7	$\Rightarrow$	Included the first denominator so 14 is now a factor of the LCD.
$2\times 7\times 2\times 3$	$\Rightarrow$	Moved on to 12. The LCD already had one of my factors of 2 so I now included the second factor of 2 and a factor of 3. 12 ($2\times 2\times 3$) is now included in the LCD.

$2\times7\times2\times3$	$\Rightarrow$	Moved on to the last denominator, (the 21) and realized both factors (the 3 and the 7) are already included in the LCD so no new factor was needed.
84	$\Rightarrow$	I used the grouping $(7\times3)\times2\times2$ to see the LCD is 84.

c) $4\frac{7}{10}$ and $\frac{25}{9}$ and $\frac{2}{25}$

$10=2\times5 \quad 9=3\times3 \quad 25=5\times5$	$\Rightarrow$	Prime factored the denominators.
2×5	$\Rightarrow$	Took the first denominator.
$2\times5\times3\times3$	$\Rightarrow$	Needed two factors of 3 to include a factor of 9 in the LCD.
$2\times5\times3\times3\times5$	$\Rightarrow$	Needed a second factor of 5 so that 25 is a factor of the LCD.
450	$\Rightarrow$	Used the grouping $(3\times3)(5)(2\times5)$ to see that nine times five is forty-five and forty-five times ten gives an LCD of 450.

Homework 1.7 Find the LCD of the following fractions.

9) $\frac{3}{4}, \frac{3}{10}$	10) $\frac{1}{12}, \frac{9}{8}$	11) $-\frac{9}{14}, -\frac{5}{6}$	12) $-\frac{1}{2}, \frac{4}{3}, \frac{5}{6}$	13) $-\frac{1}{6}, 2\frac{5}{16}, \frac{7}{12}$
14) $-2\frac{3}{4}, -\frac{5}{36}, \frac{13}{20}$	15) $\frac{5}{9}, \frac{7}{10}, \frac{1}{4}, \frac{2}{25}$	16) $-\frac{1}{6}, \frac{5}{18}, \frac{3}{4}, \frac{8}{15}$		

1.7.3 Building Equivalent Fractions

Once you've found the least common denominator, you'll need to replace each fraction that doesn't have the LCD, with an equivalent fraction that has the LCD. Two fractions are equivalent if they have the same value.

We can change a fraction's look, without changing its value, if we multiply the original fraction by 1. The form of 1 we choose will depend on the factors in our LCD that are missing from our original denominator. For example, say I started with $\frac{4}{7}$ and knew the common denominator was 35. Then the form of 1 I'd want would be $\frac{5}{5}$ since a factor of 5 is needed to take my original denominator of 7 to my common denominator of 35, $\frac{4}{7}\times1=\frac{4}{7}\times\frac{5}{5}=\frac{20}{35}$. Here's a procedure to help build equivalent fractions.

***Procedure* – Building Equivalent Fractions**
1. Find the factor, which when multiplied to the original denominator, gives the LCD as the product. (If you don't "see" the factor, you can always divide the LCD by the original denominator.) 2. Multiply both the numerator and denominator of the original fraction by the factor found in step 1.

Here's some practice building equivalent fractions.

Practice 1.7.3 Building Equivalent Fractions

Rewrite the factor of 1 and multiply, so the product is an equivalent fraction with the new denominator.

a) $\frac{4}{5} \times 1 = \frac{}{20}$

$\frac{4}{5} \times \frac{4}{4} = \frac{}{20}$ ⇒ Multiplying by 1, in the form $\frac{4}{4}$, gives an equivalent fraction with a denominator of 20.

$\frac{4}{5} \times \frac{4}{4} = \frac{16}{20}$ ⇒ Multiplied. $\frac{16}{20}$ has the same value as $\frac{4}{5}$.

b) $-\frac{7}{8} \times 1 = -\frac{}{104}$

$-\frac{7}{8} \times \frac{13}{13} = -\frac{}{104}$ ⇒ To see what form of 1 to use I divided $104 \div 8 = 13$ to see that $\frac{13}{13}$ is the form of 1 I want.

$-\frac{7}{8} \times \frac{13}{13} = -\frac{91}{104}$ ⇒ Multiplied to build the equivalent fraction.

Homework 1.7 *Rewrite the factor of 1 and multiply, so the product is an equivalent fraction with the new denominator.*

17) $\frac{5}{6} \times 1 = \frac{}{48}$ 18) $\frac{-3}{1} \times 1 = \frac{}{6}$ 19) $\frac{18}{25} \times 1 = \frac{}{75}$ 20) $\frac{11}{9} \times 1 = \frac{}{117}$ 21) $\frac{-2}{9} \times 1 = \frac{}{153}$

1.7.4 Adding and Subtracting Fractions

Now that you can find the LCD and build equivalent fractions, you're ready to add and subtract fractions that don't share a common denominator. Here's our general procedure.

***Procedure* – Adding and Subtracting Fractions**

1. Find the LCD of all the denominators.
2. Rewrite each fraction without the LCD, as an equivalent fraction with the LCD.
3. Add or subtract the numerators. Keep the common denominator.
4. Reduce if possible.

Practice 1.7.4 Adding and Subtracting Fractions

Simplify.

a) $\frac{1}{2} - \frac{1}{3}$

$2 \times 3 = 6$ ⇒ Found the LCD.

$\frac{1}{2}=\frac{3}{6}$ and $\frac{1}{3}=\frac{2}{6}$ $\Rightarrow$ Built equivalent fractions using the LCD.

$\frac{3}{6}-\frac{2}{6}=\frac{3-2}{6}=\frac{1}{6}$ $\Rightarrow$ Subtracted the numerators and kept the LCD.

b) $\frac{7}{6}-\frac{5}{14}+\frac{7}{12}$

$2\times2\times3\times7=84$ $\Rightarrow$ Found the LCD.

$\frac{7}{6}=\frac{98}{84}$ $\frac{5}{14}=\frac{30}{84}$ $\frac{7}{12}=\frac{49}{84}$ $\Rightarrow$ Built equivalent fractions using the LCD.

$\frac{98-30+49}{84}=\frac{117}{84}$ $\Rightarrow$ Added and subtracted the numerators. Kept the LCD.

$\frac{117}{84}=\frac{\cancel{3}\times3\times13}{2\times2\times\cancel{3}\times7}=\frac{39}{28}$ $\Rightarrow$ Reduced. Both 117 and 84 share a common factor of 3. (The sum of the digits of both numbers are a multiple of 3, $1+1+7=9$ and $8+4=12$.)

c) $1\frac{1}{6}-2\frac{5}{9}$

$\frac{7}{6}-\frac{23}{9}$ $\Rightarrow$ It's often easiest to do problems with mixed numbers as improper fractions.

$2\times3\times3=18$ $\Rightarrow$ Found the LCD.

$\frac{7}{6}=\frac{21}{18}$ $\frac{23}{9}=\frac{46}{18}$ $\Rightarrow$ Built equivalent fractions using the LCD.

$\frac{21-46}{18}=-\frac{25}{18}=-1\frac{7}{18}$ $\Rightarrow$ Subtracted the numerators, kept the LCD and wrote the difference as a mixed number.

Homework 1.7 Simplify.

22) $\frac{4}{15}+\frac{2}{5}$ 23) $\frac{5}{7}-\frac{3}{14}$ 24) $\frac{3}{4}+\frac{5}{16}$ 25) $3\frac{1}{2}+1\frac{2}{3}$

26) $\frac{1}{9}-2$ 27) $\frac{5}{8}-\frac{-1}{18}$ 28) $3-\frac{7}{5}$ 29) $\frac{1}{4}+4$

30) $\frac{7}{9}+1\frac{1}{12}$ 31) $\frac{3}{10}-1\frac{1}{15}$ 32) $-\frac{1}{2}+\frac{2}{3}-\frac{3}{4}$ 33) $\frac{4}{5}-\frac{3}{8}-\frac{7}{10}$

34) $2\frac{1}{6}+3-1\frac{7}{9}$ 35) $\frac{1}{7}-\frac{1}{5}+\frac{23}{35}$ 36) $\frac{3}{10}-1\frac{1}{15}+\frac{6}{5}$ 37) $\frac{-4}{9}-\frac{-5}{3}-\frac{5}{4}$

38) $6-1\frac{5}{12}-4\frac{7}{18}$ 39) $\left(\frac{4}{5}\right)\left(\frac{4}{5}\right)-\frac{4}{5}$ 40) $\frac{3}{4}-\frac{2}{3}\div\frac{4}{3}$ 41) $\left(\frac{3}{4}+\frac{3}{2}\right)\div\frac{27}{10}$

Homework 1.7 Answers

1) $\frac{15}{15}=1$ 2) $\frac{-18}{9}=-2$ 3) $\frac{24}{3}=8$ 4) $\frac{3}{4}$ 5) $\frac{0}{5}=0$ 6) $\frac{-5}{8}$

7) $\frac{-2}{11}$ 8) $\frac{-11}{40}$ 9) $2\times2\times5=20$ 10) $2\times2\times2\times3=24$ 11) $2\times3\times7=42$

12) $2\times3=6$ 13) $2\times2\times2\times2\times3=48$ 14) $2\times2\times3\times3\times5=180$

15) $2\times2\times3\times3\times5\times5=900$ 16) $2\times2\times3\times3\times5=180$ 17) $\frac{5}{6}\times\frac{8}{8}=\frac{40}{48}$

18) $\frac{-3}{1}\times\frac{6}{6}=\frac{-18}{6}$ 19) $\frac{18}{25}\times\frac{3}{3}=\frac{54}{75}$ 20) $\frac{11}{9}\times\frac{13}{13}=\frac{143}{117}$ 21) $\frac{-2}{9}\times\frac{17}{17}=-\frac{34}{153}$

22) $\frac{2}{3}$ 23) $\frac{1}{2}$ 24) $\frac{17}{16}$ 25) $5\frac{1}{6}$ 26) $-\frac{17}{9}$ 27) $\frac{49}{72}$ 28) $\frac{8}{5}$

29) $\frac{17}{4}$ 30) $1\frac{31}{36}$ 31) $-\frac{19}{30}$ 32) $-\frac{7}{12}$ 33) $-\frac{11}{40}$ 34) $3\frac{7}{18}$ 35) $\frac{3}{5}$

36) $\frac{13}{30}$ 37) $-\frac{1}{36}$ 38) $\frac{7}{36}$ 39) $\frac{-4}{25}$ 40) $\frac{25}{36}$ 41) $\frac{5}{6}$

1.8 Applying the Order of Operations

Earlier we discussed that a mathematical expression is a meaningful collection of numbers, letters, operators and grouping. In this section, we're going to begin **evaluating** an expression by substituting a number in place of a letter and then simplifying.

1.8.1 A Word About Variables

Although it's common to call any letter that stands in place of an unknown number a "variable", it's important to appreciate that sometimes we expect the value will actually "vary" and sometimes we don't. For example, in the first topic below, we'll practice checking whether the value 4 solves the equation $x+2(x+1)=11$. In this context, the variable *x* is standing in place of a single unknown number, so it's value really isn't expected to "vary".

After checking answers we'll shift our attention to functions. In a **function**, variables are used to help describe a <u>relationship</u> between two or more items and as the value of one item varies, so might the value of the other. Today, we'll concentrate on a special kind of function called a **formula** which uses mathematics to express a rule or a fact. For instance, we'll practice using the formula $F = {}^{9}\!/_{5}C+32$ to convert a Celsius temperature (represented by *C* in the formula), to a Fahrenheit temperature (represented by *F* in the formula). Notice as we vary the value of *C* then, after multiplying by $\frac{9}{5}$ and adding 32, the resulting value for *F* will also change.

1.8.2 Checking an Answer

When two expressions are separated by an **equality symbol**, =. we have an **equation**. Equations imply that the expression on the left of the equality symbol has the same value as the expression on the right. For example, the equation, $x+1=5$, implies the left side, $x+1$, and the right side, 5, are both 5. A **solution** for $x+1=5$ is any value that evaluates to 5 on the left. I hope you can see that 4 is a solution for $x+1=5$ since, after substituting 4 for *x* and simplifying, the left side also takes on the value 5. We have **solved** an equation when we find all the values that are a solution. All the solutions for an equation are collected in a **solution set**. The solution set for $x+1=5$ would be $\{4\}$.

Unlike $x+1=5$, where most people can "see" the solution is 4, it's uncommon for someone to look at $-2y-1=-3(y+5)$ and have the solution -14 pop into their head. In chapter 2 you'll learn a procedure to help infer the solution for $-2y-1=-3(y+5)$. Unfortunately, if you make a mistake carrying out the procedure, then your answer probably won't match the solution. To help convince ourselves that the answer and the solution match, we'll return to the original equation and replace each occurrence of the variable with our answer. If, after simplifying, the left expression and right expression have the same value, we assume that our answer is the solution. If instead, the left value doesn't match the right value, we assume something about our solving procedure wasn't right. This process of evaluating an answer is known as **checking**. Here's an example.

Say we started with $2(y-1)=y+4$ and, after carrying out the procedure, we came up with the answer 5. To evaluate our answer, we'd return to the original equation, replace all occurrences of *y* with 5 and simplify the left and right side to see if both expressions take on the same value. Unfortunately, in this case, the left side evaluates to 8 and the right side to 9. This tells us that our answer, 5, probably isn't the solution and we should go through our solution procedure, step by step, and look for errors.

$2(y-1)$	$y+4$
$2(5-1)$	$5+4$
$2(4)$	9
8	

Let's assume we found an error, carried out the procedure a second time and found the new answer, 6. After evaluating each expression with 6, we find the left and right expression both take on the value 10. This implies that 6 is the solution and our solution set would be $\{6\}$.

$2(y-1)$	$y+4$
$2(6-1)$	$6+4$
$2(5)$	10
10	

Please notice that I don't put an equality symbol between my expressions when I begin checking an answer. If I did, I'd be assuming the values are the same before I've shown the two expressions actually do have the same value. Here's some practice checking an answer.

Practice 1.8.2 Checking an Answer

Evaluate the supplied answer and state whether it is, or is not, a solution.

a) $x+2(x+1)=11$ with answer 4

$4+2(4+1)$	11	$\Rightarrow$	Substituted 4 for every *x* on the left side. The right side is already simplified.
$4+2(5)$ $4+10$ 14	11	$\Rightarrow$	Simplified the left side. Since the values are different, 4 is not a solution.

b) $-2y-1=-3(y+5)$ with answer -14

$-2(-14)-1$	$-3(-14+5)$	$\Rightarrow$	Substituted -14 for every *y*.
$28-1$ 27	$-3(-9)$ 27	$\Rightarrow$	Simplified each side paying particular attention to getting the signs right. Since the expressions have the same value, -14 is probably the solution.

c) $\frac{k}{2}-\frac{k}{3}=\frac{k-1}{4}$ with answer 3

$\frac{3}{2}-\frac{3}{3}$	$\frac{3-1}{4}$	$\Rightarrow$	Substituted 3 for every *k*.
$\frac{9}{6}-\frac{6}{6}$ $\frac{3}{6}$	$\frac{2}{4}$	$\Rightarrow$	Both expressions are equal to one-half so 3 is probably the solution.

Homework 1.8 Evaluate the supplied answer and state whether it is, or is not, a solution.	
1) $46 = 2L + 2(12)$ with answer 11	2) $2k - 5 = -9k - 27$ with answer -2
3) $-3(y+1) = 3(y+1)$ with answer 1	4) $\frac{2}{15} + \frac{a}{5} = \frac{1}{3}$ with answer 1
5) $\frac{x-2}{2} = \frac{x+4}{6}$ with answer 5	6) $3(x-1) + 2 = x + 4$ with answer 2
7) $\frac{y-2}{18} - \frac{y}{4} = \frac{-8}{9}$ with answer 4	8) $2(4-t) = 2t - (3t+4)$ with answer 12
9) $\frac{3+k}{9} - \frac{k}{6} = \frac{k}{36} + 2$ with answer -20	10) $-14 - 5(a+3) = -(1-2a)$ with answer -4
11) $-15 = -2(z+1) + 3(z-1)$ with answer -10	12) $x + \frac{x}{2} - \frac{1}{4} = \frac{x}{3} + \frac{1}{3}$ with answer -1

1.8.3 Applying Formulas - Temperature

Fahrenheit is the temperature scale named after physicist Daniel Fahrenheit. It's the official temperature scale for the United States. Celsius (centigrade) is the temperature scale named after astronomer Anders Celsius. It's the official scale for most of the rest of the world. The formula $F = {}^{9}\!/_{5}C + 32$ converts a Celsius temperature to its equivalent Fahrenheit temperature. The formula $C = \frac{5(F-32)}{9}$ converts a Fahrenheit temperature to its Celsius equivalent. Here's some practice using temperature formulas.

Practice 1.8.3 Applying Formulas – Temperature

Use the appropriate formula to answer the following questions.

a) Someone who's only familiar with Fahrenheit temperature is told a natural hot spring is 65° Celsius. Would it be a good idea to soak in the pool made by the hot spring?

$F = {}^{9}\!/_{5}C + 32$	$\Rightarrow$	Since I know the Celsius temperature, and I'm looking for the Fahrenheit temperature, I need the formula where I supply a value for *C* and simplify to find a value for *F*.
$F = {}^{9}\!/_{5}(65) + 32$	$\Rightarrow$	Substituted the given Celsius value into the formula.
$F = 149$	$\Rightarrow$	Simplified using the order of operations.
It's not a good idea since the temperature is too hot.	$\Rightarrow$	Answered the question in English. A temperature of 149° Fahrenheit can cause third degree burns in about 2 seconds.

b) Water freezes at 32° Fahrenheit. At what Celsius temperature does water freeze.

$C = \frac{5(F-32)}{9}$	$\Rightarrow$	Since I've been given a value for *F*, and need to find a value for *C*, I'll use the formula where I supply the Fahrenheit temperature and find the equivalent Celsius temperature.

$C = \frac{5(32-32)}{9}$	$\Rightarrow$ Substituted the Fahrenheit value into the formula.
$C = \frac{5(0)}{9} = 0$	$\Rightarrow$ Simplified.
Water freezes at 0° Celsius.	$\Rightarrow$ Made sure to answer the question using English.

Homework 1.8 Use the appropriate formula to answer the following questions.

13) Water boils at 212° F. Find the Celsius temperature at which water boils.

14) A passenger from Canada says they prefer the temperature in your car to be around 20 degrees. To accommodate their needs what Fahrenheit temperature should you use?

15) On a trip to Europe you plan to cook a dish for your relatives and you need to set the oven to 350° Fahrenheit. What Celsius temperature should you use?

16) 98.6° Fahrenheit is often used as the "normal" temperature for a person. What Celsius temperature would be considered normal?

17) At 48° Celsius it takes 5 minutes to develop a third-degree burn. At 60° Celsius it takes 5 seconds. Find the equivalent times in degrees Fahrenheit.

18) With a car, it's important to never remove a radiator cap while the engine is hot. An American automobile manufacturer currently has a sticker on the radiator warning that, when hot, the antifreeze is between 195° and 220° Fahrenheit. If they plan on selling the car in Mexico, what Celsius temperatures should the warning be changed to? Round your temperatures to the nearest whole degree.

1.8.4 Applying Formulas - Circles

Before algebra was developed, it was often difficult to exactly measure the distance around the boundary of a circle (the **circumference**). A common technique might have been to lay a rope along the boundary, straighten the rope out, and measure its straight-line length. Many cultures tried to find a way to measure the circumference of a circle using only a straight-line distance. One intuitive straight-line distance for circles is to measure from one edge to the opposite edge while going through the center. This distance is known as the **diameter**.

An interesting insight many cultures discovered was, for a circle of any size, if you used a quotient to compare the distance of the circumference to the distance of the diameter, the value was consistently close to 3. Using C for the distance of the circumference and D for the distance of the diameter, we can write a formula for this insight as $\frac{C}{D} \approx 3$. The wavy lines, $\approx$, imply the value is close, but not exact. We now know this constant value as the number pi, π, and this implies $\frac{C}{D} = \pi$. If you're interested in finding the circumference of a circle given the diameter, the quotient can be rewritten as $C = \pi D$. If you're interested in finding the diameter of a circle given the circumference, the quotient can be rewritten as $D = \frac{C}{\pi}$.

Many calculators have a button that gives a value for pi, but for this set of exercises let's all use the value 3.14 to approximate the value of pi, and, if necessary, round to the tenths place so our answers will be consistent.

Practice 1.8.4 Applying Formulas - Circles

Use the appropriate formula to answer the following questions.

a) A lake that is approximately circular has a diameter of 2.8 miles. Approximately how many miles of shore line does it have?

$C = \pi D$	$\Rightarrow$	Since I was given a diameter and asked to find a circumference (the shoreline is approximately the boundary of a circle), this is the proper formula.
$C = 3.14(2.8)$ $C = 8.8$	$\Rightarrow$	Evaluated the supplied value. Used 3.14 for pi and rounded the result to the tenths place.
The lake has approximately 8.8 miles of shoreline.	$\Rightarrow$	Answered the question.

Homework 1.8 Use the appropriate formula to answer the following questions.

19) On the tenth day of the tenth month some college students decide to build a crop circle with a diameter of $10 \times 10 = 100$ feet. What will the circumference of the crop circle be?

20) To train for a triathlon, a woman runs around a lake with a circumference of about 4 miles and then swims across the lake from shore to shore. If her swim takes her through the middle of the lake, about how far does she swim?

21) The diameter of the Earth is approximately 7,918 miles. How many miles of rope would you need to circle the Earth if the rope was lying on the ground? Round to the nearest mile.

22) A pizza restaurant decides to create a "super-sized" pizza with a 60-inch circumference. Currently their largest box is a square with 16 inch sides. Will their current box work for the super-sized pizza? If pizza boxes only come with side lengths that are natural numbers, and their current box doesn't work, what side length would work for their super-sized pizza?

1.8.5 Applying Formulas - Paying Off a Loan

The function $A = 6,500 - 250t$ predicts the amount you still owe on a \$6,500 loan, *A*, if you supply the number of months you've been paying, *t*. The function $t = 26 - 0.004A$ estimates how many months you've been paying on the loan, *t*, if you supply *A*, the amount still outstanding on the loan.

Homework 1.8 Use the appropriate function to answer the following questions.

23) How much will you still owe on the loan after paying for 1 year?

24) How long will it take to get the loan down to around \$1,000?

25) What was the original amount of the loan?

26) How long will it take to pay off the loan?

Homework 1.8 Answers

1) Both expressions simplify to 46 so 11 is a solution.

2) Both expressions simplify to -9 so -2 is a solution.

3) The left expression simplifies to -6 and the right expression to 6. 1 is not a solution.

4) Both expressions simplify to $5/15$ or $1/3$ so 1 is a solution.

5) Both expressions simplify to $9/6$ or $3/2$ so 5 is a solution.

6) The left expression simplifies to 5 and the right expression to 6. 2 is not a solution.

7) Both expressions simplify to $-32/36$ or $-8/9$ so 4 is a solution.

8) Both expressions simplify to -16 so 12 is a solution.

9) Both expressions simplify to $52/36$ or $13/9$ so -20 is a solution.

10) Both expressions simplify to -9 so -4 is a solution.

11) Both expressions simplify to -15 so -10 is a solution.

12) The left expression simplifies to $-7/4$ and the right expression to 0. -1 is not a solution.

13) $C = \frac{5(212-32)}{9} = 100$ Water boils at 100° Celsius.

14) $F = 9/5(20)+32 = 68$ You should set the temperature at 68° Fahrenheit.

15) $C = \frac{5(350-32)}{9} = 176.\bar{6}$ The oven should be set to 177° Celsius.

16) $C = \frac{5(98.6-32)}{9} = 37$ Normal temperature would be 37° Celsius.

17) $F = 9/5(48)+32 = 118.4$

$F = 9/5(60)+32 = 140$ At 140° F, it takes about 5 seconds to develop a third-degree burn.

18) $C = \frac{5(195-32)}{9} \approx 90.6$, $C = \frac{5(220-32)}{9} \approx 104.4$ The sticker should warn that the temperature is between 90 and 104 degrees Celsius.

19) $C = (3.14)(100) = 314$ The circumference of the crop circle will be 314 feet.

20) $D = 4/3.14 \approx 1.3$ She swims about 1.3 miles.

21) $C = (3.14)(7918) = 24{,}863$ I'd need about 24,863 miles of rope.

22) $D = 60/3.14 \approx 19.1$ They'll need a new square box that's 20 inches on a side.

23) $A = 6{,}500 - 250(12) = 3{,}500$ After paying on the loan for a year you'll still owe $3,500.

24) $t = 26 - 0.004(1{,}000) = 22$ It will take 22 months to reduce the loan amount to $1,000.

25) $A = 6{,}500 - 250(0) = 6{,}500$ At time 0 the original loan amount was $6,500.

26) $t = 26 - 0.004(0) = 26$, The loan amount will go to 0 in 26 months.

Section 1.9 Preparing for Linear Equations Part 1

Previously, I mentioned that in the next chapter, you'll be solving equations like $4(k-5)=-3(k+9)$ where the letter k implies that we'd like to find a single unknown number. In this section, and the next, we'll practice some of the vocabulary and skills you'll need when you solve equations.

1.9.1 Some Vocabulary for Algebraic Expressions

An expression like $3x$ is often called a **variable term.** A variable term is the product of a **coefficient** factor and a variable factor. With the variable term $3x$ the coefficient 3, tells us to multiply 3 to whatever value we choose for x. If we replace x with 2 then $3x$ has a value of 6. If we replace x with 4 then $3x$ has a value of 12.

If you don't "see" a coefficient, then the factor is a 1 or a -1. So x should be thought of as the product $1x$ and $-x$ should be thought of as the product $-1x$. Please notice that $-x$ implies opposite, not a negative value so $-x$ might be positive or negative depending on the value we use to replace x. If we replace x with positive four, $-(4)$, we get its opposite negative four, $-(4)=-1\times 4=-4$. On the other hand, if we replace x with negative four, $-(-4)$, we get its opposite positive four, $-(-4)=-1\times -4=4$. The misconception that $-x$ is necessarily a negative value, leads to all kinds of misunderstandings as students continue in algebra.

Terms whose value remains the same, regardless of the value we substitute for the variable, are known as **constant terms** or just constants. For example, the expression $3x+5$ has a constant term of 5.

When we discuss expressions with variable and constant terms, we'll always take the point of view that subtracting is adding the opposite. So, if you and I are discussing $5y-7$, we'll both agree the constant is -7 since we'll both see the difference $5y-7$ as the sum $5y+-7$.

When an expression has variable and constant terms, it's traditional to write the expression in **standard form**. For now, standard form means an expression isn't simplified until the variable term is to the left of the constant term. So, $3+4x$ is not in standard form while $4x+3$ is in standard form. Notice that with addition, changing the order of terms doesn't affect the sum. This isn't the case with subtraction though because which number is the minuend and which is the subtrahend usually does matter. So to write $-9-3n$ in standard form we first think of the difference as a sum, $-9+-3n$, then we reorder the terms, $-3n+-9$ and finally we rewrite adding a negative value as a subtraction, $-3n-9$. With a little practice people usually do these steps, "in their head".

Expressions like $4x+3$ and $-3n-9$ are examples of polynomials. Here's the definition of a polynomial.

***Definition* – Polynomial**
A polynomial in x is a single term, or a sum of terms, where each term is a variable term or a constant. Every variable term is a product where one factor is the coefficient and the other factor is a power of x where the exponent is a natural number. Example: $-3x+5$

Although it's fine to call $-3x$ a one-term polynomial and $-3x+5$ a two-term polynomial it's probably more common to call a one-term polynomial a **monomial** and a two-term polynomial a **binomial**. We'll discuss polynomials in more detail in chapter 5.

Practice 1.9.1 Some Vocabulary for Algebraic Expressions
Write the expression in standard form as a sum with explicit coefficients and describe the polynomial.

a) $2x-9$

$2x+-9$	$\Rightarrow$	Wrote the difference as a sum.
$2x+-9$	$\Rightarrow$	This binomial has a coefficient of 2 and a constant of -9.

b) $-h$

$-1h$	$\Rightarrow$	Made the coefficient explicit. This monomial has a coefficient of -1 and no constant term (or, if you'd prefer, a constant term of 0).

c) $3-y$

$3-1y$ $3+-1y$ $-1y+3$	$\Rightarrow$	Made the coefficient of y explicit, wrote the subtraction as addition of the opposite and reordered the terms into standard form.
$-1y+3$	$\Rightarrow$	The binomial has the coefficient -1 and the constant 3.

Homework 1.9 Write the expression in standard form as a sum with explicit coefficients and describe the polynomial.

1) $3+4k$ 2) $x-1$ 3) $-h-3$ 4) $-1+7c$ 5) $6-2a$ 6) $-k$

1.9.2 Beginning to Add and Subtract Polynomials

To add and subtract polynomials we first have to identify the "like" terms. For now, variable terms are like if they have the same letter. For example, $2x$, $-7x$, and x are all like terms since they all have the variable x. Notice the coefficients have nothing to do with whether terms are like. Constants are also considered like terms.

A quick way to add and subtract polynomials is to first write the polynomial as a sum with explicit coefficients, then add and subtract like variable terms left to right, and finally add and subtract constant terms left to right. By starting with the variable terms you'll automatically be writing the polynomial in standard form.

It's best to simplify the polynomial in a single line. For example, to simplify $3x+15+2x+11$ I'd first sum $3x$ and $2x$ to get $5x$. Then I'd cross out the variable terms and sum the constants to 26. Because I started with the variable terms the polynomial is already in standard form. Here's some practice.

$3x+15+2x+11$
$5x$

$\cancel{3x}+15+\cancel{2x}+11$
$5x+26$

Practice 1.9.2 Beginning to Add and Subtract Polynomials
Simplify.

a) $3y+5+y-2$

$3y+5+1y+-2$	$\Rightarrow$	Wrote the expression as a sum and made the coefficients explicit.
$4y$	$\Rightarrow$	Began by adding the variable terms left to right, $3y+1y=4y$.
$\cancel{3y}+5+\cancel{1y}+-2$ $4y+3$	$\Rightarrow$	After crossing out the variable terms I simplified the constants left to right.

b) $3-k+4-2k-7$

$3+-1k+4+-2k+-7$	$\Rightarrow$	Wrote the expression as a sum and made the coefficients explicit.
$-3k$	$\Rightarrow$	Simplified variable terms first.
$-3k$	$\Rightarrow$	The constants summed to 0.

c) $-y+2y-15-(-2y)+11-(-6)$

$-1y+2y+-15+2y+11+6$	$\Rightarrow$	Wrote the expression as a sum and made the coefficients explicit.
$3y-2$	$\Rightarrow$	Simplified the variable terms and then the constants.

Homework 1.9 Simplify.

7) $3y+5+y-2$ 8) $5h+4-9h+1$ 9) $x-4x-2x+7-x$ 10) $4a-2+a-12a$

11) $17-2r+11r+5$ 12) $2a-(-2a)-8-2$ 13) $2y-(-5y)+4y-y-9y$

14) $8-k-4+7-(-2k)$ 15) $-5-y+2y-2-(-5)$ 16) $3x+9-5x-12-x$

17) $2r+5r+4r-r-9r$ 18) $-17t-2-(-11t)-15+5t$ 19) $4-a-16-a-6-(-2a)$

20) $-7k-k-k+4-(-2k)$ 21) $-14-(-y)+y-(-3)+11$

22) $4p-11+6p+4-10p-9$ 23) $7-40q+11q-(-4)-12q-(-29q)$

24) $-(-4)-y-12+15y-11y$ 25) $-10+12w-4w+6w-8w-(-10)$

26) $6k-4k-3k+9-11-4$ 27) $7-(-8)-(-8w)-5w+6-2w$

28) $-17p-40p+11p+12p-(-5p)+29p$

Homework 1.9 Answers

1) $4k+3$ The binomial has a coefficient of 4 and the constant 3.

2) $1x+-1$ The binomial has a coefficient of 1 and the constant -1.

3) $-1h+-3$ The binomial has a coefficient of -1 and the constant -3.

4) $7c+-1$ The binomial has a coefficient of 7 and the constant -1.

5) $-2a+6$ The binomial has a coefficient of -2 and the constant 6.

6) $-1k$ The monomial has a coefficient -1 and the constant 0.

7) $4y+3$	8) $-4h+5$	9) $-6x+7$	10) $-7a-2$	11) $9r+22$
12) $4a-10$	13) y	14) $k+11$	15) $y-2$	16) $-3x-3$
17) r	18) $-t-17$	19) -18	20) $-7k+4$	21) $2y$
22) -16	23) $-12q+11$	24) $3y-8$	25) $6w$	26) $-k-6$
27) $w+21$	28) 0			

Section 1.10 Preparing for Linear Equations – Part 2

In this section we'll practice adding, subtracting and multiplying polynomials.

1.10.1 Some Vocabulary for Distribution

Earlier we used the distributive property to replace a sum with a product. Today, we'll use the distributive property the other way and replace a product with a sum.

Property **– The Distributive Property of Multiplication over Addition**
The product $a(b+c)$ can be replaced by the sum $ab+ac$.
Example: $3(x+6)=3x+3(6)$

We'll use the distributive property to simplify an expression like $3(x+6)$, where the sum inside the parentheses, the $x+6$, can't be simplified since the terms aren't like. Even though we can't add inside the parentheses, we can still use the distributive property to replace the product $3(x+6)$ with the sum $3x+3(6)$ which simplifies to $3x+18$. This replacement is often called "distributing" the three. Here's some practice with the vocabulary you'll need to discuss distribution.

Practice 1.10.1 Some Vocabulary for Distribution
Fill in the blanks using the words term, sum, factor, product, minuend, subtrahend, difference, dividend, divisor or quotient.

a) Before simplifying $3(x+4)$;
4 is a____, $x+4$ is a____, $(x+4)$ is a____, 3 is a____, and $3(x+4)$ is a____.

term, sum, factor, factor, product $\Rightarrow$ Since 4 is being added to x, 4 is a term and $x+4$ is a sum. Because of the parentheses $(x+4)$ is being multiplied to 3 so both $(x+4)$ and 3 are factors. If x where a value, then the last operation done when simplifying would be multiplication, which makes $3(x+4)$ a product.

Homework 1.10 *Fill in the blanks using the words term, sum, factor, product, minuend, subtrahend, difference, dividend, divisor or quotient*

1) Given $-1(2x-5)$;

2 is a____, x is a____, $2x$ is a____ and the____, 5 is the____, $2x-5$ is a____, $(2x-5)$ is a____, -1 is a____ and $-1(2x-5)$ is a____.

2) Given $9(y+5)-40$;

y is a____, 5 is a____, $y+5$ is a____, $(y+5)$ is a____, 9 is a____, $9(y+5)$ is both a____ and the____, 40 is the____ and $9(y+5)-40$ is a____.

3) Given $3(2+k)+5(k+4)$;

2 is a____, k is a____, $2+k$ is a____, $(2+k)$ is a____, 3 is a____, $3(2+k)$ is a____ and a____, k is a____, 4 is a____, $k+4$ is a____, 5 is a____, $(k+4)$ is a____, $5(k+4)$ is a____ and a____ and $3(2+k)+5(k+4)$ is a____.

1.10.2 Multiplying Using the Distributive Property

Here's some practice where we distribute the common factor and then use our skills from the last section to add and subtract like terms.

Practice 1.10.2 Multiplying Using the Distributive Property.
Simplify. Write all answers in standard form.

a) $3(x+6)$

$3(x)+3(6)$ ⇒	Distributed the factor of 3. Usually the first product is written without the parentheses as $3x$.
$3x+18$ ⇒	Simplified.

b) $5(y-4)+8y+12$

$5y-20+8y+12$ ⇒	Distributed the factor of 5 to both terms inside the parentheses. Made sure the 5 didn't multiply beyond the parentheses to either the $8y$ or the 12.
$13y-8$ ⇒	Collected like terms and wrote in standard form.

c) $5(y-4+8y+12)$

$5(9y+8)$ ⇒	Followed the order of operations and simplified inside parentheses first.
$45y+40$ ⇒	Distributed the common factor of 5.

d) $-3(-4r+1)+-2(r+7)$

$12r-3+-2r-14$ ⇒	Distributed with both terms $12r+-3+-2r+-14$, and wrote adding a negative as subtraction.
$10r-17$ ⇒	Simplified.

Homework 1.10 Simplify. Write all answers in standard form.

4) $3(2x+3)+5x$ 5) $3(2w)+6+3(2w+6)$ 6) $2(4+z)+7(z-2)$

7) $4(x-4+3x-2)$ 8) $12m+9(m-4)-30$ 9) $-7k+3(2-k)+5(k-3)$

10) $9(3t+5-11t)+12(4+6t+1)$ 11) $-2(k+3)-2(3k)+4(k+3)-3(2)$

12) $-15x+5(x+7)+7(x-5)$ 13) $-1(-q+4)+-2(4q+1)$

14) $-3(2y+8)+-9(-2y+1)$ 15) $-2(-9k+5)+-7(-2k+1)$

16) $-2(a+2)+6(2a-11)+-3(-4a+2)$ 17) $8(-2y-3)+-6(-2y+3)+-3(6y+4)$

1.10.3 Simplifying Subtraction as Adding the Opposite

When simplifying an expression like $2p-3(p-12)$ students often ask, "Is that a negative three or a subtract three?". As I mentioned earlier, the answer depends on the point of view you wish to take. Currently, it's a subtraction, but if you decide to think of the subtraction as adding the opposite, $2p+-3(p-12)$, it becomes a factor of negative three. In practice, most people take the point of view they're adding the opposite. Whether you change the subtraction in your mind, or on paper, I'd suggest changing subtractions to adding the opposite before distributing.

Practice 1.10.3 Simplifying Subtraction as Adding the Opposite

Simplify by adding the opposite.

a) $2p-3(p-12)$

$2p+-3(p+-12)$	$\Rightarrow$	Changed all subtractions to adding the opposite.
$2p+-3p+(-3)(-12)$ $2p+-3p+36$	$\Rightarrow$	Distributed and multiplied.
$-p+36$	$\Rightarrow$	Combined like terms.

b) $-6(2b-3)-4(2-b)$

$-6(2b+-3)+-4(2+-b)$	$\Rightarrow$	Changed all subtractions to adding the opposite.
$-12b+18+-8+4b$ $-8b+10$	$\Rightarrow$	Distributed and combined like terms.

c) $-8(k-4)-(k-2)$

$-8(k+-4)+-1(k+-2)$	$\Rightarrow$	Included a factor of 1 when rewriting the second difference.
$-8k+32+-k+2$ $-9k+34$	$\Rightarrow$	Distributed and combined like terms.

Homework 1.10 Simplify by adding the opposite

18) $5(-2z+8)-(2-z)$

19) $15x-3(3-x)-3x$

20) $-2(3x)-2(3-x)$

21) $4x-3x-(4-x-2x)+4$

22) $6(y+1)-3(y-4)-(-2y-3)$

23) $3a-(3-12a)-2+a$

24) $-8x-(2-x)-5(2x-9)$

25) $-3(2k)-3[-2-(k-2)]$

26) $-6(x-1)-(x-6)-(6-x+6)$

27) $x-[x-5(5-x)-5x]$

Homework 1.10 Answers

1) factor, factor, product, minuend, subtrahend, difference, factor, factor, product

2) term, term, sum, factor, factor, product, minuend, subtrahend, difference

3) term, term, sum, factor, factor, product, term, term, term, sum, factor, factor, product, term, sum

4) $11x+9$ 5) $12w+24$ 6) $9z-6$ 7) $16x-24$ 8) $21m-66$

9) $-5k-9$ 10) 105 11) $-4k$ 12) $-3x$ 13) $-7q-6$

14) $12y-33$ 15) $32k-17$ 16) $22a-76$ 17) $-22y-54$ 18) $-9z+38$

19) $15x-9$ 20) $-4x-6$ 21) $4x$ 22) $5y+21$ 23) $16a-5$ 24) $-17x+43$

25) $-3k$ 26) $-6x$ 27) 25

Chapter 2

Linear Equations and Inequalities

2.1 An Introduction to Linear Equations

In this chapter, we'll practice solving, and then applying, **one-variable linear equations**. The words, "one-variable", tell us that any variable terms in our equation will have the same letter. The word "linear" tells us that the exponent on any variable will be a 1. We'll discuss exponents in detail in a later chapter. In practice, people tend not to include the, one-variable, and just use linear equation. An important point about linear equations is that, if there is a solution, there is usually only one.

2.1.1 The Idea of an Equivalent Equation

Sometimes, you can look at an equation and see the solution. For example, if you look at $x+3=8$ you should be able to see the solution is 5. On the other hand, you probably can't see that 5 is also the solution for $4(x+3)=3x+17$. The equations $x+3=8$ and $4(x+3)=3x+17$ are **equivalent equations** because they have the same solution set, $\{5\}$.

In this chapter, we'll discuss how to transform an equation like $4(x+3)=3x+17$, where it's difficult to see the solution, into a simpler but equivalent equation like $x=5$, where it's easy to see the solution. Be aware that it's common for people to stop at $x=5$ and say they've solved the equation assuming anyone can see that the solution set would be $\{5\}$.

We'll use two strategies to create equivalent equations. First, we'll use your skills from chapter 1. Second, we'll use the property of equality.

Property **– The Property of Equality**
If two expressions are equal, then operating on both in the same way results in two expressions that are equal.

The property of equality is asking us to view an equation like $2x+1=5$ as two expressions, both of which have the value 5. The right expression is obviously 5 while the left expression will be 5 as soon as we replace *x* with the solution value (which in this case is 2). The property of equality says that since both the left and right expression have the value 5, subtracting 1 from each $2x+1-1=5-1$, and simplifying $2x=4$, might change the way the expressions look, but they'll remain equal. Most importantly, the resulting equation will still have the same solution set. (Notice 2 is still the solution for $2x=4$.)

Before moving on I want to make a couple of important points about the property of equality. First, we're not allowed to divide both expressions by 0. Recall that dividing by 0 results in an expression that's undefined. Second, it isn't helpful to multiply the left expression and the right expression by 0 since the result isn't an equivalent equation. For example, if we start with $x=5$, which has the solution set $\{5\}$, and multiply both expressions by 0, $0(x)=0(5)$ then, by the zero-product rule, we'll get $0=0$ which clearly doesn't have the solution set $\{5\}$.

2.1.2 *A Rational for the General Procedure*

When I look at the equation $x=-12$, the solution, -12, just pops into my head. If I use the property of equality to make a small change, like adding 8 to both expressions, then, after simplifying on the right side, an answer doesn't "pop" as easily. With a little thought though, I can see that -12 is still the solution to $x+8=-4$.

$$\begin{aligned} x &= -12 \\ x+8 &= -12+8 \\ x+8 &= -4 \end{aligned}$$

As I increase the number of operators affecting *x*, the solution gets harder and harder to "see". For example, if I multiplied both expressions by 3 and simplified on the right, and then subtracted 10 from both expressions and again simplified on the right it takes me significantly longer to "see" that the solution to $3(x+8)-10=-22$ is still -12.

$$\begin{aligned} 3(x+8) &= (3)(-4) \\ 3(x+8) &= -12 \\ 3(x+8)-10 &= -12-10 \\ 3(x+8)-10 &= -22 \end{aligned}$$

The remedy for this increase in operations would be to "undo" the subtraction of 10, the multiplication by 3 and the addition of 8 to get back to the original equation, $x=-12$. The purpose of the general procedure is to make sure we undo operations in the right order.

2.1.3 *The General Procedure*

The general procedure helps us reduce the number of operations affecting our variable in a way that keeps the resulting equations equivalent.

***Procedure* – Solving One-Variable Linear Equations**
1. If fractions are present multiply every term by the LCD of all the terms and reduce (if possible). 2. Simplify the left expression. 3. Simplify the right expression. 4. If both expressions have a variable term compare the coefficients and use addition (or subtraction) to isolate the variable term on the side that had the **greater** coefficient. 5. If both expressions have a constant term, use addition (or subtraction) to isolate the constant term on the side that's opposite the variable term. 6. If the coefficient isn't a 1, use division to reduce the coefficient to 1. 7. If an answer results from steps 1 – 6, check to make sure it's a solution.

Steps 1 through 3 help us reduce operators by individually simplifying the left expression and the right expression. We'll leave steps 1 through 3 for later. Steps 4 – 6 help us reduce operations by building terms of 0 and a factor of 1. Steps 4 – 6 will be the focus of this lesson. Step 7 doesn't help us solve the equation, instead, it improves the chances the answer we found using steps 1 – 6 is actually the solution. Let's begin with steps 4 and 5 of the general procedure which are both based on the **additive identity**.

***Property* – The Additive Identity**
0 is the additive identity. The sum of an expression and 0 can be replaced by the expression itself. Example: $a+0=a$

2.1.4 Step 5 of the General Procedure

If I look at the equation $x+8=-4$ the solution doesn't instantly pop into my head because 8 is being added to the variable. A common strategy to "undo" the addition of 8 is to use the property of equality and subtract 8 from both expressions. Since adding 8 and subtracting 8 are inverse operations the sum on the left will go from $x+8$ to $x+0$. Then, using the additive identity, I'm able to replace $x+0$ with just x. The steps are worked out to the right.

$$x+8=-4$$
$$x+8\mathbf{-8}=-4\mathbf{-8}$$
$$x+0=-12$$
$$x=-12$$

Please notice that solving equations has nothing to do with "things" being "moved" from one side to the other. There is no "moving" property in algebra. Instead, solving equations is all about working with the equality of two individual expressions.

Practice 2.1.4 Step 5 of the General Procedure

Solve by building a sum with a term of 0. Check your answer.

a) $y+10=4$		
$y+10\mathbf{-10}=4\mathbf{-10}$	$\Rightarrow$	Since I want to "undo" an addition of 10, I'll use the property of equality to subtract 10 from both expressions.
$y+0=-6$ $y=-6$	$\Rightarrow$	On the left, the inverse operations sum to 0. On the right, I simplified the integers. Next, using the additive identity, I can replace $y+0$ with y. My answer is -6.
$y+10$ 4 $-6+10$ 4	$\Rightarrow$	Checking the answer shows both expressions have the value 4 so I'm confident the solution set is $\{-6\}$.
b) $-11=p-9$		
$-11\mathbf{+9}=p-9\mathbf{+9}$	$\Rightarrow$	To undo the subtraction of 9, I used the property of equality and added 9 to both expressions.
$-2=p+0$ $-2=p$	$\Rightarrow$	After simplifying, and using the additive identity, a good answer would be -2.
-11 $p-9$ $-2-9$ -11	$\Rightarrow$	Checking the answer shows both expressions have the value -11 so I'm confident the solution set is $\{-2\}$.

Homework 2.1 Solve by building a sum with a term of 0. Check your answer.

1) $x-5=17$	2) $-9=m-3$	3) $-15+y=-4$	4) $-5=y+2$	5) $12=-4+k$

2.1.5 Step 4 of the General Procedure

Now let's use step 4 to help with equations that have variable terms in both expressions. One reason it's difficult to see the solution for $4x = 3x + 12$ is that both the left expression and the right expression have a variable term. It would be easier to see the solution if we just had one variable term. Using step 4 to build a sum of zero helps make this happen.

To start the process, I'd first decide whether I'd like to have the variable term on the left side or the right side. There isn't a "right" answer to this question, but my preference is to keep the variable term on the side with the greater coefficient. Since 4 is greater than 3, I'd prefer to have my variable term on the left. I now switch my attention to the right side and add or subtract in a way that will make the variable terms sum to 0. Since $3x$ is currently being added on the right, I can force a sum of 0 by also subtracting $3x$ on the right. Of course, by the property of equality, I'll also need to subtract $3x$ from the left side. After simplifying, and using the additive identity, I have a single variable term on the left and it's easy to see the answer is 12. Checking the answer shows that 12 is the solution, $\begin{array}{cc} 4(12) & 3(12)+12 \\ 48 & 36+12 \\ & 48 \end{array}$.

$$4x = 3x + 12$$
$$4x - \mathbf{3x} = 3x + 12 - \mathbf{3x}$$
$$x = 0 + 12$$
$$x = 12$$

Practice 2.1.5 Step 4 of the General Procedure

Solve by building a sum with a term of 0. Check your answer.

a) $6k = 5k - 4$

$6k = 5k - 4$	$\Rightarrow$	My variable term will be on the left since 6 is greater than 5.
$6k - \mathbf{5k} = 5k - 4 - \mathbf{5k}$	$\Rightarrow$	Isolating the variable term on the left means summing the variable terms on the right to 0. Used the property of equality and subtracted $5k$ from both the left expression and the right expression.
$k = -4$	$\Rightarrow$	After simplifying, the single variable term is isolated on the left. My answer is -4.
$\begin{array}{cc} 6(-4) & 5(-4)-4 \\ -24 & -20-4 \\ & -24 \end{array}$	$\Rightarrow$	Returned to the original equation and checked the answer. The check shows -4 is the solution. The solution set would be $\{-4\}$.

b) $15 + 3y = 4y$

$15 + 3y = 4y$	$\Rightarrow$	This time I'll isolate the variable term on the right since 4 is greater than 3.
$15 + 3y - \mathbf{3y} = 4y - \mathbf{3y}$ $15 = y$	$\Rightarrow$	To sum the variable terms on the left to 0, I subtracted $3y$ from both expressions. After simplifying, the single variable term is on the right. My answer is 15.

$15+3(15)$ $4(15)$ $15+45$ 60 60	$\Rightarrow$ Returned to the original equation and checked the answer. The check shows 15 is the solution. The solution set would be $\{15\}$.
c) $-7x=-8x-1$	
$-7x=-8x-1$	$\Rightarrow$ This time I'll isolate the variable term on the left since -7 is greater than -8.
$-7x+\mathbf{8x}=-8x-1+\mathbf{8x}$ $x=-1$	$\Rightarrow$ To sum the variable terms on the right to 0, I added $8x$ to both expressions. After simplifying, the single variable term is on the left. My answer is -1.
$-7(-1)$ $-8(-1)-1$ 7 $8-1$ 7	$\Rightarrow$ Returned to the original equation and checked the answer. The check shows -1 is the solution. The solution set would be $\{-1\}$.

Homework 2.1 Solve by building a sum with a term of 0. Check your answer.

6) $x-5=2x$ 7) $-2w=-3w+6$ 8) $-15+3y=4y$ 9) $-2k-5=-k$ 10) $-5t=-6t+5$

2.1.6 Combining Steps 4 and 5

Now let's practice combining steps 4 and 5. First, we'll use step 4 and isolate the variable term on the side with the larger coefficient. Then, we'll use step 5 to isolate the constant on the opposite side. Please realize that mathematically, there are many "right" ways to solve these equations. I'm showing you the way that, in general, I believe is the easiest.

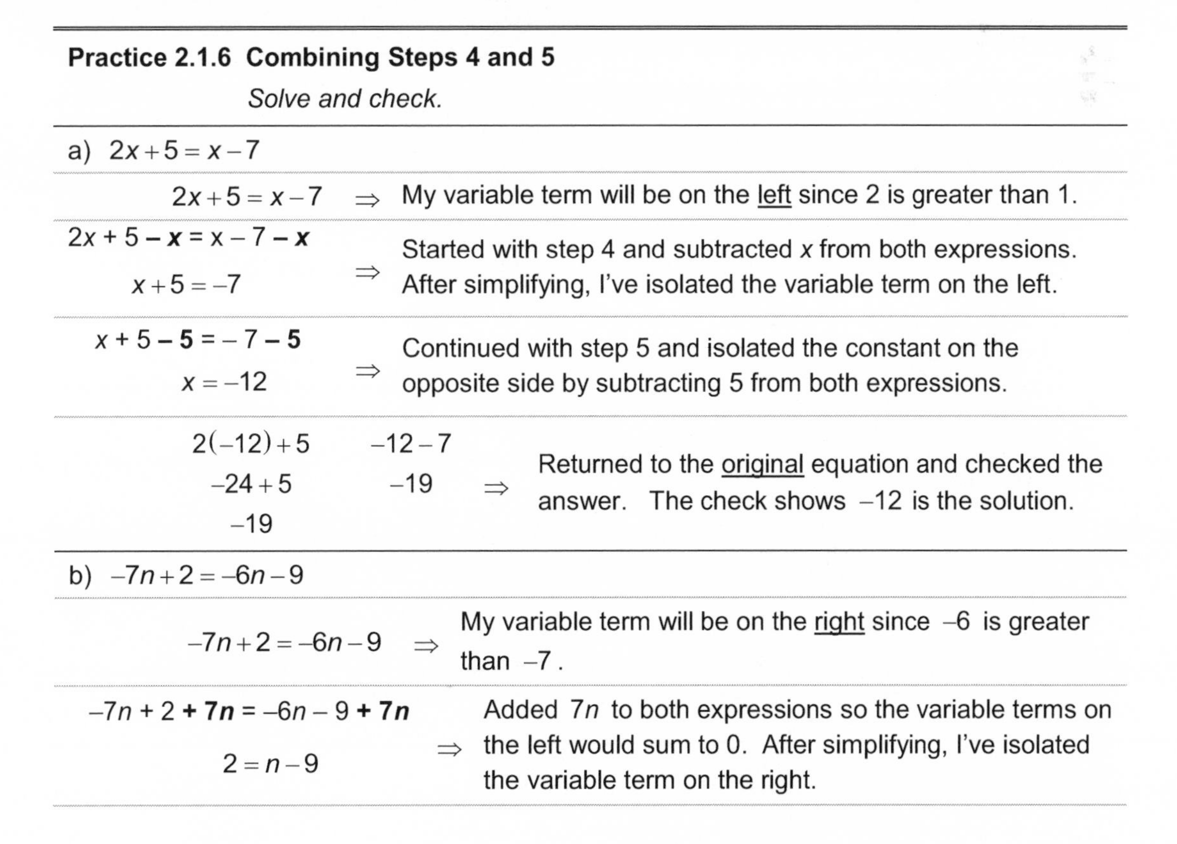

Practice 2.1.6 Combining Steps 4 and 5

Solve and check.

a) $2x+5=x-7$	
$2x+5=x-7$	$\Rightarrow$ My variable term will be on the left since 2 is greater than 1.
$2x+5-\mathbf{x}=x-7-\mathbf{x}$ $x+5=-7$	$\Rightarrow$ Started with step 4 and subtracted x from both expressions. After simplifying, I've isolated the variable term on the left.
$x+5-\mathbf{5}=-7-\mathbf{5}$ $x=-12$	$\Rightarrow$ Continued with step 5 and isolated the constant on the opposite side by subtracting 5 from both expressions.
$2(-12)+5$ $-12-7$ $-24+5$ -19 -19	$\Rightarrow$ Returned to the original equation and checked the answer. The check shows -12 is the solution.
b) $-7n+2=-6n-9$	
$-7n+2=-6n-9$	$\Rightarrow$ My variable term will be on the right since -6 is greater than -7.
$-7n+2+\mathbf{7n}=-6n-9+\mathbf{7n}$ $2=n-9$	$\Rightarrow$ Added $7n$ to both expressions so the variable terms on the left would sum to 0. After simplifying, I've isolated the variable term on the right.

$2 + \mathbf{9} = n - 9 + \mathbf{9}$ $11 = n$	$\Rightarrow$ Added 9 to both expressions. After simplifying, the constant is isolated on the side opposite the variable term.
$-7(11)+2$ $-6(11)-9$ $-77+2$ $-66-9$ -75 -75	$\Rightarrow$ Returned to the original equation and checked the answer. The check shows 11 is the solution.

c) $14-17a=-18a-12$

$14+-17a=-18a-12$	$\Rightarrow$ Thought of the left side as a sum to help realize that the variable term will be on the left since -17 is greater than -18.
$14 - 17a + \mathbf{18a} = -18a - 12 + \mathbf{18a}$ $14+a=-12$	$\Rightarrow$ Added $18a$ to both expressions and simplified. The variable term is now isolated on the left.
$14 + a - \mathbf{14} = -12 - \mathbf{14}$ $a=-26$	$\Rightarrow$ Subtracted 14 from both expressions and simplified so the constant terms on the left would sum to 0. The constant term is now isolated on the side opposite the variable term.
$14-17(-26)$ $-18(-26)-12$ $14-(-442)$ $468-12$ $14+442$ 456 456	$\Rightarrow$ Returned to the original equation and checked the answer. The check shows . . is the solution. Noticed how important the work with integer signs was when simplifying both expressions.

Homework 2.1 Solve and check.

11) $3x-5=2x+17$	12) $5x+7=6x-3$	13) $15+7y=-1+6y$
14) $2k+18=3k+18$	15) $-9+x=2x-3$	16) $-7m-13=-12-8m$
17) $11-15h=-14h+4$	18) $-5-11a=14-12a$	

2.1.7 Step 6 of the General Procedure

Steps 4 and 5 involved working with sums and using opposites to build a term of 0. This allowed us to use the additive identity and replace an expression like $x+0$ with x. Step 6 involves working with products so our strategy will be to build a product where one of our factors is 1. This will allow us to use the multiplicative identity to reduce operators by replacing the product $1x$ with x.

***Property* – The Multiplicative Identity**
1 is the multiplicative identity. The product of 1 and another factor can be replaced by the other factor. Example: $1a=a$

If we look at the equation $2k = -40$, the factor of 2 on the left makes it a little difficult to see what the value of k should be. Notice we can't "solve for" k using adding or subtracting because 2 is being multiplied to the variable and adding or subtracting doesn't "undo" multiplication. On the other hand, multiplying and dividing are inverse operations so we'll use division to isolate k.

The strategy is to use the property of equality and divide both expressions by 2 (the coefficient of the variable term). Reducing the common factors of 2 on the left leaves $1k$ and simplifying on the right gives -20. Now, on the left, the multiplicative identity allows us to replace $1k$ with k and we've isolated the variable.

$$2k = -40$$
$$\frac{\cancel{2}k}{\cancel{2}} = \frac{-40}{2}$$
$$1k = -20$$
$$k = -20$$

Practice 2.1.7 Step 6 of the General Procedure

Solve and check.

a) $-2y = 14$

$\frac{-2y}{-2} = \frac{14}{-2}$	$\Rightarrow$	The coefficient of the variable term is -2 so I used the property of equality and divided both expressions by -2.
$\frac{\cancel{-2}y}{\cancel{-2}} = \frac{14}{-2}$ $1y = -7$	$\Rightarrow$	Reduced the coefficient to positive one on the left and simplified on the right.
$y = -7$	$\Rightarrow$	Used the multiplicative identity to replace $1y$ with y.
$2(-7) \quad -14$ -14	$\Rightarrow$	Checked the answer. The solution set is $\{-7\}$.

b) $3 = -k$

$\frac{3}{-1} = \frac{(-1)k}{-1}$	$\Rightarrow$	Thought of $-k$ as $-1 \times k$ and divided both expressions by the coefficient of the variable.
$-3 = 1k$ $-3 = k$	$\Rightarrow$	Reduced the common factors of -1 on the right and simplified on the left. Used the multiplicative identity to replace $1k$ with k.
$3 \quad -1(-3)$ 3	$\Rightarrow$	Checked the answer. The solution is -3.

Homework 2.1 Solve and check.

19) $2a = -24$ 20) $27 = 9p$ 21) $-8v = -16$ 22) $30 = -5h$ 23) $40 = -k$

2.1.8 Combining Steps 4 – 6

Now let's practice putting steps four through 6 together.

Practice 2.1.8 Combining Steps 4 – 6

Solve and check.

a) $x+6=4x-18$

$x+6-x=4x-18-x$

$6=3x-18$

⇒ Started with step 4 and decided to isolate the variable term on the right since 4 is greater than 1.

$6+18=3x-18+18$

$24=3x$

⇒ Moved to step 5 and added 18 to both expressions. Isolated the constant term on the side opposite the variable term.

$\frac{24}{3}=\frac{3x}{3}$

$8=1x$

$8=x$

⇒ Moved to step 6 and used the property of equality to divide both expressions by 3. After simplifying on the left and reducing on the right I'm able to use the multiplicative identity on the right to replace $1x$ with x. My answer is 8.

$8+6$	$4(8)-18$
14	$32-18$
	14

⇒ Checking the answer confirms the solution set is $\{8\}$.

b) $4+11h=-22-2h$

$4+11h+2h=-22-2h+2h$

$4+13h=-22$

⇒ Started with step 4 and decided to have a single variable term on the left since 11 is greater than -2. Added $2h$ to both expressions and simplified.

$4+13h-4=-22-4$

$13h=-26$

⇒ Moved to step 5 and subtracted 4 from both expressions to isolate the constant term on the side opposite the variable term.

$\frac{13h}{13}=\frac{-26}{13}$

$h=-2$

⇒ Moved to step 6 and used division, and the multiplicative identity, to make it easy to "see" an answer.

$4+11(-2)$	$-22-2(-2)$
$4+-22$	$-22+4$
-18	-18

⇒ Checking the answer confirms that -2 is the solution.

Homework 2.1 Solve and check.

24) $3x-5=x+5$	25) $t-4=6t+16$	26) $2y+5=-3y+20$
27) $7+11p=4p-21$	28) $-9-7m=m+31$	29) $8y-25=3y-25$
30) $14n-14=7n+14$	31) $-1-9k=-3-11k$	32) $-12y+25=12y+1$
33) $3x-15=-2x+20$	34) $-7-m=m-3$	35) $2y-4=-4+3y$
36) $-B+12=3-4B$		

Homework 2.1 Answers

1) {22}	2) {–6}	3) {11}	4) {–7}	5) {16}	6) {–5}
7) {6}	8) {–15}	9) {–5}	10) {5}	11) {22}	12) {10}
13) {–16}	14) {0}	15) {–6}	16) {1}	17) {7}	18) {19}
19) {–12}	20) {3}	21) {2}	22) {–6}	23) {–40}	24) {5}
25) {–4}	26) {3}	27) {–4}	28) {–5}	29) {0}	30) {4}
31) {–1}	32) {1}	33) {7}	34) {–2}	35) {0}	36) {–3}

2.2 Continuing with Linear Equations

In the last section, we practiced decreasing the number of operators using steps 4 – 6 of the general procedure. In this section, we'll practice decreasing the number of operators using steps 1 – 3. Steps 2 and 3 reduce the number of operators by simplifying the left expression and the right expression individually. Here's some practice.

2.2.1 Steps 2 and 3 of the General Procedure

In steps 2 and 3 we use the techniques you practiced in chapter 1 to simplify the left expression and the right expression individually. Correctly simplifying the left and right expression will always result in an equivalent equation.

Practice 2.2.1 Steps 2 and 3 of the General Procedure

Solve and check.

a) $2(y-5)+6y=-34$	
$2y-10+6y=-34$ $8y-10=-34$	$\Rightarrow$ Started with step 2 and simplified the left expression.
$8y-10=-34$	$\Rightarrow$ Moved to step 3. The right expression is already simplified.
$8y-10=-34$	$\Rightarrow$ Moved to step 4. The variable term is already isolated.
$8y-10+10=-34+10$ $8y=-24$	$\Rightarrow$ Moved to step 5 and used addition to isolate the constant term on the side opposite the variable term.
$\frac{8y}{8}=\frac{-24}{8}$ $y=-3$	$\Rightarrow$ Moved to step 6, divided both expressions by 8 and simplified to isolate the variable. The answer is -3.
$2(-3-5)+6(-3) \quad -34$ $2(-8)-18$ -34	$\Rightarrow$ Moved to step 7 and evaluated the answer using the <u>original</u> equation. Since both expressions simplify to -34, -3 is the solution.
b) $3y+7-4y=3y+4-5$	
$7-y=3y+4-5$	$\Rightarrow$ Began with step 2 and simplified the left side
$7-y=3y-1$	$\Rightarrow$ Moved to step 3 and simplified the right side
$7-y+y=3y-1+y$ $7=4y-1$	$\Rightarrow$ Moved to step 4 and isolated the variable term on the right side (since 3 is greater than -1).
$7+1=4y-1+1$ $8=4y$	$\Rightarrow$ Moved to step 5 and isolated the constant term on the side opposite the variable term.
$\frac{8}{4}=\frac{4y}{4}$ $2=y$	$\Rightarrow$ Moved to step 6 and divided both expressions by 4. My answer is 2.

$3(2)+7-4(2)$ $\quad 3(2)+4-5$ $6+7-8$ $\quad 6+4-5$ 5 $\quad 5$	$\Rightarrow$	Moved to step 7 and evaluated 2 in the <u>original</u> equation. Both expressions simplified to 5 so I'm confident the solution set is $\{2\}$.
c) $2(5+x)=-3(x+6)+x$		
$10+2x=-3x-18+x$ $10+2x=-2x-18$	$\Rightarrow$	Began with steps 2 and 3 and simplified each expression.
$10+2x+2x=-2x-18+2x$ $10+4x=-18$ $10+4x-10=-18-10$ $4x=-28$	$\Rightarrow$	Moved to steps 4 and 5. Isolated the variable term on the left and then isolated the constant term on the right (the side opposite the variable term).
$\frac{4x}{4}=\frac{-28}{4}$ $1x=-7$ $x=-7$	$\Rightarrow$	Moved to step 6 and divided both expressions by 4. The answer is -7.
$2(5+-7)$ $\quad -3(-7+6)+-7$ $2(-2)$ $\quad -3(-1)+-7$ -4 $\quad -4$	$\Rightarrow$	Returned to the original equation and evaluated the answer. Since both expressions simplify to -4 the solution set is $\{-7\}$.

Homework 2.2 Solve and check.

1) $3(k+1)-9=-3$
2) $-7t-11=-6(t+1)$
3) $-2(y-2)+3y-6=0$
4) $6(-2-2m)=-13m-4$
5) $5=45(3-x)+55x$
6) $-3(y-1)=15-5y$
7) $-4(k-5)-3(k+9)=0$
8) $-8=-6(y+6)+5(1+y)$
9) $c=3-5(c+3)+4(3-2c)$
10) $-3-1=2(t+1)-7t-6$
11) $-6+2(5-3k)+7k=2(-6+k)$
12) $5x-(3-x)=14x-(9x-12)$
13) $3(1+p)+5(2-p)=-3(1+p)+4$
14) $3(x-5)-2(x-4)+7=-5(-2)$
15) $4x+3(x+1)-3-6x=-4x+4(x+6)$
16) $29-(4-2k)-k=12-2k-(1-k)$

2.2.2 Reviewing the LCD

If our equation has denominators other than 1, we can simplify the equation using the LCD and the property of equality. Let's review finding the LCD before we begin step 1 of the general procedure.

Practice 2.2.2 Reviewing LCD

Find the LCD of the following fractions.

a) $\frac{-9}{14}$ and $1\frac{5}{12}$ and $\frac{2}{21}$

$14 = 2\times7$ $12 = 2\times2\times3$ $21 = 7\times3$	$\Rightarrow$	Prime factored each denominator.
2×7	$\Rightarrow$	Took the first denominator. 14 is now a factor of the LCD.
$2\times7\times\mathbf{2}\times\mathbf{3}$	$\Rightarrow$	For 12 to be a factor of the LCD needed a second factor of 2 and a factor of 3.
$2\times\mathbf{7}\times2\times\mathbf{3}$	$\Rightarrow$	21 is already included in the LCD. No new factor is needed.
84	$\Rightarrow$	The LCD is the resulting product.

Homework 2.2 Find the LCD of the following fractions.

17) $\frac{3}{4}, \frac{3}{10}$ 18) $-\frac{1}{20}, \frac{4}{9}, \frac{5}{6}$ 19) $\frac{5}{27}, \frac{7}{10}, \frac{8}{45}, \frac{1}{18}$ 20) $-\frac{1}{6}, \frac{5}{18}, \frac{3}{4}, \frac{8}{15}$

2.2.3 Step 1 of the General Procedure

Whenever possible, mathematicians like to reuse previous ideas to handle new situations. That's the purpose of step 1 in the general procedure. To solve an equation like $\frac{x}{2}+\frac{3}{4}=\frac{2x}{3}-\frac{1}{4}$, where we have a number of denominators not equal to 1, we'll use the LCD to quickly reduce denominators to 1 so we can reuse the techniques we've already practiced.

To begin, I find the LCD of all the terms which in this case is 12. Please notice that I'm considering the left terms and the right terms at the same time. Next, I use the property of equality to multiply both expressions by the LCD. Then I'd distribute the LCD to the numerator of each term.

The LCD is 12

$$\frac{12}{1}\left(\frac{x}{2}+\frac{3}{4}\right)=\frac{12}{1}\left(\frac{2x}{3}-\frac{1}{4}\right)$$

$$\frac{12x}{2}+\frac{12(3)}{4}=\frac{12(2x)}{3}-\frac{12(1)}{4}$$

Now comes the crucial step. Since I used the LCD of all the terms (both on the left and the right) it has to be the case that all the denominators can be reduced to a denominator of 1.

$$\frac{\overset{6}{\cancel{12}}x}{\cancel{2}}+\frac{\overset{3}{\cancel{12}}(3)}{\cancel{4}}=\frac{\overset{4}{\cancel{12}}(2x)}{\cancel{3}}-\frac{\overset{3}{\cancel{12}}(1)}{\cancel{4}}$$

$$6x+3(3)=4(2x)-3(1)$$

$$6x+9=8x-3$$

After reducing, I move on to line 2 and begin reusing my previous skills. Make sure to return to the original equation to check your answer.

$$6x+9-6x=8x-3-6x$$

$$9=2x-3$$

$$12=2x$$

$$6=x$$

Now let's solve $\frac{x+1}{5}-2=\frac{x-4}{3}+\frac{7}{15}$ so I can point out a couple of the common issues you'll need to be aware of when you use this technique. After determining 15 is the LCD we'd multiply, and then distribute, 15 to the numerator of each term. The first issue I want you to notice are the parentheses with both $15(x+1)$ and $15(x-4)$. These help to make sure you'll multiply 15 to both terms in the numerator.

$$\frac{15}{1}\left(\frac{x+1}{5}-2\right)=\frac{15}{1}\left(\frac{x-4}{3}+\frac{7}{15}\right)$$

$$\frac{15\boldsymbol{(x+1)}}{5}-15(2)=\frac{15\boldsymbol{(x-4)}}{3}+\frac{15(7)}{15}$$

The second issue you should notice is how we needed to distribute the 15 to the term of 2. Sometimes students miss this product.

$$\frac{15(x+1)}{5}-\mathbf{15(2)}=\frac{15(x-4)}{3}+\frac{15(7)}{15}$$

Now we reduce denominators and continue with the solution procedure.

$$\frac{\overset{3}{\cancel{15}}(x+1)}{\cancel{5}}-30=\frac{\overset{5}{\cancel{15}}(x-4)}{\cancel{3}}+\frac{\overset{1}{\cancel{15}}(1)}{\cancel{15}}$$

$$3(x+1)-30=5(x-4)+1$$

$$3x+3-30=5x-20+1$$

$$3x-27=5x-19$$

$$-27=2x-19$$

$$-8=2x$$

$$-4=x$$

Practice 2.2.3 Step 1 of the General Procedure

Solve and check.

a) $\frac{x}{5}+\frac{2}{15}=\frac{1}{3}$

$5\times3=15$ ⇒ The LCD for all the fractions is 15.

$$\frac{15}{1}\left(\frac{x}{5}+\frac{2}{15}\right)=\frac{15}{1}\left(\frac{1}{3}\right)$$

$$\frac{15x}{5}+\frac{15(2)}{15}=\frac{15(1)}{3}$$

⇒ Used the property of equality to multiply both expressions by the LCD. Then distributed the LCD to the numerator of each term.

$$\frac{\overset{3}{\cancel{15}}x}{\cancel{5}}+\frac{\overset{1}{\cancel{15}}(2)}{\cancel{15}}=\frac{\overset{5}{\cancel{15}}(1)}{\cancel{3}}$$

$$3x+2=5$$

⇒ Reduced the denominators to 1.

$$3x+2-2=5-2$$

$$3x=3$$

$$x=1$$

⇒ Solved using steps 2 – 6.

b) $\frac{x}{8}-\frac{4}{3}=\frac{x-1}{6}-2$

$2\times2\times2\times3=24$	$\Rightarrow$	The LCD for all the fractions is 24.
$\frac{24}{1}\left(\frac{x}{8}-\frac{4}{3}\right)=\frac{24}{1}\left(\frac{x-1}{6}-2\right)$ $\frac{24x}{8}-\frac{24(4)}{3}=\frac{24(x-1)}{6}-24(2)$	$\Rightarrow$	Used the property of equality and distributed 24 to all the terms.
$\frac{\overset{3}{\cancel{24}}x}{\cancel{8}}-\frac{\overset{8}{\cancel{24}}(4)}{\cancel{3}}=\frac{\overset{4}{\cancel{24}}(x-1)}{\cancel{6}}-48$ $3x-32=4(x-1)-48$	$\Rightarrow$	Reduced the denominators to 1.
$3x-32=4x-4-48$ $3x-32=4x-52$ $-32=x-52$ $20=x$	$\Rightarrow$	Solved using steps 2 through 4.

Homework 2.2 Solve and check.

21) $\frac{w}{3}+1=\frac{w}{4}$ 22) $4=\frac{-2}{7}x$ 23) $\frac{2}{15}+\frac{a}{5}=\frac{1}{3}$ 24) $\frac{k+1}{4}=\frac{8-k}{5}$ 25) $\frac{1}{14}x+\frac{6}{7}=1$

26) $\frac{x+1}{12}=\frac{x}{9}+3$ 27) $\frac{t}{2}-\frac{t}{4}=\frac{t}{6}-1$ 28) $x+\frac{x-9}{2}=\frac{x+5}{6}$ 29) $\frac{2}{3}-\frac{m}{12}=\frac{m+4}{6}$

30) $\frac{y-2}{18}-\frac{y}{4}=\frac{-8}{9}$ 31) $x+\frac{x}{2}-\frac{7}{6}=\frac{x}{3}$ 32) $1+\frac{x}{4}-\frac{x-8}{6}=0$ 33) $\frac{p-2}{14}+3=\frac{p-2}{21}-2$

34) $\frac{x+1}{2}-\frac{x-1}{3}=\frac{x}{4}$ 35) $\frac{2-n}{30}=\frac{n-1}{3}-\frac{49}{15}$ 36) $\frac{k+1}{9}+\frac{k+1}{6}=1+k$

Homework 2.2 Answers

1) $\{1\}$ 2) $\{-5\}$ 3) $\{2\}$ 4) $\{8\}$ 5) $\{-13\}$ 6) $\{6\}$

7) $\{-1\}$ 8) $\{-23\}$ 9) $\{0\}$ 10) $\{0\}$ 11) $\{16\}$ 12) $\{15\}$

13) $\{-12\}$ 14) $\{10\}$ 15) $\{24\}$ 16) $\{-7\}$ 17) $2\times2\times5=20$

18) $2\times2\times3\times3\times5=180$ 19) $2\times3\times3\times3\times5=270$ 20) $2\times2\times3\times3\times5=180$

21) $\{-12\}$ 22) $\{-14\}$ 23) $\{1\}$ 24) $\{3\}$ 25) $\{2\}$ 26) $\{-105\}$

27) $\{-12\}$ 28) $\{4\}$ 29) $\{0\}$ 30) $\{4\}$ 31) $\{1\}$ 32) $\{-28\}$

33) $\{-208\}$ 34) $\{10\}$ 35) $\{10\}$ 36) $\{-1\}$

2.3 Concluding Linear Equations

When we solve a linear equation like $2x+1=2(x+2)$ we begin by assuming there is a single value, which when substituted for *x*, will make the value of the left and right expression equal. This assumption is often, but not always, true.

2.3.1 Identities, Contradictions and Conditional Equations

The sentence, “This month has thirty-one days.” is conditionally true. Its truth is a condition of the month we’re considering. If someone tells us that, “This month” is July, then the sentence is true. If someone tells us that, “This month” is April, then the sentence is false. Like a conditional sentence, the truth of a **conditional equation** depends on the value we’re considering. For instance, $x+1=2x$ is true if we replace *x* with 1 and it’s false if we replace *x* with a value other than 1. Until now, all our equations have been conditional equations.

The sentence, “January has thirty-one days.” is always true. There are linear equations which are also always true. For example, the equation $x+2=2+x$ is always true, regardless of the value we choose for *x*. An equation where any value is a solution is known as an **identity**. For now, if your equation is an identity, you’ll need to make a statement like, “Any value is a solution.” or “All real numbers.” implying every real number is a solution.

The sentence, “February has thirty-one days.” is always false. There are linear equations which are also always false. For example, the equation $x=x+2$ is always false regardless of the value we choose for *x*. An equation where no value is a solution is known as a **contradiction**. For now, if your equation is a contradiction, you’ll need to make a statement like, “There isn’t a solution.”

One characteristic of identities and contradictions is that, as you try to solve the equation, you’re only left with constants.

Practice 2.3.1 Identities, Contradictions and Conditional Equations
Solve. Make sure you include an answer.

a) $2x+1=2(x+2)$		
$2x+1=2x+4$	$\Rightarrow$	Began with step 3 and simplified the right side.
$2x+1-2x=2x+4-2x$ $0+1=0+4$ $1=4$	$\Rightarrow$	Moved to step 4 and subtracted $2x$ from both expressions. Noticed my equation no longer has a variable term.
$1=4$	$\Rightarrow$	Since I’m left with an equation that’s always false, my original equation was a contradiction.
There is no solution.	$\Rightarrow$	Made sure to include an answer.

b) $6(2-x)+3(2x+5)=27$

$12-6x+6x+15=27$ $27=27$	$\Rightarrow$	Began with step 2 and simplified the left side. Noticed there are no longer any variable terms and the equation I'm left with is always true. The original equation was an identity.
All real numbers.	$\Rightarrow$	Made sure to include an answer.

Homework 2.3 Solve. Make sure you include an answer.

1) $6p-4-11p=14-5p$ 2) $-3y+2(y-1)=-2-y$ 3) $4k+9-2(2k-1)=11$

4) $\frac{h-2}{3}-2=\frac{1}{6}+\frac{h-5}{3}$ 5) $-3(y+1)-3(y-1)=-6(y+1)$

2.3.2 Mixed Practice with Linear Equations

The following equations might be a conditional equation, a contradiction or an identity. Also, until today's problems, conditional equations had integer solutions. Now, the solution set for conditional equations is open to all rational numbers.

Homework 2.3.2 Solve and when possible, check.

6) $4t+6+4(1-t)=10$ 7) $-2(x-4)=8(1-x)$ 8) $3m-(m-1)=4(2+2m)$

9) $3t-5(3-5)=3+3t$ 10) $\frac{3x}{5}=\frac{1}{15}+\frac{2(x+1)}{3}$ 11) $7-2(5-3x)-(5x-3)=x$

12) $\frac{h-2}{3}-2=\frac{1}{6}+\frac{h}{2}$ 13) $\frac{7+2k}{6}+\frac{1}{6}=\frac{k}{3}$ 14) $4-(r-7)-2(5-r)=3r+14$

15) $\frac{2}{9}k-\frac{1}{3}=\frac{7}{3}k-\frac{7}{9}$ 16) $3x-7=15x+9(x-1)+2$ 17) $4b-7=6\left(\frac{b}{3}+5\right)$

18) $2[-6(y+8)+45]=-1-(7y-15)$ 19) $\frac{t+12}{12}+\frac{t-4}{4}=\frac{t}{3}$

20) $6x=40(x+3)-30(x+4)$ 21) $k-\frac{k+1}{6}=\frac{5k}{6}$

22) $18\left(\frac{x}{9}-1\right)=12\left(\frac{1}{6}-\frac{2x}{3}\right)$ 23) $y-[y-(y-5)+38]=5[y-8]$

24) $\frac{x}{4}+\frac{36+x}{12}=3+\frac{x}{3}$

Homework 2.3 Answers

1) No solution 2) All real numbers 3) All real numbers 4) No solution 5) No solution

6) All real numbers 7) $\{0\}$ 8) $\left\{-\frac{7}{6}\right\}$ 9) No solution 10) $\{-11\}$

11) All real numbers 12) $\{-17\}$ 13) No solution 14) $\left\{-\frac{13}{2}\right\}$ 15) $\left\{\frac{4}{19}\right\}$

16) $\{0\}$ 17) $\left\{\frac{37}{2}\right\}$ 18) $\{-4\}$ 19) All real numbers 20) $\{0\}$

21) No solution 22) $\{2\}$ 23) $\left\{-\frac{3}{4}\right\}$ 24) All real numbers

2.4 An Introduction to Word Problems

Your work in previous sections has mostly involved skills that a computer can do faster and more accurately. Solving real problems, which involve translating from English to algebra, is something a computer can't do. A computer can't translate English because it's currently not possible to write a step by step process that deals with the fluidity and nuance of language. Thankfully, you were born to process language. This section begins your practice with problem solving skills, skills that are uniquely human.

2.4.1 A General Template for Problem Solving

I'd like to start with a word problem where most students instantly see an answer without needing algebra at all;

> To help prepare for a half-marathon a woman runs 7 miles per day. If her morning run is 4 miles further than her evening run, how far does she run in the evening?

If you're like me the answer is obvious, it's three miles. If you didn't try answering the question earlier, please read it now and see how naturally the answer three miles comes to mind.

Unfortunately, as you might suspect, if we check 3 miles, it doesn't work. If she runs 3 miles in the evening, then her morning run must be 7 miles (4 miles further), but 7 miles is the total after combining both runs, so something is wrong.

In this section you'll practice some strategies for moving forward algebraically when your intuition fails. In general, the idea is to translate the English problem into an algebraic equation, solve the algebraic equation, and then translate any algebraic solution back to English in a way that's helpful for others. Please notice these are strategies, not a true step by step process. If there were a step by step process that solved all applied problems, we'd simply code the process and turn all applied problems over to computers.

Solving applied problems successfully is the result of practice. You should copy these steps to a note card and refer to them until they become automatic.

Some General Strategies for Solving Word Problems
1. Write goal statement(s) describing what the unknown values will count when you finish the problem. Leave a space for these values. 2. Build an English equation that relates the unknown values. a) Look for sums or differences and work backwards to terms. 3. Translate the English equation to an algebraic equation. 4. Solve the algebraic equation and find values for all the unknowns. 5. Check your answer(s) using the <u>original</u> English problem. 6. Finish your goal statement(s).

Although computers can't solve a general word problem, they are able to do steps 4 – 6. It's steps 1 through 3, (and especially step 2), that only people can do. As you work on solving word problems please keep this important idea in mind;

Think like an expert
Experts spend most of their time on steps 1 and 2 of the process. Once an expert can build an English equation they simply translate to algebra and solve. Novices on the other hand, try to read the problem and jump to step 3. Think like an expert, spend time on steps 1 and 2 so that you "understand" the problem.

To start learning the process for solving word problems we'll begin with "number" problems which only involve steps 3 through 5. Computers are able to do step 3 with number problems because there's a very limited number of words. You'll find number problems useful since they practice translating the language for operations into algebra.

2.4.2 Using Steps 3 through 5 to Solve "Number" Problems

An important part of using algebra is translating from English to the four basic operators. For instance, "four more than a number" translates to $n+4$ (assuming n represents the number). Here are some common English words that describe our basic operations.

Addition – more, more than, increased by, sum, plus, a gain of

Subtraction – difference, decreased by, reduced by, minus, less, less than, fewer than, declined

Multiplication – times, twice, product, double, triple

Division – quotient, divided by, per, divide

When you translate operations it's important to keep in mind that addition and multiplication are commutative but subtraction and division are not. For instance, whether you translate, "four more than a number", correctly as $n+4$ or incorrectly as $4+n$, the sum is the same. On the other hand, if you translate, "four less than a number" it has to have the correct order $n-4$ since $4-n$ gives the opposite value.

Let's practice solving some number problems.

Practice 2.4.2 Using Steps 3 through 5 to Solve "Number" Problems

Use steps 3-5 to find the unknown number.

a) The sum of four and twice a number is negative ten. Find the number.

$\underbrace{\text{The sum of 4 and twice a number}}_{4+2n}\ \underset{=}{\text{is}}\ \underbrace{\text{negative ten}}_{-10}$ $4+2n=-10$	$\Rightarrow$	Step 3 asked me to translate from English to algebra. Sum implies addition. Twice implies two times. Notice, in this context, is implies equals.

$4+2n=-10$ $2n=-14$ $n=-7$	$\Rightarrow$ Moved to step 4 and solved the equation.
$\underbrace{\text{The sum of 4 and twice a number}}_{4+2(-7)}$ is $\underbrace{\text{negative ten}}_{-10}$ $4+-14$ -10	$\Rightarrow$ Following step 5 I returned to the <u>original problem</u> and substituted –7 for "a number". The check looks good.

b) Three less than the quotient of a number and four is two. Find the number.

$\underbrace{\text{Three less than the quotient of a number and four}}_{n/4 - 3}$ is two. $= 2$ $n/4 - 3 = 2$	$\Rightarrow$ Started with step 3 and translated the English to algebra. Noticed that less than is written in a different order than it is spoken.
$n/4 - 3 = 2$ $n = 20$	$\Rightarrow$ Moved to step 4 and solved the equation.
$\underbrace{\text{Three less than the quotient of twenty and four}}_{20/4 - 3}$ is two. 2 $5-3$ 2	$\Rightarrow$ Moved to step 5, returned to the original English and substituted 20 for "a number". The check looks good.

Homework 2.4 Use steps 3-5 to find the unknown number.

1) The sum of a number and fifty is twenty-eight.
2) Six is fifteen less than a number.
3) Three times a number added to ten is forty-three.
4) The quotient of a number and negative eight is negative five.
5) A number decreased by sixty is forty-seven.
6) Six is twelve plus the product of a number and three.
7) A number reduced by negative nine is twice the number.
8) The sum of four and the quotient of a number and five is negative six.
9) Seventeen minus a number is one less than the number.
10) After being divided by two a number is four more than itself.

2.4.3 Step 1 – Writing Goal Statements

Now that you've practiced doing what a computer can do, let's start working on the steps that only people can do.

Step 1 of the template asks you to determine what value(s) you'll need to solve the problem and write goal statements that identify this information. For example, in this situation;

To help raise money, a school sells 200 tickets to a play. Some of the tickets were child tickets and some were adult tickets. If 84 more child tickets than adult tickets were sold, how many of each type of ticket were sold?

the last sentence, "how many of each type of ticket were sold?" indicates that we'll have solved the problem when we can count the number of child tickets sold and the number of adult tickets sold. One way to write goal statements for this problem would be;

The number of child tickets sold is __. The number of adult tickets sold is __.

Another would be;

I'd sell ____ child tickets and _____ adult tickets.

Please notice that each goal statement is written as a statement, not as a question and that I've left a blank for the values I don't know. Here's some practice writing goal statements.

Practice 2.4.3 Step 1 – Writing Goal Statements

Write goal statements for each problem.

a) Together a washer and dryer cost $1000. If the washer was $600 more than the dryer, what was the price of each?

The price of the dryer is $___. The price of the washer is $___.	⇒	The expression, "what was the price of each?" tells me the values I'll need to finish the problem. Made sure to include a unit of dollars.

b) Two items are consecutive, if one follows the other in order. If the sum of two consecutive integers is 14, find the integers.

The smaller integer is ___. The larger integer is ___.	⇒	Since the expression, "find the integers." Is plural I know I'm looking for the value of more than one integer. Used the idea of smaller and larger to keep the integers separate.

c) A student spends $385 buying textbooks for three classes. Find the cost of each textbook if the biology textbook is $60 more than the humanities textbook and twice the cost of the math textbook.

The biology text costs $_____. The humanities text costs $_____. The math text costs $_____.	⇒	The expression, "Find the cost of each textbook…" identifies the goal. Needed three goal statements since the problem has three items. Remembered to include units.

Homework 2.4 Write goal statements for each problem.

11) After reducing the price by 10% a T.V. sold for $500. What was the original price?

12) Two people go out for dinner. One person's dinner was half as much as the other. If together they paid a total of $48 find the price of each dinner?

13) A family drives to a cabin which is 300 miles away. After driving for a while they took a break and then finish the trip. If the distance before the break was 50 miles less than the distance after the break, how far did they drive before and after the break?

14) A community center splits $1,200 into three prizes. The second prize is $600 less than the first prize and the third prize is half the amount of the second prize. Find the amount for each prize.

15) The difference of two consecutive integers is -63 , find the integers.

16) If the sum of two consecutive **odd** integers is 113, find the integers.

17) To help prepare for a triathlon a woman runs 7 miles per day. If her morning run is 4 miles further than her evening run how far does she run in the evening?

2.4.4 Step 2 – Writing an English Equation From a Sum

Now let's move to step 2, the most difficult step in the process. Today we'll concentrate on one of the most helpful strategies for building the English equation, identifying a sum and working backwards to the terms. Let's apply this strategy to our problem about the child and adult tickets;

> To help raise money, a school sells 200 tickets to a play. Some of the tickets were child tickets and some were adult tickets. If 84 more child tickets than adult tickets were sold, how many of each type of ticket were sold?

The first constraint," a school sells 200 tickets to a play", is a sum. It comes from adding together the number of child tickets sold and the number of adult tickets sold. Here's an example of an effective English equation which shows the terms that build the total number of tickets;

The total tickets sold = the number of child tickets + the number of adult tickets

Here's some practice with building an English equation using a sum.

Practice 2.4.4 Step 2 – Writing an English Equation from a Sum

Build an English equation from the sum.

a) Together a washer and dryer cost $1,250. If the dryer was $600 less than the washer, what was the price of each?

The total cost	⇒	Identified the sum.
total cost = the washer cost + the dryer cost.	⇒	Identified the terms that built the sum.

b) To help prepare for a triathlon a woman runs 7 miles per day. If her morning run is 4 miles further than her evening run how far does she run in the evening?

The total miles	⇒	Identified the sum.
total miles = the morning miles + the evening miles.	⇒	Identified the terms.

c) A community center splits $1,600 into three prizes. The third prize is $800 less than the first prize and the second prize is twice the amount of the third prize. Find the amount for each prize.

Total prizes	⇒	Identified the sum
The total dollars = the 1st prize dollars + the 2nd prize dollars + the 3rd prize dollars.	⇒	Identified the terms that built the sum.

Homework 2.4 Build an English equation from the sum (or difference).

18) A $52.50 cable bill has a fixed monthly fee and a charge for a premium move channel. If the movie channel is $37 less than the fixed fee what is the cost for the premium movie channel?

19) Two people go out for dinner. One person's dinner was half as much as the other. If together they paid a total of $48 find the price of each dinner?

20) After decreasing the price by 20% a T.V. sold for $810. What was the original price?

21) A hiker splits their 40-mile trip into three parts. The first part was 8 miles further than the second part and the third part was twice as long as the second part. How long was each part of the trip?

22) A large container of juice costs $5. If the juice alone costs $4 more than the empty container, find the cost of the empty container.

23) Two things are consecutive if one follows the other in order. If the sum of two consecutive <u>even</u> integers is -334, find the integers.

24) A gas station sells a certain number of gallons of regular, one third as much super as regular and 8,000 less gallons of premium as regular. Find the gallons of each type of gasoline sold if 39,600 gallons were sold all together.

2.4.5 Step 3 – Translating to Algebra

Once you're happy with step 2 the next step is to translate the English equation to an algebraic equation by translating each English term to an equivalent algebraic term. As an example let's return to our problem about the tickets.

> To help raise money, a school sells 200 tickets to a play. Some of the tickets were child tickets and some were adult tickets. If 84 more child tickets than adult tickets were sold, how many of each type of ticket were sold?

Which we translated to the English equation;

$$\text{the total number of tickets} = \text{the number of child tickets} + \text{the number of adult tickets}$$

Although I like to write the <u>number</u> of child tickets and the <u>number</u> of adult tickets, people often "shorthand" the information using, for example, "the child tickets" or just "child tickets".

To translate the English equation, I'll start on the left side which translates to 200.

$$\underbrace{\text{the total number of tickets}}_{200} = \text{the number of child tickets} + \text{the number of adult tickets}$$

Translating the right side will take some insight. Since we don't know the number of child tickets or the number of adult tickets both will have to be represented by a variable. One approach is to use two different variables, say c for the number of child tickets and a, for the number of adult tickets. This is a good approach, just not the one we'll take in this section. Today, we'll pick a single variable to represent one of the unknown values and then build a second expression, <u>using this same variable</u>, to represent the other unknown value.

It's common at this point for students to worry whether a or c is the "right" variable to use. Fortunately, one isn't "right" and the other "wrong", instead, depending on how the problem is worded, one might seem more natural to you than the other. For instance, say you decided to find a, the number of adult tickets. Then your translation so far should look like this;

$$\underbrace{\text{The total number of tickets}}_{200} = \underbrace{\text{the number of child tickets}}_{} + \underbrace{\text{the number of adult tickets}}_{a}$$

To complete the translation, you'll need to build an expression, using the variable a, to represent the number of child tickets. Returning to the original problem we're told that, "84 more child tickets were sold". This implies that adding 84 to the number of adult tickets, gives us the number of child tickets. Here's the final translation;

$$\underbrace{\text{The total number of tickets}}_{200} = \underbrace{\text{the number of child tickets}}_{a+84} + \underbrace{\text{the number of adult tickets}}_{a}$$

What would have happened if we'd decided to find the number of child tickets instead? Well, if we let c represent the number of child tickets then our translation would be;

$$\underbrace{\text{The total number of tickets}}_{200} = \underbrace{\text{the number of child tickets}}_{c} + \underbrace{\text{the number of adult tickets}}_{}$$

Now, we'd have to realize that if there are 84 more child tickets then there has to be 84 less adult tickets. This implies that $c-84$ would be the number of adult tickets. Here's the final translation from the point of view of the child tickets.

$$\underbrace{\text{The total number of tickets}}_{200} = \underbrace{\text{the number of child tickets}}_{c} + \underbrace{\text{the number of adult tickets}}_{c-84}.$$

Both these approaches are "right" in that either one will give you the correct answers. Personally, I find the first approach more natural but that doesn't make it "better". If you find the second approach more natural then that's the better approach for you.

Practice 2.4.5 Step 3 – Translating to Algebra

Translate the English equation to an algebraic equation.

a) Together a washer and dryer cost \$1,000. If the washer was \$600 more than the dryer translate the English equation, the total cost = washer cost + dryer cost , to algebra. Let d stand for the cost of the dryer.

$$\underbrace{\text{The total cost}}_{1{,}000} = \underbrace{\text{washer cost}}_{d+600} + \underbrace{\text{dryer cost}}_{d}$$ $\Rightarrow$ Translated to algebra.

b) Each month you pay \$52.50 for a cable bill. The bill has a fixed fee plus a charge for a premium move channel. If the premium movie channel is \$37 less than the fixed fee translate the English equation, the total bill = the fixed fee + the movie channel cost to algebra. Let f stand for the cost of the fixed fee.

$$\underbrace{\text{the total bill}}_{52.50} = \underbrace{\text{the fixed fee}}_{f} + \underbrace{\text{the movie channel cost}}_{f-37}$$

$\Rightarrow$ Translated to algebra.

c) Two things are consecutive if one follows the other in order. If the sum of two consecutive odd integers is 116, translate the English equation, the sum = the smaller odd integer + the larger odd integer , to algebra. Let s stand for the value of the smaller integer.

$$\underbrace{\text{the sum}}_{116} = \underbrace{\text{the smaller odd integer}}_{s} + \underbrace{\text{the larger odd integer}}_{s+2}$$

$\Rightarrow$ The larger integer is consecutive and odd so it's two more than the smaller integer.

d) Three people, with a combined height of 199 inches, are lined up from shortest to tallest. If the tallest person is 7 inches taller than the shortest person and the middle person is 4 inches shorter than the tallest person translate the English equation, total inches = short inches + middle inches + tallest inches., to algebra. Let s stand for the height of the shortest person.

$$\underbrace{\text{total inches}}_{199} = \underbrace{\text{short inches}}_{s} + \underbrace{\text{middle inches}}_{(s+7)-4} + \underbrace{\text{tallest inches}}_{s+7}$$

$\Rightarrow$ Translated to algebra.

Homework 2.4 Translate the English equation to an algebraic equation.

25) After reducing the price by 10% a T.V. sold for \$500. Translate the English equation, the final price = the original price − the discount to algebra. Let p stand for the original price.

26) Two people go out for dinner. One person's dinner was half as much as the other. If together they paid a total of \$48 translate the English equation, total cost = the cheaper dinner + the expensive dinner to algebra. Let e stand for the cost of the expensive dinner.

27) A hiker split their 40-mile trip into three parts. The first part was 8 miles further than the second part and the third part was twice as long as the second part. Translate the English equation, the total distance = the 1st distance + the 2nd distance + the 3rd distance to algebra. Let s stand for the distance of the second part.

28) Two things are consecutive if one follows the other in order. If the sum of two consecutive even integers is -334, Translate the English equation, the sum = the smaller even integer + the larger even integer , to algebra. Let s stand for the value of the smaller integer.

29) A man started with \$800 and after paying an initiation fee and the first month's membership he had \$300 left. If the initiation fee was three times the monthly membership Translate the English equation, original total = amount left + initiation fee + monthly charge , to algebra. Let m stand for the monthly charge.

30) To help prepare for a triathlon a woman runs 7 miles per day. If her morning run is 4 miles further than her evening run Translate the English equation, total distance $=$ morning distance $+$ evening distance, to algebra. Let e stand for the distance run in the evening.

31) A gas station sells a certain number of gallons of regular, one third as much super as regular and 8,000 less gallons of premium as regular. Translate the English equation, total gallons $=$ gallons of regular $+$ gallons of super $+$ gallons of premium to algebra if 39,600 gallons were sold all together. Let r stand for the gallons of regular that are sold.

2.4.6 Applying the General Template

Now that you've practiced each step, let's put everything together.

A General Template for Solving Word Problems
1. Write goal statement(s) using items and their units (if they have units) to describe any unknown values. 2. Build an English equation that relates the unknown values. a) Look for sums or differences and work backwards to terms. 3. Translate the English equation to an algebraic equation. 4. Solve the algebraic equation, find values for all the unknowns. 5. Check your answer(s) using the <u>original</u> English problem. 6. Finish your goal statement(s).

Practice 2.4.6 Applying the General Template

Solve the following problems using the general template.

a) To help prepare for a triathlon a woman runs 7 miles per day. If her morning run is 4 miles further than her evening run how far does she run in the evening?

Step 1 $\Rightarrow$ Her evening run is _____ miles.

Step 2 $\Rightarrow$ the total distance = the morning distance $+$ the evening distance.

Step 3 $\Rightarrow$ $\underbrace{\text{the total distance}}_{7} = \underbrace{\text{the morning distance}}_{e+4} + \underbrace{\text{the evening distance}}_{e}$.

Step 4 $\Rightarrow$

$$7 = e + 4 + e$$
$$7 = 2e + 4$$
$$3 = 2e$$
$$1.5 = e$$

Step 5 $\Rightarrow$ If her evening run is 1.5 miles than her morning run is 4 miles further or 5.5 miles and together her daily run is 7 miles.

Step 6 $\Rightarrow$ Her evening run is <u>1.5</u> miles.

b) Some students rent a room for spring break. They pay a one-time "housekeeping" fee and also \$197.50 per day for 5 days. If the total cost was \$1,082.50, how much was the housekeeping fee?

Step 1 $\Rightarrow$	The housekeeping fee was \$ ____.
Step 2 $\Rightarrow$	total cost = housekeeping fee + the 5 day fee.
Step 3 $\Rightarrow$	$\underbrace{\text{Total cost}}_{1,082.50} = \underbrace{\text{the housekeeping fee}}_{h} + \underbrace{\text{the 5 day fee}}_{197.50(5)}$.
Step 4 $\Rightarrow$	$1,082.50 = h + 197.50(5)$ $1,082.50 = h + 987.50$ $95 = h$
Step 5 $\Rightarrow$	The housekeeping fee and the total daily fee should be \$1,082.50. $95 + 197.50(5) = 1,082.50$, the check looks good.
Step 6 $\Rightarrow$	The housekeeping fee was \$95.

Before you start on this homework I want to caution you about some common mistakes students make. First, don't use arithmetic to solve these problems. Being comfortable with arithmetic is good but being good at both arithmetic and algebra is better. You're using arithmetic if you find yourself operating on numbers instead of using numbers and letters.

Second, as I mentioned earlier, spend your time on steps 1 through 3. When students spend time on these steps their comments go from, "I just can't do word problems" to "I can't figure out how to write the relationship between the morning run and the evening run."

Homework 2.4 Solve the following problems using the general template.

32) The first night of a high school play, 40 more adult tickets than child tickets were sold. If they sold 276 tickets, how many of each type of ticket were sold?

33) Together a washer and dryer costs \$1,000. If the dryer was \$400 less than the washer, what was the cost of the dryer?

34) Two people go out for dinner. One person's order was half as much as the other. If together they paid a total of \$48 what was the price of the **least** expensive dinner?

35) Two items are consecutive if one follows the other in order. If the sum of two consecutive integers is 113, find the integers.

36) A hiker splits their 40-mile trip into three parts. The first part was 8 miles further than the second part and the third part was twice as long as the second part. How long was each part of the trip?

37) Two consecutive **odd** integers have a sum of -20 . Find the integers.

38) A man started with \$800 and, after paying an initiation fee and the first month's membership, he had \$300 left. If the initiation fee was three times the monthly membership, how much was the initiation fee?

39) On a monthly mobile phone bill, a student pays a constant fee and a variable data fee. If their total bill was \$63 and the data fee was one-sixth their constant fee how much was each fee?

40) The sum of the angle measures of a triangle always equals 180 degrees. If two of the angles are equal and the third is 42 degrees less than either of the other angles find the measure of all the angles.

41) After decreasing the price by 20% a T.V. sold for \$810. What was the original price?

42) A family drives to a cabin which is 300 miles away. They drive a while at 55 mph, take a break, and then finish the trip at 60 mph. If they drove 50 fewer miles before the break, how far was each part of the trip?

43) The perimeter of a rectangle is the distance of the boundary and can be found by adding together the distance of the two lengths and the distance of the two widths. The length of a rectangular park is half a mile longer than the width. If the perimeter of the park is 3.4 miles find the dimensions (the length and the width) of the park.

44) A community center splits \$2,000 into three prizes. The second prize is \$600 less than the first prize and the third prize is \$400 less than the second prize. How much was each prize?

45) A student makes a \$700 deposit to rent a moving van. It costs \$79.99 a day to rent the van. After returning the van the student receives back \$399.95 of their original deposit. How many days did they rent the van?

46) A student rented a car. They paid \$19.99 regardless of the miles driven and twenty-five cents for every mile driven. If their total cost was \$173.24, how many miles did they drive?

47) To help raise money, a school sells tickets to a play and earns \$932.25. Some of the tickets were \$3.75 child tickets and some were \$5.25 adult tickets. If 75 child tickets were sold, how many adult tickets were sold?

Homework 2.4 Answers

1) $n=-22$ 2) $n=21$ 3) $n=11$ 4) $n=40$ 5) $n=107$

6) $n=-2$ 7) $n=9$ 8) $n=-50$ 9) $n=9$ 10) $n=-8$

11) Originally the TV cost $__.

12) The cheaper dinner costs $____.
The expensive dinner costs $___.

13) The family drove ____ miles before the break.
The family drove ____ miles after the break.

14) The first prize was $____.
The second prize was $_____.
The third prize was $____.

15) The smaller integer is ______.
The larger integer is ______.

16) The smaller odd integer is _____.
The larger odd integer is ______.

17) Her evening run is ______ miles.

18) The total bill = the fixed monthly fee + the movie channel charge

19) The total cost = cost of the expensive dinner + cost of cheaper dinner

20) The final price = original price - 20% of the original price

21) The total miles = miles for the 1st part + miles for the 2nd part + miles for the 3rd part.

22) The total cost = the cost of the juice + the cost of the container.

23) The sum = the smaller even integer + the larger even integer

24) The total gallons = the gallons of regular + the gallons of super + the gallons of premium

25) $\underbrace{\text{the final price}}_{500} = \underbrace{\text{the original price}}_{p} - \underbrace{\text{the discount}}_{0.10p}$

26) $\underbrace{\text{total cost}}_{48} = \underbrace{\text{the cheaper dinner}}_{\frac{1}{2}e} + \underbrace{\text{the expensive dinner}}_{e}$

27) $\underbrace{\text{the total distance}}_{40} = \underbrace{\text{the 1st distance}}_{s+8} + \underbrace{\text{the 2nd distance}}_{s} + \underbrace{\text{the 3rd distance}}_{2s}$

28) $\underbrace{\text{the sum}}_{-334} = \underbrace{\text{the smaller even integer}}_{s} + \underbrace{\text{the larger even integer}}_{s+2}$

29) $\underbrace{\text{original total}}_{800} = \underbrace{\text{amount left}}_{300} + \underbrace{\text{initiation fee}}_{3m} + \underbrace{\text{monthly charge}}_{m}$

30) $\underbrace{\text{total distance}}_{7} = \underbrace{\text{morning distance}}_{e+4} + \underbrace{\text{evening distance}}_{e}$

31) $\underbrace{\text{total gallons}}_{39{,}600} = \underbrace{\text{gallons of regular}}_{r} + \underbrace{\text{gallons of super}}_{\frac{1}{3}r} + \underbrace{\text{gallons of premium}}_{r-8{,}000}$

32)

Step 3 ⇒ $\underbrace{\text{Tickets sold}}_{276} = \underbrace{\text{adult tickets sold}}_{c+40} + \underbrace{\text{child tickets sold}}_{c}.$

Step 6 ⇒ There were 158 adult tickets. There were 118 child tickets.

33)

Step 3 ⇒ $\underbrace{\text{total cost}}_{1{,}000} = \underbrace{\text{the cost of the washer}}_{w} + \underbrace{\text{the cost of the dryer}}_{w-400}.$

Step 6 ⇒ The cost of the dryer was $300.

34)

Step 3 ⇒ $\underbrace{\text{the cost for both}}_{48} = \underbrace{\text{cheap dinner cost}}_{\frac{1}{2}e} + \underbrace{\text{expensive dinner cost}}_{e}.$

Step 6 ⇒ The price for the least expensive dinner was $16.

35)

Step 3 ⇒ $\underbrace{\text{the integer sum}}_{113} = \underbrace{\text{the smaller integer}}_{s} + \underbrace{\text{the larger integer}}_{s+1}.$

Step 6 ⇒ The smaller integer is 56. The larger integer is 57.

36)

Step 3 ⇒ $\underbrace{\text{total miles}}_{40} = \underbrace{\text{miles of 1rst part}}_{s+8} + \underbrace{\text{miles of 2nd part}}_{s} + \underbrace{\text{miles of 3rd part}}_{2s}.$

Step 6 ⇒ Part 1 was 16 miles. Part 2 was 8 miles. Part 3 was 16 miles.

37)

Step 3 ⇒ $\underbrace{\text{the integer sum}}_{-20} = \underbrace{\text{the smaller odd integer}}_{s} + \underbrace{\text{the larger odd integer}}_{s+2}.$

Step 6 ⇒ The smaller odd integer is –11. The larger odd integer is –9.

38)

Step 3 ⇒ $\underbrace{\text{money left}}_{300} = \underbrace{\text{starting amount}}_{800} - \underbrace{\text{initiation fee}}_{3m} - \underbrace{\text{first month's membership}}_{m}.$

Step 6 ⇒ The initiation fee was $375.

39)

Step 3 ⇒ $\underbrace{\text{monthly fee}}_{63} = \underbrace{\text{constant fee}}_{c} + \underbrace{\text{variable data fee}}_{\frac{1}{6}c}.$

Step 6 ⇒ The constant fee was $54 and the variable data fee was $9.

40)

Step 3 $\Rightarrow$ $$\underbrace{\text{the total degrees}}_{180} = \underbrace{\text{angle 1 degrees}}_{d} + \underbrace{\text{angle 2 degrees}}_{d} + \underbrace{\text{angle 3 degrees}}_{d-42}.$$

Step 6 $\Rightarrow$ The two equal angles are each $\underline{74}°$ and the smaller angle is $\underline{32}°$.

41)

Step 3 $\Rightarrow$ $$\underbrace{\text{final price}}_{810} = \underbrace{\text{original price}}_{p} - \underbrace{\text{20\% of original price}}_{0.20p}.$$

Step 6 $\Rightarrow$ The original price was $\$\underline{1,012.50}$.

42)

Step 3 $\Rightarrow$ $$\underbrace{\text{total distance}}_{300} = \underbrace{\text{distance before break}}_{a-50} + \underbrace{\text{distance after break}}_{a}.$$

Step 6 $\Rightarrow$ They drove $\underline{125}$ miles before the break and $\underline{175}$ miles after the break.

43)

Step 3 $\Rightarrow$ $$\underbrace{\text{the perimeter}}_{3.4} = \underbrace{\text{the length}}_{w+0.5} + \underbrace{\text{the length}}_{w+0.5} + \underbrace{\text{the width}}_{w} + \underbrace{\text{the width}}_{w}$$

Step 6 $\Rightarrow$ The width is $\underline{0.6}$ miles and the length is $\underline{1.1}$ miles.

44)

Step 3 $\Rightarrow$ $$\underbrace{\text{total prize money}}_{2,000} = \underbrace{\text{the 1st prize}}_{f} + \underbrace{\text{the 2nd prize}}_{f-600} + \underbrace{\text{the 3rd prize}}_{(f-600)-400}$$

Step 6 $\Rightarrow$ The 1st prize is \$1,200, the 2nd prize is $\$\underline{600}$ and the 3rd prize is $\$\underline{200}$.

45)

Step 3 $\Rightarrow$ $$\underbrace{\text{the dollars back}}_{399.95} = \underbrace{\text{the original deposit}}_{700} - \underbrace{\text{the van dollars}}_{79.99d}$$

Step 6 $\Rightarrow$ The student rented the van for $\underline{7}$ days.

46)

Step 3 $\Rightarrow$ $$\underbrace{\text{the total cost}}_{173.24} = \underbrace{\text{the fixed amount}}_{19.99} + \underbrace{\text{the amount for the miles driven}}_{0.25m}$$

Step 6 $\Rightarrow$ The student drove $\underline{613}$ miles.

47)

Step 3 $\Rightarrow$ $$\underbrace{\text{the total dollars}}_{932.25} = \underbrace{\text{child tickets dollars}}_{75(3.75)} + \underbrace{\text{adult tickets dollars}}_{a(5.25)}.$$

Step 6 $\Rightarrow$ $\underline{124}$ adult tickets were sold.

2.5 The Solution Set for a Linear Inequality

Unlike an equation, which says two expressions are equal, an inequality says one expression is greater than or less than another. In this section, we'll begin working with linear inequalities.

2.5.1 The Inequality Symbol

To discuss an inequality, we'll accept that our set of numbers are ordered with one number being "greater than" another, if it's further to the right on the number line. For instance, 1 is greater than -2. Since the number to the right is greater it's natural to think of the number to the left as "less than" so -2 would be less than 1.

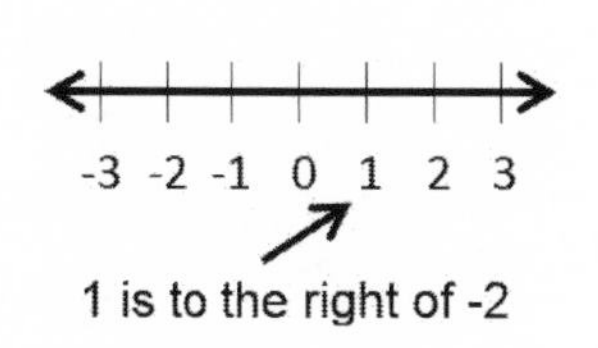

To convey the idea of greater than or less than, we use the inequality symbol, >. The "larger" end of the symbol is closest to the number which is greater while the "smaller" end is closest to the number that is less. To say negative two is greater than negative five we would write $-2 > -5$. To say negative five is less than negative two we would write $-5 < -2$.

When it comes to inequalities one of the most harmful misconceptions students bring to class, is that the symbol is an arrow that is "pointing" in a certain direction on the number line. Please notice that both $-2 > -5$ and $-5 < -2$ are describing the same relationship between the values -2 and -5 even though the inequality symbol "points" first to the right and then to the left.

Inequalities like $-2 > -5$ are known as strict inequalities because they don't include the option that the expressions might be equal. A second type of inequality symbol allows for one expression to be greater than or equal to another $\geq$, or less than or equal to another $\leq$. For example, $7 \leq 7$ says that 7 is less than, or equal to, 7. Here's some practice using the inequality symbol.

Practice 2.5.1 The Inequality Symbol
Translate an inequality written in English to mathematics and an inequality written in mathematics to English.

a) $14 > 9$

"Fourteen is greater than nine." $\Rightarrow$ The larger part of the symbol is closest to the number that's greater.

b) "Negative one is less than zero."

$-1 < 0$ $\Rightarrow$ The smaller part of the symbol is closest to the number which is less.

c) $3 \geq 3$

Three is greater than or equal to three. $\Rightarrow$ Noticed the line under the inequality symbol.

Homework 2.5 Translate an inequality written in English to mathematics and an inequality written in mathematics to English.

1) $12 > -2$ 2) Negative two is greater than or equal to negative seven. 3) $-2 \le 0$

4) Five is less than six. 5) $1 \ge 1$ 6) Negative eight is less than negative one.

Practice 2.5.1 The Inequality Symbol

Fill in the box with the symbol that makes the inequality true.

a) $14 \square 9$

$14 \boxed{>} 9$ $\Rightarrow$ Fourteen is to the right of nine on the number line so it's greater.

b) $-3\frac{1}{2} \square -3\frac{3}{4}$

$-3\frac{1}{2} \boxed{>} -3\frac{3}{4}$ $\Rightarrow$ Negative three and three-quarters is further to the left on the number line.

c) $7 \square 7$

$7 \boxed{\le} 7$ or $7 \boxed{\ge} 7$ $\Rightarrow$ Either symbol makes the inequality true.

Homework 2.5 Fill in the box with the symbol that makes the inequality true.

7) $12 \square -2$ 8) $-8 \square -12$ 9) $0 \square -2$ 10) $-9 \square -9$

11) $-2.1 \square -1.9$ 12) $0 \square 0$

2.5.2 The Solution Set for an Inequality

To solve a linear inequality, we find all values that make the inequality true. Unlike a linear equation, where the solution set usually had a single value, the solution set of an inequality is usually an infinite set of values. There are three common ways to represent this infinite set of values, set notation, a graph and interval notation. Let's begin with set notation.

2.5.3 Using Set Notation to Represent a Solution Set

With linear equations, we often represented the solution using a solution set. For example, the solution set for $x = 2$ was $\{2\}$. Using set notation to represent the solution set of a linear inequality is an extension of this idea. To show the solution set for a linear inequality we again use braces and then include a description of the values which belong in the set. For instance, the solution set for $x > 2$ would be $\{x \mid x > 2\}$. The leftmost *x* represents the set of real numbers, the vertical line, |, is often read "such that" and the third part, $x > 2$ gives us a way of deciding whether a specific real number belongs in the set or not. In this case, real numbers greater than 2 belong in the set while those less than or equal to 2 do not. To describe this solution set we might say, "The set of all real numbers that are greater than two." Here's some practice using set notation to describe the solution set for an inequality.

Practice 2.5.3 Using Set Notation to Represent a Solution Set
Express the solution of the inequality using set notation.

a) $m < -1$

$\{m \mid m < -1\}$ $\Rightarrow$ The solution set includes all values less than -1.

b) $t \geq 8/3$

$\{t \mid t \geq 8/3\}$ $\Rightarrow$ $8/3$ and all values to the right of $8/3$ should be included in the solution set.

c) $4 \geq p$

$\{p \mid p \leq 4\}$ $\Rightarrow$ The tradition with set notation is to put the variable on the left side of the inequality symbol. If 4 is greater than the values we're interested in, then those numbers are all less than 4.

Homework 2.5 Express the solution of the inequality using set notation.

13) $t \leq 7$ 14) $-4 \leq c$ 15) $\frac{1}{2} > q$ 16) $x > 0$ 17) $-1 \geq a$

2.5.4 Using a Graph to Represent a Solution Set

A second way to express the solution set is with a graph. To draw the graph, we start with the number line and use the value we've been given to divide the line into intervals. At the value itself we draw either a parenthesis or a bracket. A parenthesis implies the value itself is not included in the solution set. A bracket implies the value itself is included in the solution set. Last, we darken the interval where the values make the inequality true. For $x > 2$ the graph would be

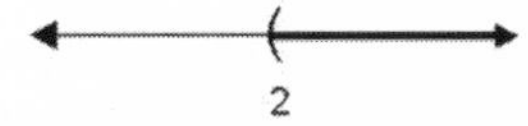

. Notice the line is darker for values to the right of 2. Also, since the values can't equal 2, there's a parenthesis at 2. Here's some practice expressing the solution set of an inequality as a graph.

Practice 2.5.4 Using a Graph to Represent a Solution Set
Express the solution set as a graph.

a) $m < -1$

-1 $\Rightarrow$ Split the number line at -1 and shaded to the left since the solution set includes values less than -1. Used a parenthesis since -1 is not included in the solution set.

b) $4 \geq p$

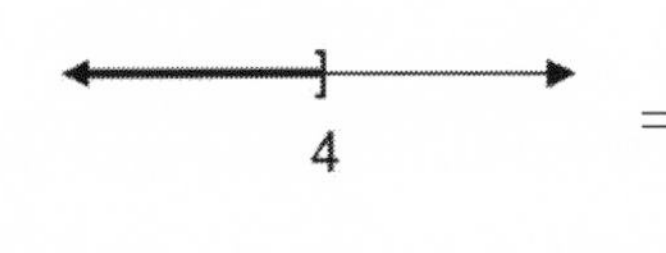

$\Rightarrow$ Split the number line at 4 and shaded to the left since the solution set is values less than or equal to 4. Remembered the inequality symbol doesn't have anything to do with "pointing". Used a bracket since 4 is included in the solution set.

c) $t \geq \frac{8}{3}$

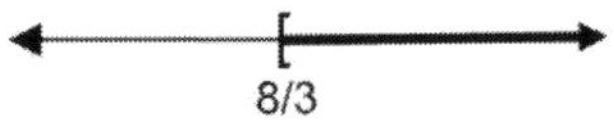

$\Rightarrow$ Split the number line at $\frac{8}{3}$ and shaded to the right since the solution set is values greater than or equal to $\frac{8}{3}$. Used a bracket since $\frac{8}{3}$ is included in the solution set.

d) $-12 < y$

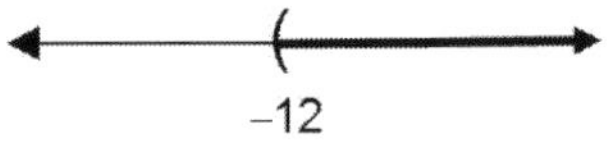

$\Rightarrow$ Split the number line at -12 and shaded to the right. Used a parenthesis since it's a strict inequality so -12 is not included in the solution set.

Homework 2.5 Express the solution set as a graph.

18) $t \leq 7$ 19) $-4 \leq c$ 20) $-\frac{1}{2} > q$ 21) $x > 0$ 22) $-1 \geq a$

2.5.5 Using Interval Notation to Represent a Solution Set

A third way to represent the solution set is using interval notation. With interval notation, we again use a parenthesis for strict inequalities and a bracket when the solution set includes the value itself. To express the idea of including all values to the right of our number we use the positive infinity symbol ∞. To express the idea of including all values to the left we use the symbol for negative infinity $-\infty$. We always use a parenthesis with negative infinity or positive infinity since there isn't a "last" real number either to the left or to the right on the number line.

Practice 2.5.5 Using Interval Notation to Represent a Solution Set
Express the solution set using interval notation.

a) $\{x \mid x \leq 3\}$

$(-\infty, 3]$ $\Rightarrow$ The solution set is values to the left of 3 and 3 itself. Used a bracket since 3 is included in the solution set.

b) 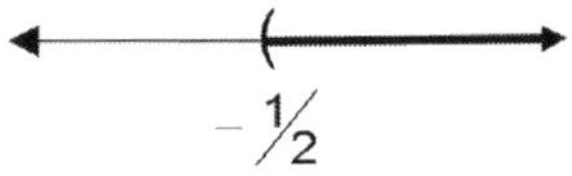

$\left(-\frac{1}{2}, \infty\right)$ $\Rightarrow$ Used parentheses for both the left and right side.

c) $-5 > t$

$(-\infty, -5)$ $\Rightarrow$ I'm interested in real numbers that are less than -5.

Homework 2.5 Express the solution set using interval notation.

23) $-1.3 \geq r$ 24) 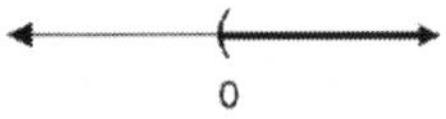

25) $\{x \mid x < 0.5\}$ 26) $x \geq -15$

27) 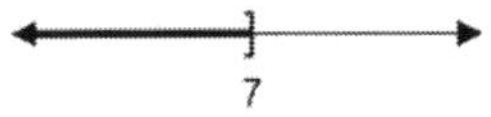

28) $0 < x$ 29) 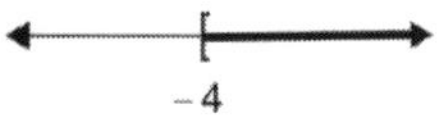

30) $\{k \mid k > -3.3\}$

2.5.6 Mixed Practice with Solution Sets

Now let's put together our different ways of expressing the solution set for an inequality.

Practice 2.5.6 Mixed Practice with Solution Sets

Express the solution set that's shown, in the two ways that are not shown.

a) $\{x \mid x \le 3\}$

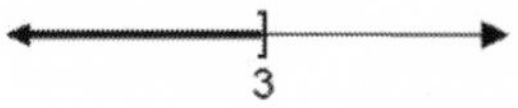

$\Rightarrow$ Set notation is given so I'll need to express the solution set as a graph and in interval notation. Expressed the solution as a graph.

$(-\infty, 3]$ $\Rightarrow$ Expressed the solution in interval notation.

b) 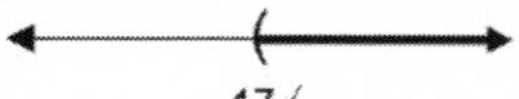

$\{x \mid x > -17/18\}$ $\Rightarrow$ Since a graph is given I'll need to expression the solution set in set notation and in interval notation. Used set notation first. It's common to use *x* as the variable.

$(-17/18, \infty)$ $\Rightarrow$ Used curved brackets for both the left and right side.

Homework 2.5 Express the solution set that's shown, in the two ways that are not shown.

31) $[5,\infty)$

32) 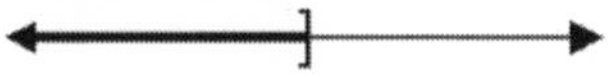

33) $\{x \mid x < 0\}$

34) $\{x \mid x \ge -15\}$

35) $(-\infty,7]$

36) 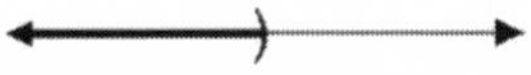

37) 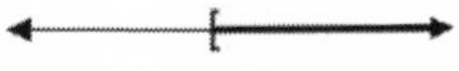

38) $\{x \mid x > -3.75\}$

39) $(-17/18, \infty)$

40) $(-\infty,0]$

41) 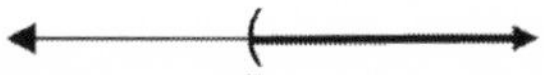

42) $\{x \mid x \le -1\}$

2.5.7 Inequalities with Special Solution Sets

Just like linear equalities, linear inequalities might be a conditional inequality, an identity or a contradiction. So far all of our examples have been conditional linear inequalities, that is, they were true for some real numbers, but not for others. As you might expect an inequality that's always false will be a contradiction and an inequality that's always true will be an identity. Like with linear equalities, you should expect you have a contradiction or identity if during the solution procedure, you're left with only constants.

We'll use the symbol $\varnothing$ when the inequality is a contradiction and has no solution. The symbol $\varnothing$ means a set (in this case the solution set) has no elements. For identities, we'll use $(-\infty,\infty)$ to imply the solution set is all the real numbers from negative infinity to positive infinity.

Practice 2.5.7 Inequalities with Special Solution Sets
Write the appropriate solution.

a) $15 < -14$

$\varnothing \Rightarrow$ The inequality is a contradiction. The solution set will be empty.

b) $3 \geq 1$

$(-\infty, \infty) \Rightarrow$ The inequality is an identity so the solution set contains all real numbers.

Homework 2.5 Write the appropriate solution.

43) $5 \geq 6$	44) $1 \leq 2$	45) $3 \leq 2$	46) $0 > -1$	47) $6 \geq 6$	48) $-8 < -9$

Homework 2.5 Answers

1) Twelve is greater than negative two.
2) $-2 \geq -7$
3) Negative two is less than, or equal to, zero.
4) $5 < 6$
5) One is greater than or equal to one.
6) $-8 < -1$
7) $12 \boxed{>} -2$
8) $-8 \boxed{>} -12$
9) $0 \boxed{>} -2$
10) $-9 \boxed{\leq} -9$ or $-9 \boxed{\geq} -9$
11) $-2.1 \boxed{<} -1.9$
12) $0 \boxed{\geq} 0$ or $0 \boxed{\leq} 0$
13) $\{t \mid t \leq 7\}$
14) $\{c \mid c \geq -4\}$
15) $\left\{q \mid q < \frac{1}{2}\right\}$
16) $\{x \mid x > 0\}$
17) $\{a \mid a \leq -1\}$
18) 7
19) -4
20) -1/2
21) 0
22) -1
23) $(-\infty, -1.3]$
24) $(0, \infty)$
25) $(-\infty, 0.5)$
26) $[-15, \infty)$
27) $(-\infty, 7]$
28) $(0, \infty)$
29) $[-4, \infty)$
30) $(-3.3, \infty)$
31) $\{x \mid x \geq 5\}$ 5
32) $\{x \mid x \leq 15\}$ $(-\infty, 15]$
33) $(-\infty, 0)$ 0
34) $[-15, \infty)$ -15
35) $\{x \mid x \leq 7\}$ 7
36) $\{x \mid x < -1\}$ $(-\infty, -1)$
37) $\left\{x \mid x \geq -\frac{1}{3}\right\}$ $\left[-\frac{1}{3}, \infty\right)$
38) $(-3.75, \infty)$ -3.75
39) $\left\{x \mid x > -\frac{17}{18}\right\}$ $-\frac{17}{18}$
40) $\{x \mid x \leq 0\}$ 0
41) $(0, \infty)$ $\{x \mid x > 0\}$
42) -1 $(-\infty, -1]$
43) $\varnothing$
44) $(-\infty, \infty)$
45) $\varnothing$
46) $(-\infty, \infty)$
47) $(-\infty, \infty)$
48) $\varnothing$

2.6 Solving Linear Inequalities

To solve linear inequalities we reuse the procedure for solving linear equations with a slight change in step 4.

Procedure **– Solving One-Variable Linear Equations**
1. If fractions are present use the LCD to reduce the denominators. 2. Simplify the left expression and the right expression individually. 3. Use additive properties to isolate the variable term and the constant on opposite sides of the inequality symbol. 4. Use multiplicative properties so the variable term has a coefficient of 1. <u>If you multiply or divide by a negative value "flip" the inequality symbol.</u>

Notice that if you multiply or divide by a <u>negative</u> value during step 4, you have to "flip" the inequality symbol. This compensates for the fact that although 2 is to the left of 3, -2 is to the right of -3. If we start with a true inequality like $2 < 3$, and multiply both sides by -1 without changing the inequality symbol, $-2 < -3$, we get a false inequality. If we change the inequality symbol, $-2 > -3$ at the same time that we multiply or divide by -1, the inequality stays true.

2.6.1 Solving Linear Inequalities

Here's some practice solving linear inequalities.

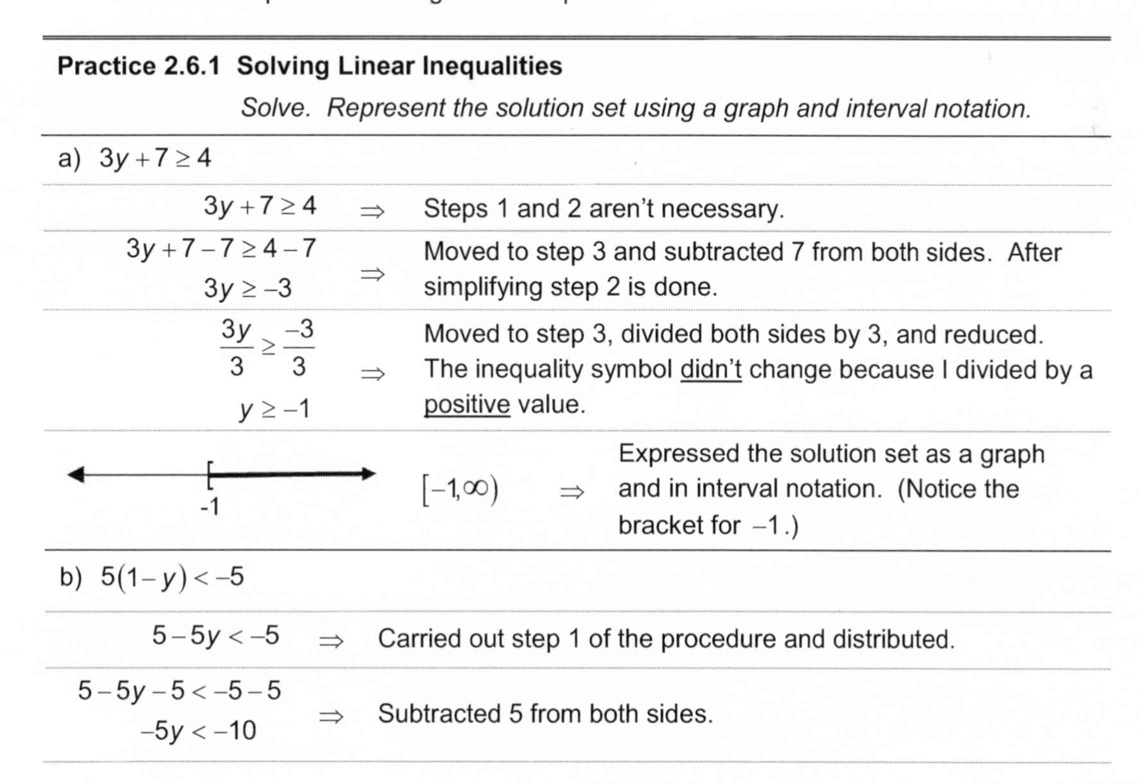

Practice 2.6.1 Solving Linear Inequalities

Solve. Represent the solution set using a graph and interval notation.

a) $3y + 7 \geq 4$

$3y + 7 \geq 4$	$\Rightarrow$	Steps 1 and 2 aren't necessary.
$3y + 7 - 7 \geq 4 - 7$ $3y \geq -3$	$\Rightarrow$	Moved to step 3 and subtracted 7 from both sides. After simplifying step 2 is done.
$\frac{3y}{3} \geq \frac{-3}{3}$ $y \geq -1$	$\Rightarrow$	Moved to step 3, divided both sides by 3, and reduced. The inequality symbol <u>didn't</u> change because I divided by a <u>positive</u> value.
(graph) -1 $[-1,\infty)$	$\Rightarrow$	Expressed the solution set as a graph and in interval notation. (Notice the bracket for -1.)

b) $5(1 - y) < -5$

$5 - 5y < -5$	$\Rightarrow$	Carried out step 1 of the procedure and distributed.
$5 - 5y - 5 < -5 - 5$ $-5y < -10$	$\Rightarrow$	Subtracted 5 from both sides.

$$\frac{-5y}{-5} > \frac{-10}{-5}$$
$$y > 2$$
$\Rightarrow$ Divided both sides by -5 and reduced. Remembered to change the inequality symbol since I divided by a negative value.

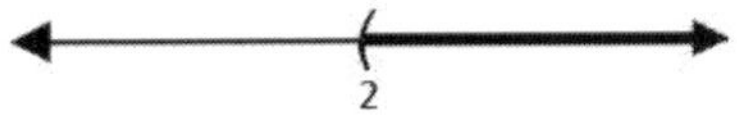

$(2,\infty)$ $\Rightarrow$ Expressed the solution set as a graph and in interval notation.

c) $5(1-y) < -5$

$5-5y < -5$ $\Rightarrow$ Distributed on the left as in the previous problem.

$$5-5y+5y < -5+5y$$
$$5 < -5+5y$$
$\Rightarrow$ Another strategy is to add $5y$ to both sides. This avoids having a negative coefficient.

$$5+5 < -5+5y+5$$
$$10 < 5y$$
$\Rightarrow$ Continued with step 2.

$$\frac{10}{5} < \frac{5y}{5}$$
$$2 < y$$
$\Rightarrow$ Divided and reduced. It wasn't necessary to change the inequality symbol this time since 5 was positive.

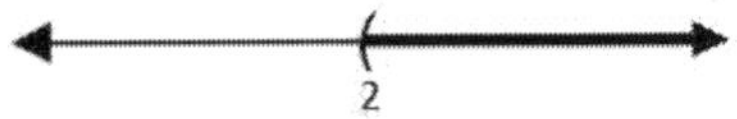

$(2,\infty)$ $\Rightarrow$ The solution remains the same as before.

d) $\frac{n}{4}+1 \ge \frac{n}{3}$

$$\frac{12}{1}\left(\frac{n}{4}+1\right) \ge \frac{12}{1}\left(\frac{n}{3}\right)$$
$$3n+12 \ge 4n$$
$\Rightarrow$ Began with step 1 of the general procedure. The LCD is $2\times2\times3=12$. Multiplied all the numerators by 12 and reduced.

$$3n+12-3n \ge 4n-3n$$
$$12 \ge n$$
$\Rightarrow$ Completed the general procedure.

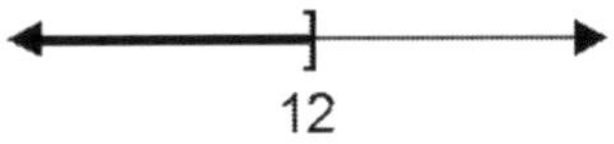

$(-\infty,12]$ $\Rightarrow$ Expressed the solution set.

Homework 2.6 Solve. Represent the solution set using a graph and interval notation.

1) $x-1<11$ 2) $-3x \ge 12$ 3) $-2 > 6-x$ 4) $x/2 - 1 \ge x/3$

5) $2p-9 < p-5$ 6) $\frac{y-4}{4} \le \frac{y-7}{7}$ 7) $-6(3-7) \ge 4(3-x)$

8) $4(1-x)+3(x-2) > 1$ 9) $\frac{4x}{5} - \frac{13}{15} > \frac{5x}{3}$ 10) $3(k-8)-12 > k-8$

2.6.2 Inequalities with Special Solution Sets

Recall the solution set of a linear inequality might be empty (for a contradiction) or contain all the real numbers (for an identity). We'll use the symbol $\varnothing$ when the solution set is empty and $(-\infty,\infty)$ when the solution set contains all the real numbers.

Practice 2.6.2 Inequalities with Special Solution Sets

Solve. Make sure you include an answer.

a) $y+15<2(y-7)-y$

$y+15<2y-14-y$ $y+15<y-14$	$\Rightarrow$	Simplified the right side.
$15<-14$	$\Rightarrow$	Subtracted *y* from both sides. The resulting inequality is a contradiction.
$\varnothing$	$\Rightarrow$	The solution set for a contradiction contains no elements.

b) $3-k\geq -1(k-1)$

$3-k\geq -k+1$	$\Rightarrow$	Simplified the right side.
$3\geq 1$	$\Rightarrow$	Tried to get a single variable term. The resulting inequality is an identity.
$(-\infty,\infty)$	$\Rightarrow$	The solution set for an identity contains all real numbers.

Homework 2.6 Solve. Make sure you include an answer.

11) $4y+1\leq 2(6+2y)$	12) $2(w-7)\leq 7(w+3)$	13) $-16>y-6(1+y)$
14) $\frac{a-2}{3}\geq\frac{a}{9}+2$	15) $x-11>12x-11(x+1)$	16) $-12(x-1)+3(4x-2)\geq 0$
17) $2c+3<c-6(1+c)$	18) $x+\frac{x}{2}\leq 3+\frac{x}{3}$	19) $-p-2(p+1)-(2-3p)\leq 0$
20) $x+2x+x+5\leq -1+4x$	21) $\frac{y+1}{9}-\frac{y-1}{6}\geq 0$	22) $k-2k+3(k-2)>-18$
23) $\frac{3k}{4}<1-\frac{1-k}{2}$	24) $0\geq 4k+2-4(k+4)$	25) $\frac{p}{4}+\frac{p+4}{3}\leq\frac{7p+16}{12}$
26) $0\geq 2(t-4)+4-(8t-6)$	27) $\frac{(r+1)}{2}\geq -1+\frac{r}{3}$	

Homework 2.6 Answers

1) 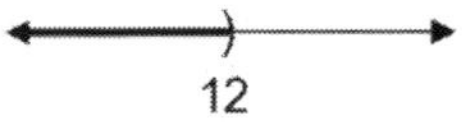

$(-\infty, 12)$

2) 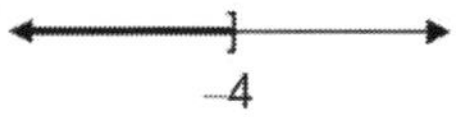

$(-\infty, -4]$

3) 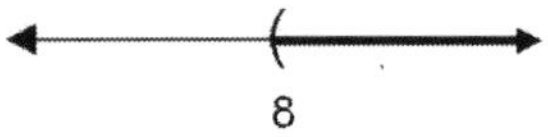

$(8, \infty)$

4) 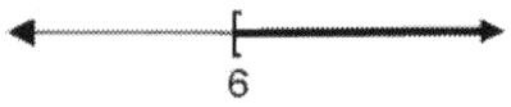

$[6, \infty)$

5) 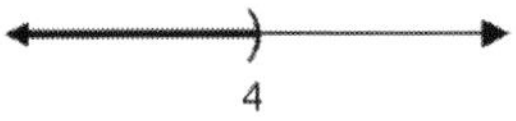

$(-\infty, 4)$

6) 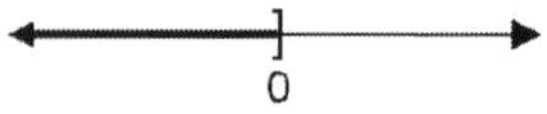

$(-\infty, 0]$

7) 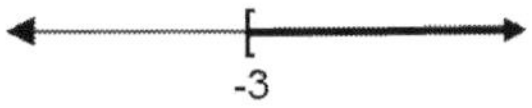

$[-3, \infty)$

8) 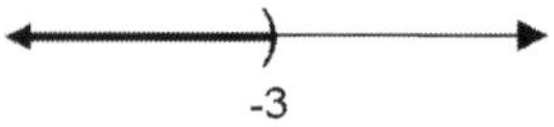

$(-\infty, -3)$

9) 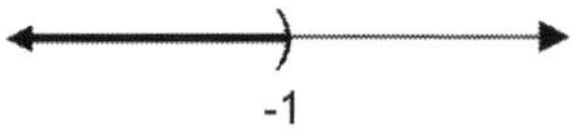

$(-\infty, -1)$

10) 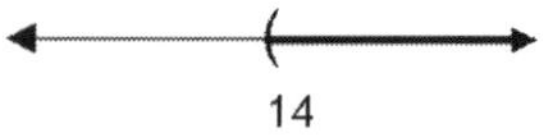

$(14, \infty)$

11) $(-\infty, \infty)$

12) 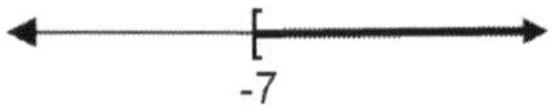

$[-7, \infty)$

13) 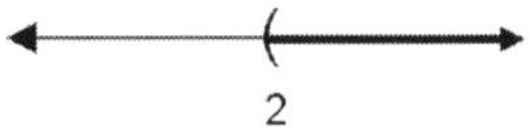

$(2, \infty)$

14) 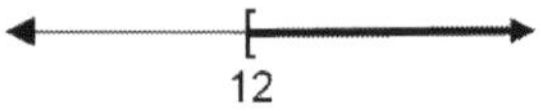

$[12, \infty)$

15) $\varnothing$

16) $(-\infty, \infty)$

17) 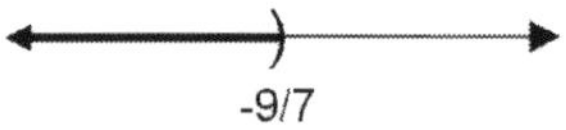

$\left(-\infty, -\frac{9}{7}\right)$

18) 

$\left(-\infty, \frac{18}{7}\right]$

19) $(-\infty, \infty)$

20) $\varnothing$

21) 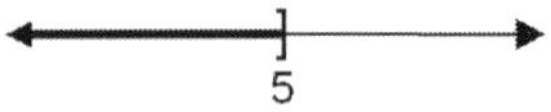

$(-\infty, 5]$

22) 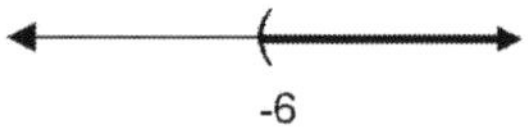

$(-6, \infty)$

23) 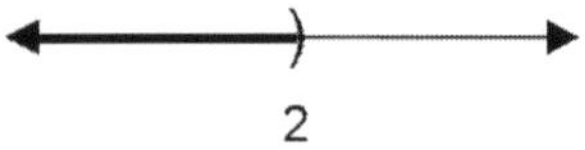

$(-\infty, 2)$

24) $(-\infty, \infty)$

25) $(-\infty, \infty)$

26) 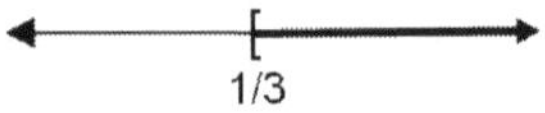

$\left[\frac{1}{3}, \infty\right)$

27) 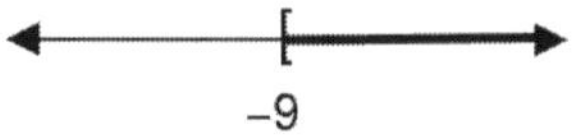

$[-9, \infty)$

Chapter 3

Functions

3.1 An Introduction to Functions

In this chapter, we'll begin working with mathematical functions. Functions are a powerful tool for translating patterns from the "real world" into algebra. It's important to keep the assumptions for functions abstract so we're able to use the same mathematics for similar patterns across fields as diverse as child development, economics or astrophysics. Unfortunately, this abstractness sometimes makes students uncomfortable with functions. In this section, I'll use some "concrete" examples to help develop ideas about building and using functions.

3.1.1 Some Beginning Examples

All of us, instinctively, look for patterns in our environment. One common pattern is believing that a feature found in one item is related to a feature found in a second item. A mathematical function uses numbers, variables, operators and grouping to describe this relationship between the two items.

Here's a mathematical function, $y = 4{,}500 - 150x$, that describes a relationship between the values for our first item (represented by *x*) and the values of our second item (represented by *y*). Although students often say to themselves that this function has a "*y* side" (the left side) and an "*x* side" (the right side), this isn't the case. Instead, the equality symbol implies that both sides are "*y* sides". The right side is simply showing us how to operate on an *x* value to find the value of *y*.

So, what's the relationship that $y = 4{,}500 - 150x$ describes? Well, here's where the beauty of abstractness becomes evident. Maybe the function describes a relationship between the number of cars remaining in a parking lot, *y*, given how many minutes it's been since a concert ended, *x*. In that case the function is telling us that when the concert ended (0 minutes) there were 4,500 cars in the parking lot, and afterwards, 150 cars left the lot every minute. To predict the number of cars remaining in the lot 15 minutes after the concert ends, we'd replace *x* with 15, $y = 4{,}500 - 150(15)$, and simplify the right expression, $y = 2{,}250$, to predict that 15 minutes after the concert ends there will still be 2,250 cars in the parking lot.

Or, say you bought a used car for $4,500 and agreed to pay $150 a month until the loan was paid off. Then $y = 4{,}500 - 150x$ (the same function we used above) would describe the amount still owed on the loan, *y*, given the number of months you've been making payments, *x*. To know how much you still owed after paying for 15 months, you'd replace *x* with 15, $y = 4{,}500 - 150(15)$, simplify the right expression, $y = 2{,}250$, and know that after paying for 15 months you'll still owe $2,250.

Or, say that a National Forest originally had a population of around 4,500 deer but since the introduction of a tick-borne illness, 150 deer have been dying per month. Then $y = 4{,}500 - 150x$ would estimate the current population, *y*, given the number of months since the illness began, *x*. To predict the population 15 months after the illness was introduced, you'd replace *x* with 15, $y = 4{,}500 - 150(15)$, and simplify the right expression, $y = 2{,}250$, to predict that after 15 months the deer population is expected to be 2,250 deer.

The point I'm trying to make is, whether the pattern we're observing comes from civil engineering, finance or forestry management, the same mathematical function can be used. By keeping the concepts of a mathematical function "abstract" we're able to use the same mathematics regardless of the context we're working in.

3.1.2 The Idea of a Data Table

When you're building a function, it's common to begin by collecting data in a **data table**. A data table uses columns (which run up /down) and rows (which run left / right) to organize numerical data. The left column holds **domain** values. The domain is the set of numbers that are considered "independent" in the sense that we're free to choose numbers from the set. In the function $y = 4{,}500 - 150x$, the variable *x* is holding the place for domain values. In fact, domain values (independent values) are commonly called, "x" values. The right column in our data table holds **range** values. The range is a set of numbers that are considered "dependent" since their value depends on the value you selected from the domain. In the function $y = 4{,}500 - 150x$, the variable *y* is holding the place for range values. Notice that finding a *y* value requires us to replace *x* with a value and simplify the expression on the right. As you might suspect, range values (dependent values) are commonly called "y" values.

Columns

Rows

Domain description	**Range description**
Domain element 1	Range element 1
Domain element 2	Range element 2
Domain element 3	Range element 3

The first row in the table describes the domain and range items. Each row, following the first, will show one domain/range pairing. The left cell shows the domain value and the right cell shows the range value that is paired with that domain value. These pairings, which are often called "**ordered pairs**", can be written individually with parentheses like this (domain value, range value) where the domain value is generally called the, **x-coordinate** and the range value is generally called the **y-coordinate**.

Here's a data table that holds some of the information for our relationship between the number of minutes since the concert ended and the number of cars still in the parking lot. Notice that the second row, $(0{,}4500)$, tells us there were 4,500 cars in the lot when the concert ended (after 0 minutes) and that the last row, $(30{,}0)$, tells us that the lot will be empty one-half hour after the concert ends.

Minutes since concert ended	Number of cars still in the lot
0	4,500
5	3,750
15	2,250
30	0

Months paying on the loan	Amount remaining ($)
0	4,500
5	3,750
15	2,250
30	0

Here's a data table with some information about the car loan. Notice the second row, $(0,4500)$ now tells us our original loan amount was $4,500 and the last row, $(30,0)$, tells us that the loan will be paid off after 30 months (two and a half years).

Height (inches)	Weight (pounds)
65	132
67	132
65	140

I need to mention that a two-item relationship is only a function if each domain element is paired with a unique range element. So far each of our data tables has represented a function but to the right I've began recording the height and weight of a group of people. This data table is no longer a function because of what has happened with rows 2 and 4 where the domain value 65 has been paired with two different range values, 132 and 140. If someone asked us to predict the weight for someone that's 65 inches tall, we'd no longer be able to give them a single answer.

Sometimes students overgeneralize this idea and think that rows 2 and 3 are a problem because they both have the weight 132 pounds. This isn't a problem though because in row 2, 65 is paired with 132 (so 65 has a unique range value) and in row 3, 67 is paired with 132 (so 67 has a unique range value). It's not until we move to row 4, and the domain value 65 has taken on two different range values 132 and 140, that we no longer have a function.

Practice 3.1.2 The Idea of a Data Table

Use the data table to answer the questions.

a)

Years after 1954	Cases of nonparalytic polio (100's)
0	181
2	81
4	37
6	16

1. What are the domain values counting?
2. What are the range values counting?
3. How many polio cases were there in 1958?
4. What year were there 18,100 cases of polio?
5. Would you predict more or less than 1,600 cases in 1962?
6. Explain whether the data table does or does not represent a function.

---

a)

1. The values in the left column count the number of years after 1954.
2. The values in the right column count the hundreds of cases of nonparalytic polio.
3. Four years after 1954 there were 3,700 cases of nonparalytic polio.
4. In 1954 (year 0) there were 18,100 cases of polio.
5. Since there were 1,600 cases in 1960, and the number of cases is continually declining, we should expect fewer cases in 1962 (year 8).
6. The data table does represent a function since each domain element is paired with a unique range element.

Homework 3.1 Use the data table to answer the question.

1)

Age of the hybrid vehicle (years)	Value of the hybrid vehicle ($)
1	25,533
3	16,777
5	11,023
8	5,871

a) What are the domain values counting?

b) What are the range values counting?

c) How old is a car with a value of $11,023?

d) After owning the car for eight years, what is its value?

e) Do you suspect the car originally cost more or less than $26,000?

f) Explain whether the data table does or does not represent a function.

2)

Hours after 6 p.m.	Wedding reception attendance
2	320
4	360
5	350
7	320

a) What are the domain values counting?

b) What are the range values counting?

c) How many people were attending the reception at 11 p.m.?

d) At what time was the attendance the same as at 8 p.m.?

e) To the closest hour, around what time do you suspect the attendance was greatest?

f) Explain whether the data table does or does not represent a function.

3)

Age over 35	Monthly cost of $100,000 worth of life insurance ($)
0	8
10	12
25	28
40	155

a) What are the domain values counting?

b) What are the range values counting?

c) How much will it cost you per month if you're 45 years old?

d) How old are you if your monthly cost is $155?

e) Compare the yearly cost at age 60 to the yearly cost at age 75.

f) Explain whether the data table does or does not represent a function.

4)

Persons age (Years)	Gallons of gas purchased
42	10
22	8
59	16
22	16

a) What are the domain values counting?

b) What are the range values counting?

c) If someone purchased 16 gallons of gas, how old were they?

d) If someone is 22 years old, how much gas did they purchase?

e) Explain whether the data table does or does not represent a function.

3.1.3 *Translating Data Tables to Graphs*

For most people, patterns that aren't apparent in a data table become obvious when the same information is presented as a graph (a picture of the data). For instance, both graphs below have five data points but notice how intuitively you "see" a pattern for the graph on the left.

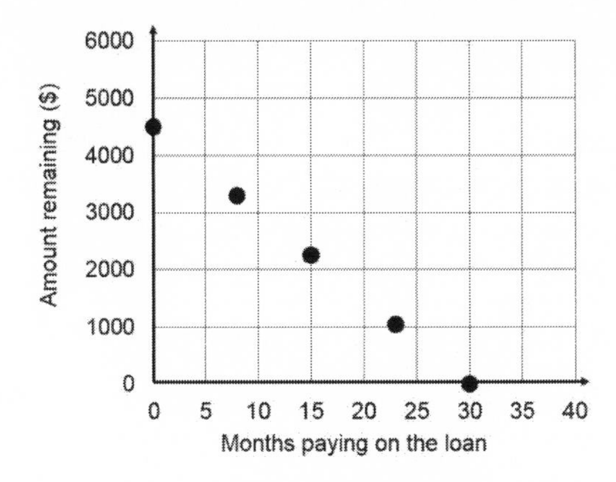

In a graph, domain information and range information are both represented by their own number line. Here's a domain number line with a point at each month value. And here's a range number line with a point at each dollar value. Please notice that each number line is "scaled" so the distances between axis values stays the same. (5 months for the first number line and $1,000 for the second.)

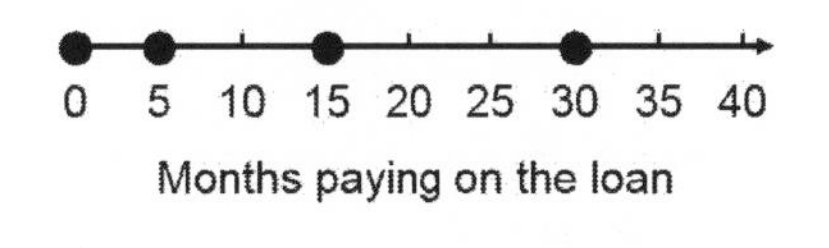

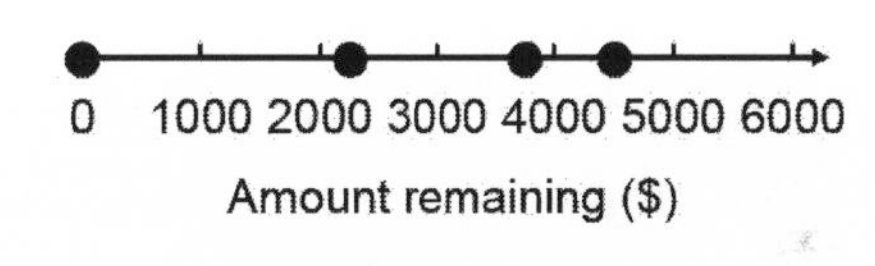

Although each individual number line has all the original data from the data table, it's not possible to see any relationship between the data. That's because each number line only has one dimension, and since both number lines are running left to right, both sets of information are in the same dimension.

To show the relationship I'll turn the range number line perpendicular to the domain number line and intersect the lines at the point (0,0) which is known as the **origin**. The domain number line (in this case information about the months) is generally called the ***x*-axis** and the range number line (in this case information about the dollars) is generally called the ***y*-axis**. Notice how my information is now in two dimensions. The "months" information runs left/right and the "dollars" information runs up/down.

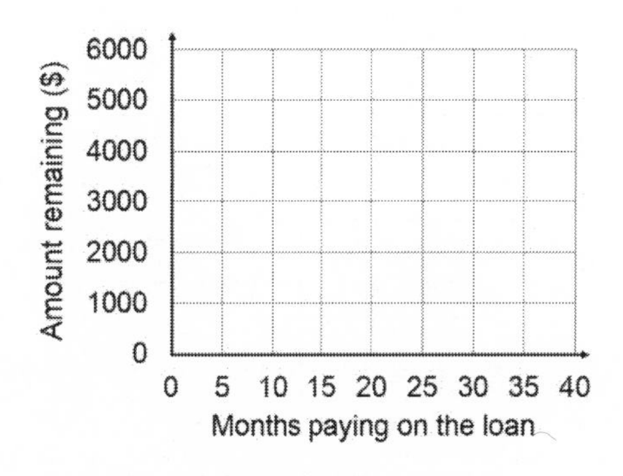

In two dimensions, each point is an ordered pair of the form, (months, dollars), where the first coordinate tells me how far to move left or right from the origin and the second coordinate tells me how far to move up or down from the origin.

In two dimensions, month 5 is a <u>line</u> that starts at 5 on the x-axis and is parallel to the *y*- axis. In the same way 3,750 is a <u>line</u> that starts at 3,750 on the *y*- axis and is drawn parallel to the *x*-axis. The point where these two lines intersect, $(5,3750)$, now carries information about both months and dollars and tells me that in month 5, the outstanding loan amount was $3,750. The graph to the right contains the ordered pair information (the row information) for the entire data table.

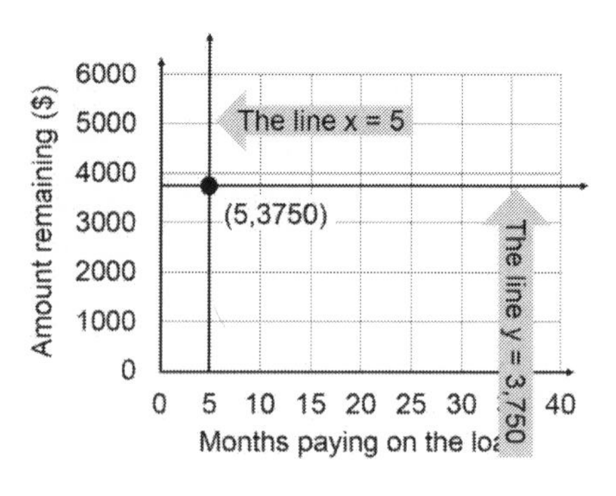

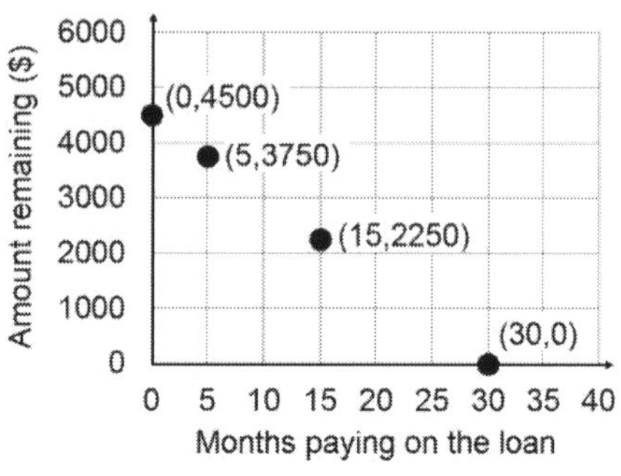

Notice if any vertical line had intersected more than one point, this wouldn't be the graph of a function. (A domain value would have had more than one range value.) This idea, that we only have the graph of a function if each vertical line from a domain value intersects a single range value, is known as the **vertical line test** for functions.

In the example above, the values for the *x*-axis and the *y*-axis were all positive or 0. If our *x* and *y* axes are general number lines, with both positive and negative values, we divide the two dimensions into four quadrants which are labeled counterclockwise from 1 to 4. The graph to the right shows the numbering. Let's practice graphing some data tables that have points in all four quadrants and then discuss the meaning of some assorted points in two dimensions.

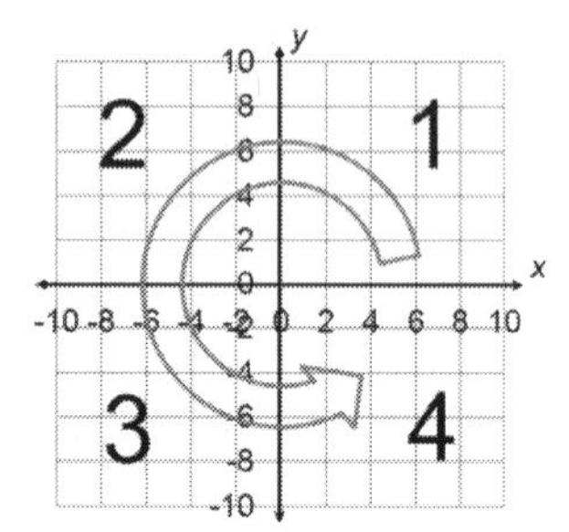

Practice 3.1.3 Translating Data Tables to Graphs

Graph the data table. After graphing, use the vertical line test to decide if you have the graph of a function.

---

a)

x	y
–8	6
–4	–1
–1	3
4	–9
8	–4

⇒

This is the graph of a function since a vertical line drawn from each domain value intersects only one range value. The graph passes the vertical line test.

Homework 3.1 Graph the data table. After graphing, use the vertical line test to decide if you have the graph of a function.

5)

x	y
-4	3
0	-2
8	-2
-8	-3
0	3

6)

x	y
-4	-2
-1	-1
0	0
1	2
2	6

7)

x	y
-3	-9
-2	-5
0	1
2	3
6	-5

8)

x	y
8	-8
4	-4
3	0
4	6
1	1

Now let's practice translating a data point into English. To accurately translate a data point, it's important to consider all available information.

Think like an expert

Novices tend to passively view a graph as a static picture. They put most of their attention on the shape of the graph and very little attention on the outside information. Experts combine the shape of the graph, and the outside labels and scaling, to actively create a story about the relationship between "*x*" and "*y*".

Practice 3.1.3 Translating Ordered Pairs to English.

Identify the coordinates of each point and discuss its meaning.

a)

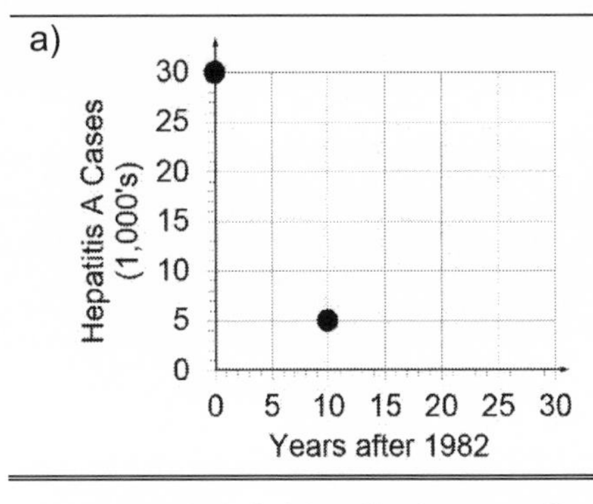

a) The point $(0,30)$ implies that in 1982 there were 30,000 cases of Hepatitis A. The point $(10,5)$ implies that in 1992 there were 5,000 cases of Hepatitis A.

Homework 3.1 Identify the coordinates of each point and discuss its meaning.

9)

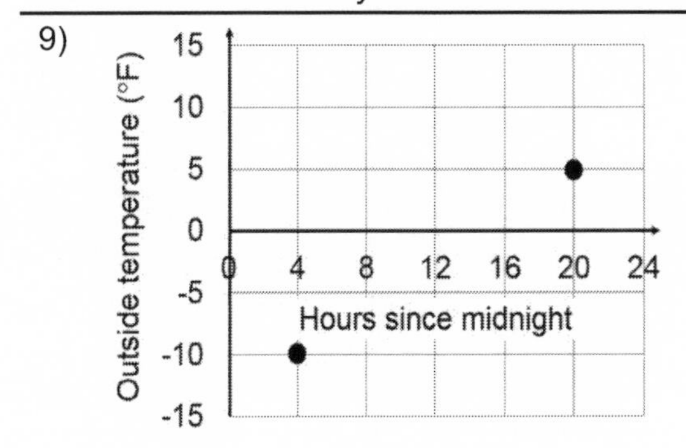

10)

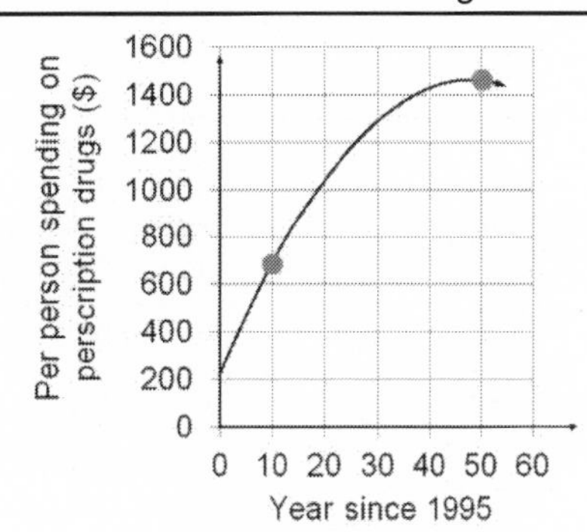

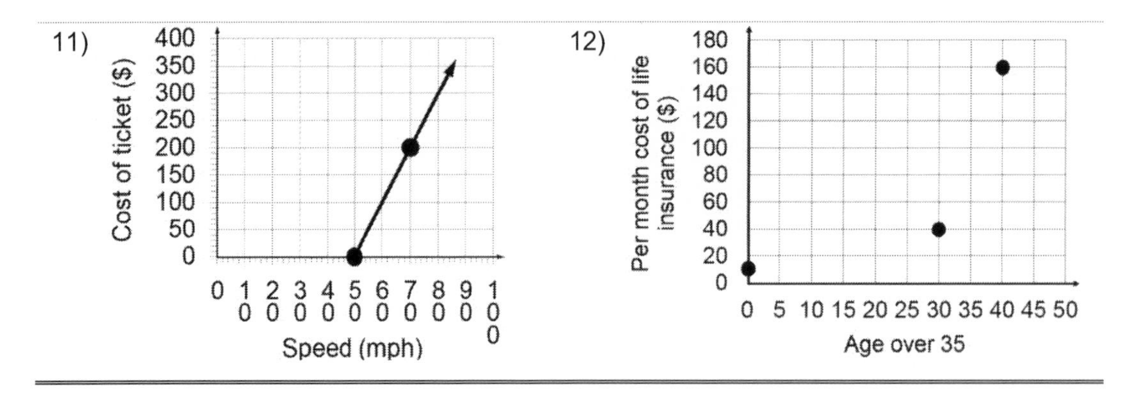

3.1.4 Increasing, Decreasing or Constant Functions

After making a graph, the notion of increasing, decreasing or constant can help describe any trends you see in the data. A function is increasing, decreasing or constant depending on what's happening to range values (y) as domain values (x) become greater (move left to right on the x-axis).

For instance, the graph to the right shows the relationship between time and distance for a family that drove for two hours to a birthday party, stayed at the party for two hours and then drove home for 2 hours. From 0 hours to 2 hours the values of y (the distance from home) are increasing so the function is increasing. From 2 hours to 4 hours the y values are neither increasing nor decreasing so the function is constant. From 4 hours to 6 hours the values of y are decreasing so the function is decreasing. Using intervals on x, we'd say the function is increasing from $x=0$ to $x=2$, is constant from $x=2$ to $x=4$, and is decreasing on $x=4$ to $x=6$.

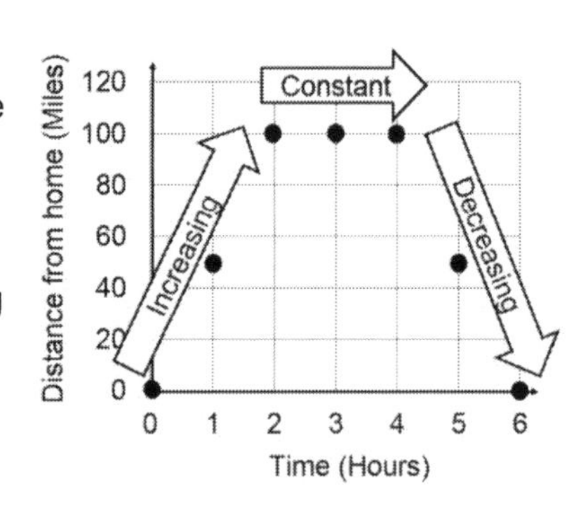

Practice 3.1.4 Increasing, Decreasing or Constant Functions

Moving left to right on the x-axis describe where the function is increasing, decreasing or constant.

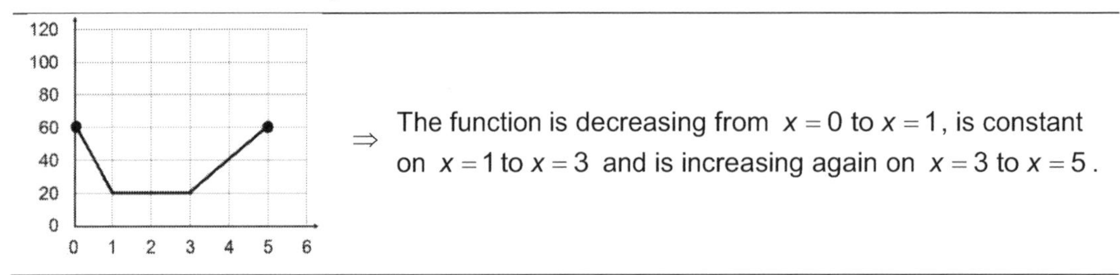

$\Rightarrow$ The function is decreasing from $x=0$ to $x=1$, is constant on $x=1$ to $x=3$ and is increasing again on $x=3$ to $x=5$.

Homework 3.1 Moving left to right on the x-axis describe where the function is increasing, decreasing or constant.

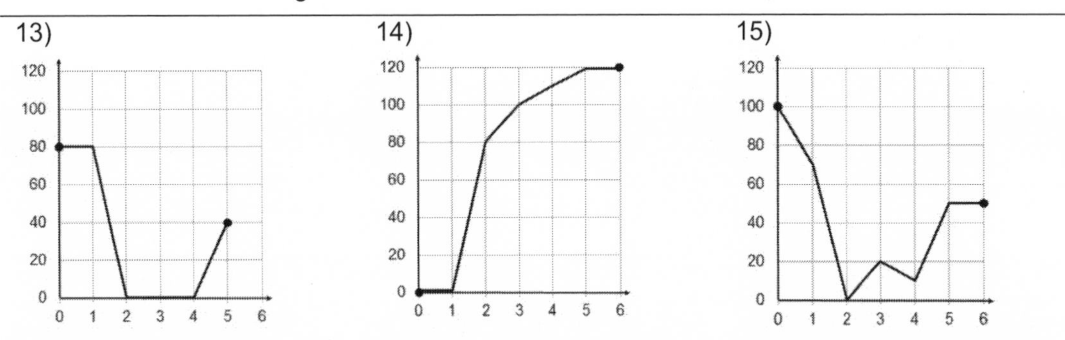

3.1.5 Describing a Trend in the Data

To describe the graph of an applied function it's important to include the labels and scaling along with the shape of the graph.

For example, to work with the graph to the right I use the labels to recognize this relationship will be about the change in the number of Hepatitis A cases over time. As I continue moving my attention inside I see that the domain values cover the 20-year period from 1995 to 2015, and the range values show from 0 to 40,000 cases of Hepatitis A. Notice each change of 1 on the *y*-axis is actually a change of 1,000 cases. Last, I'd note the points are falling over time. One narrative for this graph might be, "The cases of Hepatitis A were falling from around thirty-two thousand cases in nineteen ninety-five to around two-thousand cases in two-thousand ten."

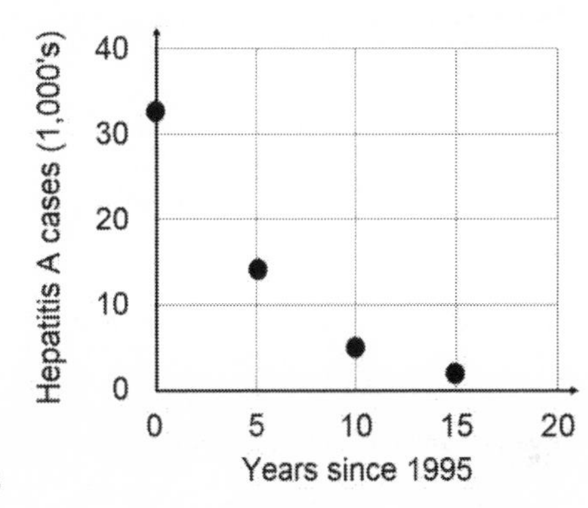

Practice 3.1.5 Describing a Trend in the Data

Using the graph, describe the relationship between x and y.

a) 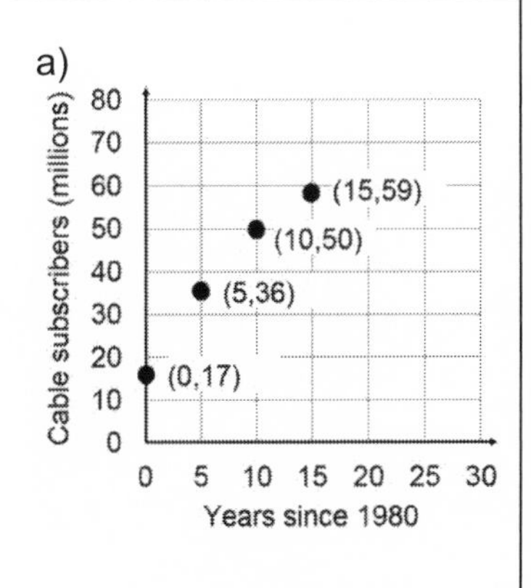	From 1980 to 1995 the number of cable subscribers increased by around 40 million.	⇒	Began with the labels and noticed this function is describing the relationship between the number of cable subscribers and the year since 1980. Next, I realized as the values of *x* increased, the values of *y* increased from 17 to 59. Noticed each 1 unit change in *y* was actually a change of 1 million.

Homework 3.1 Using the graph, describe the relationship between x and y.

16)

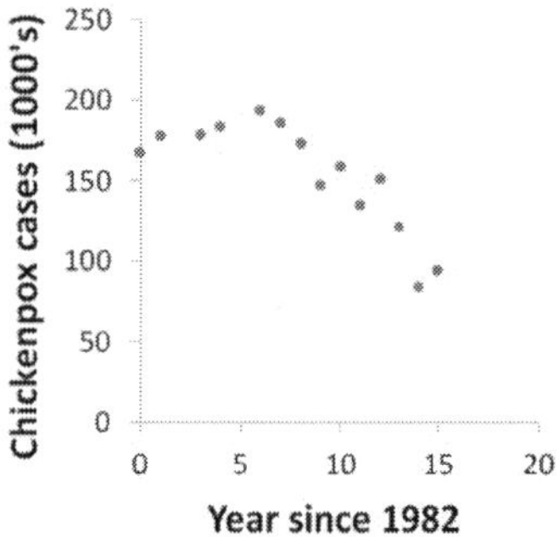
Chickenpox cases (1000's)
250
200
150
100
50
0
0 5 10 15 20
Year since 1982

17)

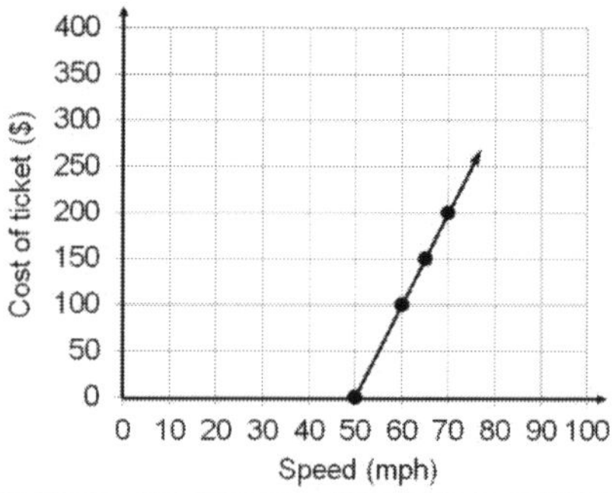
Cost of ticket ($)
400
350
300
250
200
150
100
50
0
0 10 20 30 40 50 60 70 80 90 100
Speed (mph)

18)

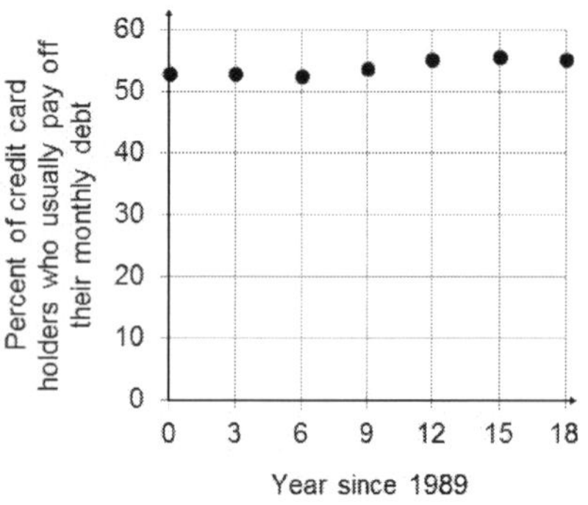
Percent of credit card holders who usually pay off their monthly debt
60
50
40
30
20
10
0
0 3 6 9 12 15 18
Year since 1989

19)

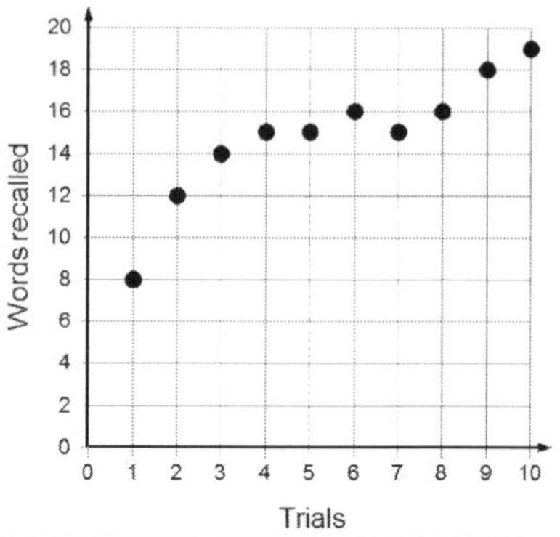
Words recalled
20
18
16
14
12
10
8
6
4
2
0
0 1 2 3 4 5 6 7 8 9 10
Trials

3.1.6 Using a Graph to Predict Values

Even though a data table and its graph represent the same information, it's more intuitive for us to find patterns using a graph. If you see a pattern in your graphed data then continuing the shape beyond the given points, allows us to predict ordered pairs we don't currently have. For example, we can start with an *x* value not in the data table and use the graph to predict the *y* value that would go with it. Or we can start with a *y* value not in the data table and use the graph to predict the *x* value it came from. Before using a function to make predictions it's essential you actively understand the relationship the function describes. Here's some practice using the relationship between *x* and *y* to estimate unknown values.

Practice 3.1.6 Using a Graph to Predict Values

Graph the function and then use your graph to answer the questions.

a)

Years after 1960	U.S. population (millions)
0	176
20	228
35	267
45	293

1. Graph the function. Scale your *x*-axis from 0 to 80 and your *y*-axis from 0 to 400.
2. Describe the trend in the data.
3. Use your graph to estimate the U.S. population in 2017.
4. Use your graph to estimate when the U.S. population first reached 250 million.
5. Use your graph to estimate when the U.S. population will be 150 million more than it was in 1970.

1.

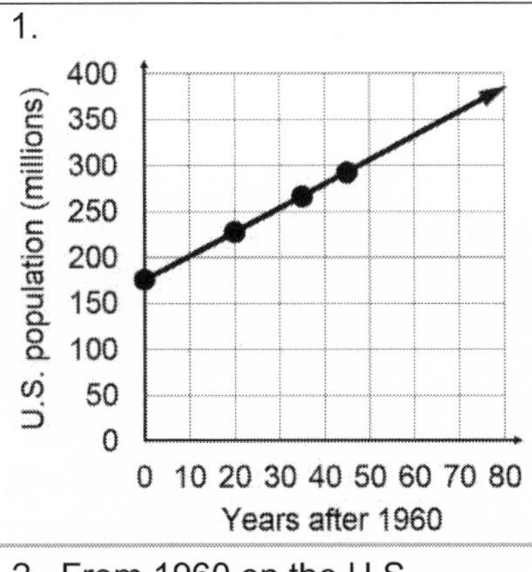

⇒ Plotted the points and connected them with a line. Made sure to correctly label both axes.

2. From 1960 on the U.S. population has been increasing. ⇒ The labels helped me describe the trend.

3. About 325 million.

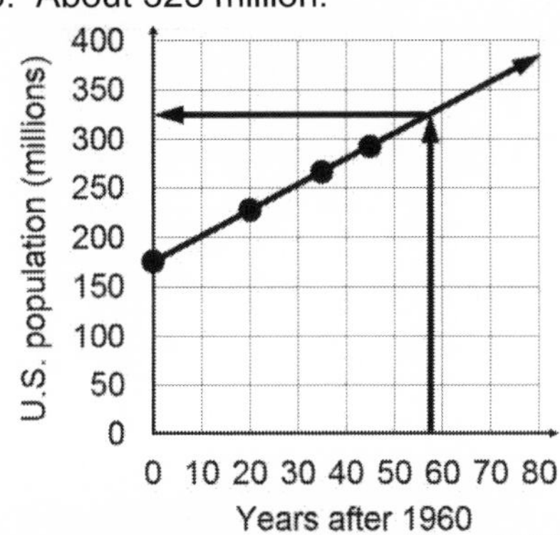

⇒ Started at year 57, (since 2017 is 57 years after 1960) moved up to the line and then left to find the appropriate range value was around 325 which would be 325 million. This is known as **extrapolation** since we're estimating beyond the original data values.

4. Around 1989.

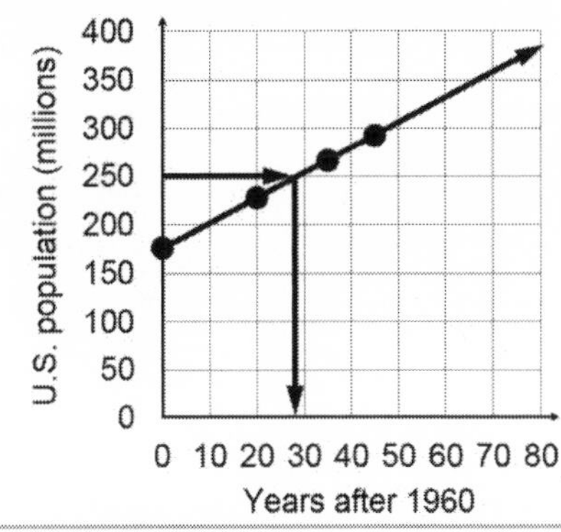

⇒ Started at 250 in the range values, moved right to the line and then moved down to see that 250 million pairs with about 29 in the domain. 29 years after 1960 would be 1989. The population actually first reached 250 million around March of 1990. This is known as **interpolation** since we're estimating within the original data values.

5. Around 2025.

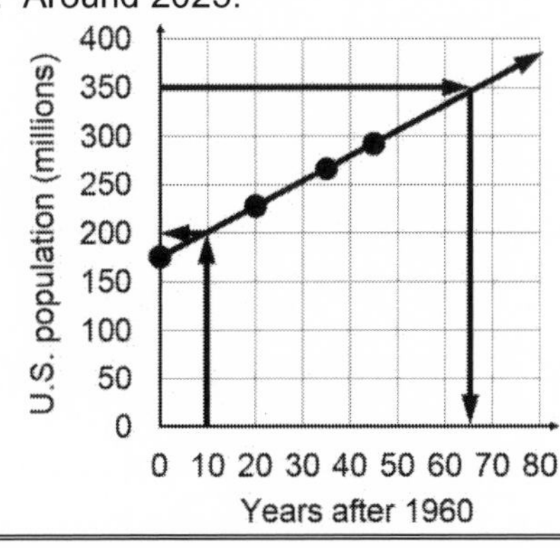

⇒ Started at year 10 (1970) moved up to the line, and then left to find the population in 1970 was about 200 million. 150 million more than 200 million would be 350 million so I started at 350 on the population axis, moved right to the line and then down to find 350 pairs with 65 on the years axis. 65 years after 1960 would be 2025.

Homework 3.1 Graph the function and then use your graph to answer the questions.

20) a) Graph the function. Scale your x-axis from 0 to 30 and your y-axis from 0 to 3,000.
b) Discuss the trend in the data.
c) Predict when $1,200 was still owed on the loan.
d) Predict the amount owed during month 0. What is the meaning of this amount?
e) Predict the number of months it will take to pay off the loan. (This occurs when the amount owed is $0.)

Months paying on a loan	Outstanding Amount ($)
5	2,000
10	1,500
15	1,000

21) a) Graph the function. Scale the x-axis from 0 to 60 and the y-axis from 0 to 80.
b) Discuss the trend in the data.
c) Predict the median income in 2030.
d) Predict the first year median income was (or will be) $60,000.
e) Predict when median income was twice the amount in 1985.

Year since 1980	Median Income ($1,000)
10	29.2
20	39.9
30	50.5

22) a) Graph the function. Scale the x-axis from 0 to 60 and the y-axis from 0 to 400.
b) Discuss the trend in the data.
c) What's the minimum number of cars that need to be washed to raise a profit of $350?
d) How much profit will washing 50 cars raise?
e) After raising $150 someone wonders what their profit will be if they wash an additional 15 cars?

Number of cars washed	Profit ($)
10	33
20	98
35	195.5

23) a) Graph the function. Scale the x-axis from 0 to 30 and the y-axis from 0 to 400.
b) Discuss the trend in the data.
c) Predict where the line touches the y-axis. What is the meaning of this y value?
d) Predict the number of cigarettes sold in 2026.
e) Predict the last year two-hundred billion cigarettes will be sold?

Year since 2001	Cigarettes sold (Billions)
4	349
8	305
12	261

Homework 3.1 Answers

1) a) Domain values count the number of years since the vehicle was purchased.

   b) Range values count how much the vehicle is worth.

   c) If the car's worth $11,023 then it's 5 years old.

   d) After 8 years, the car is worth $5,871.

   e) Between years 1 and 3 the value dropped about $9,000 or about $4,500 per year. Since in year 1 the cost was around $25,500, and $26,000 is only $500 more than the year 1 value, I suspect the car originally cost more than $26,000.

   f) The table does represent a function since each domain value has only one range value.

2) a) Domain values count the number of hours after 6 p.m.

   b) Range values count the number of people at the reception.

   c) 11 p.m. is 5 hours after 6 p.m. so it's 350 people.

   d) At 8 p.m. (hour 2) there were 320 people. There was also 320 people at hour 7 which is 1 a.m.

   e) Attendance was greatest around hour 4 which is 10 p.m.

   f) The table does represent a function since each domain value has only one range value.

3) a) Domain values count the number of years beyond age 35.

   b) Range values count the monthly cost for $100,000 worth of life insurance.

   c) If you're 45 (year 10) the cost is $12 per month.

   d) If you're paying $155 per month you're 75 years old (year 40).

   e) The cost for a year of insurance when you're 60 is $336. At 75 one year of insurance costs $1,860. You're paying an additional $1,524 per year when you're 75 years old.

   f) The table does represent a function since each domain value has only one range value.

4) a) Domain values count the number of years someone's been alive.

   b) Range values count how many gallons of gas the person purchased.

   c) If someone purchased 16 gallons they might be 59 years old or 22 years old.

   d) If someone is 22 years old they might have purchased 8 gallons or 16 gallons of gas.

   e) The data table does <u>not</u> represent a function because the domain value 22 is paired with two range values, 8 and 16. The fact that both a 59-year-old and a 22-year-old both purchased 16 gallons is fine and <u>doesn't</u> imply the data table isn't a function.

5) This is not a function. The points $(0,-2)$ and (0,3) fail the vertical line test.

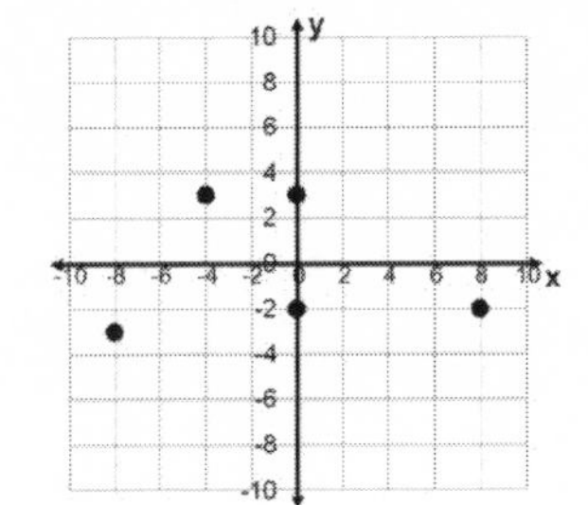

6) This is a function. The graph passes the vertical line test.

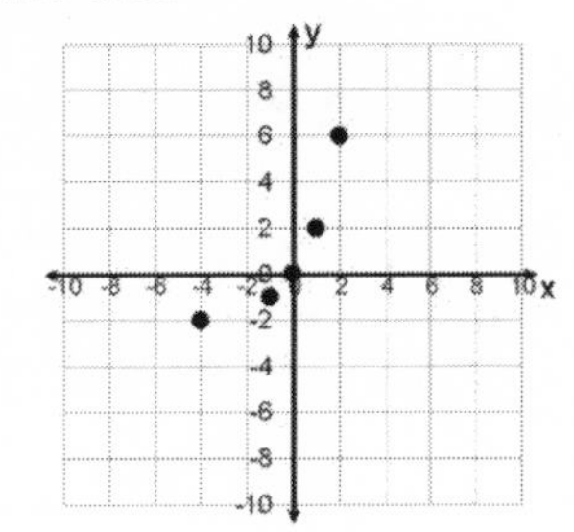

7) This is a function. The graph passes the vertical line test.

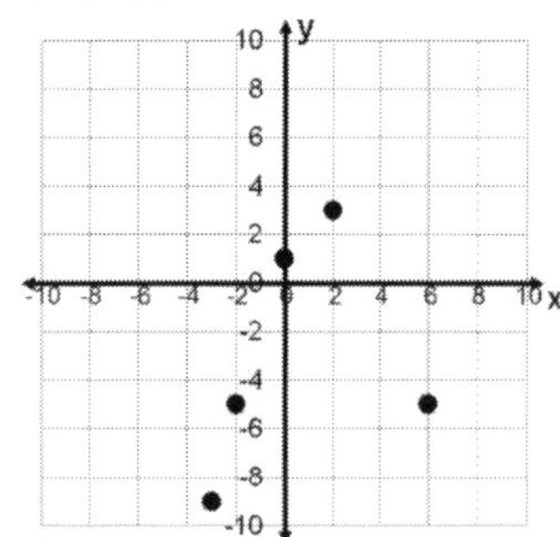

8) This is not a function since the points (4,–4) and (4, 6) fail the vertical line test.

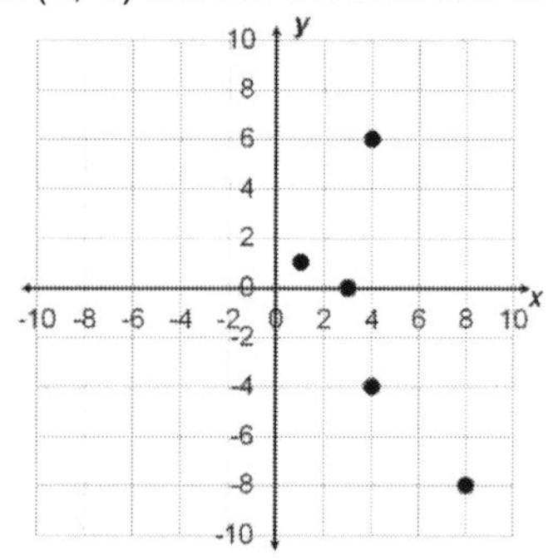

9) The point $(4,-10)$ tells us that at 4 a.m. the temperature was negative ten degrees Fahrenheit. The point (20, 5) tells us that at 8 p.m. the temperature was five degrees Fahrenheit.

10) The point (10,700) implies that in 2005 the per-person spending on drugs was \$700. The point (50, 2045) predicts that, if the pattern continues, the per-person spending will be around \$1,450 in 2045.

11) The point (0, 50) tells us that if someone is going 50 mph they don't pay a fine. The implication is that the speed limit is 50 mph. The point (70, 200) tells us that someone caught driving at 70 mph will pay a \$200 fine.

12) The point (0,10) implies that at 35 your monthly cost for life insurance would be about \$10. The point (30,40) implies that at 65 your monthly cost would be about \$40. The point (40,160) implies that at 75 your monthly cost would be around \$160.

13) The function is constant for the interval $x=0$ to $x=1$, is decreasing for the interval $x=1$ to $x=2$, is again constant for the interval $x=2$ to $x=4$ and is increasing for the interval $x=4$ to $x=5$.

14) The function is constant for the interval $x=0$ to $x=1$, is increasing for the interval $x=1$ to $x=5$, and is again constant for the interval $x=5$ to $x=6$.

15) The function is decreasing for the interval $x=0$ to $x=2$, is increasing for the interval $x=2$ to $x=3$, is decreasing for the interval $x=3$ to $x=4$, is increasing for the interval $x=4$ to $x=5$ and is constant for the interval $x=5$ to $x=6$.

16) The number of cases of Chickenpox increased between 1982 and 1987by around 40,000 cases and then decreased from 1987 through at least 1997 by around 100,000 cases.

17) As your speed increases from 50 to 70 mph, the cost of your speeding ticket increases by around \$200.

18) Between 1989 and 2007 the percent of credit card holders who didn't carry a debt month to month remained approximately constant at around 53%.

19) The number of words recalled increased from the first to the fourth trial by around 6 words, then remained relatively constant through the seventh trial and then increased again through the tenth trial by an additional 3 words.

20) a)

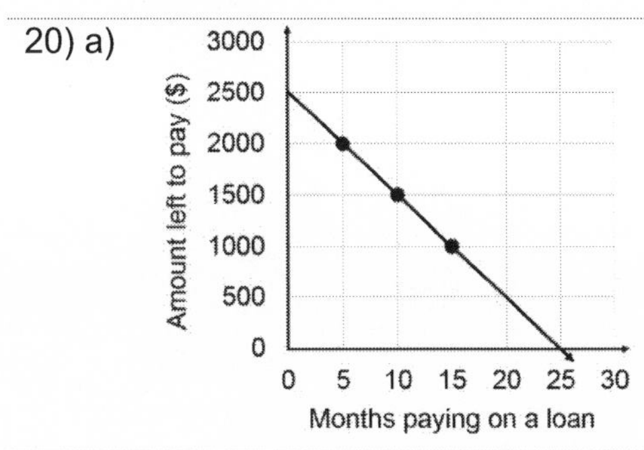

b) Over time the outstanding loan amount is declining.

c) $1,200 was owed after paying for about 12 months.

d) $2,500 was owed during month 0. This represents the original amount of the loan.

e) About 25 months.

21) a)

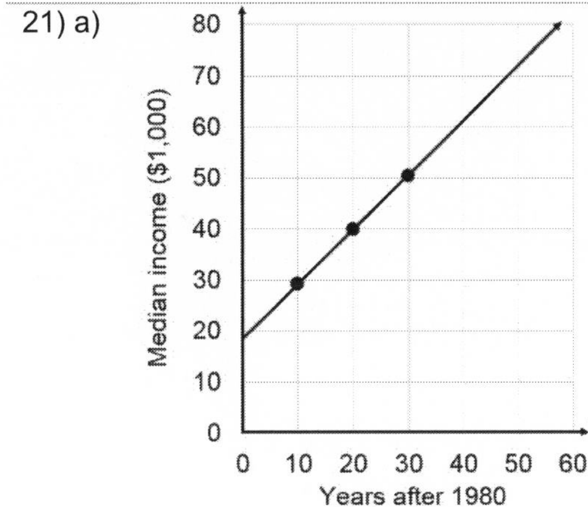

b) Median income was rising from 1980 through 2010.

c) Median income would be around $72,000.

d) Around 2019-2020.

e) In 1985 (year 5) income was around $25,000. Twice that would be $50,000 which occurred around 2010.

22) a)

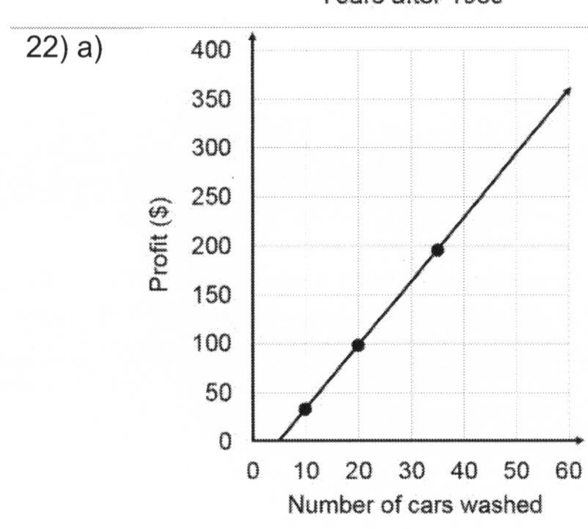

b) Profit increases as more cars are washed.

c) Around 58-59 cars.

d) Washing 50 cars will raise around $300.

e) A profit of $150 implies around 28 cars have been washed. 15 more cars will be 43 cars. Washing 43 cars will lead to a profit of around $250.

23) a)

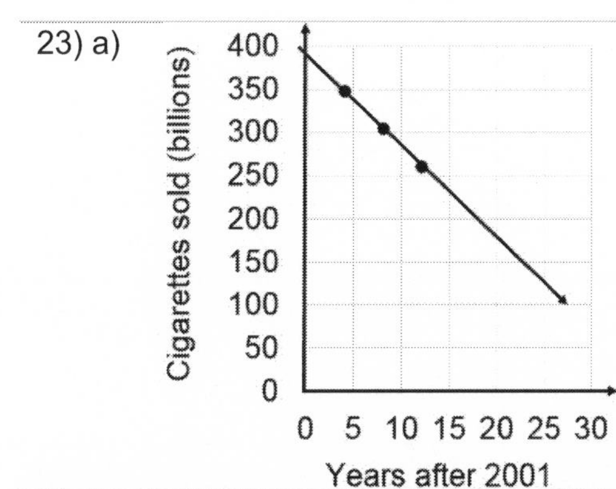

b) Cigarette sales have been decreasing.

c) In 2001 around 395 billion cigarettes were sold.

d) In 2026 about 125 billion cigarettes will be sold.

e) Around 2018.

3.2 Intercepts

Finding intercepts, and interpreting their meaning, sometimes helps us "make sense" of a function.

3.2.1 Finding and Interpreting Intercepts

Frequently, when you graph a function, the line will intersect the *x*- axis and/or the *y*-axis. These points of intersection are known as intercepts. The **y-intercept** of a function is the ordered pair that results when the value of *x* is 0. An **x-intercept** of a function is any ordered pair that results when the value of *y* is 0. Let's go through a couple examples so you'll be able to identify, and discuss the meaning of, intercepts.

If you look at the graph below, you'll see we have a relationship between how many cars were washed and the profit. As we should expect, as the number of cars washed increases, so does the profit. Notice that when the value of *x* is 0 the value of *y* is -60 so the *y*-intercept is $(0,-60)$. The meaning of $(0,-60)$ would be that when we haven't washed any cars (*x* is 0) we're $60 in debt (*y* is -60). The debt might be for things like buying supplies or advertising.

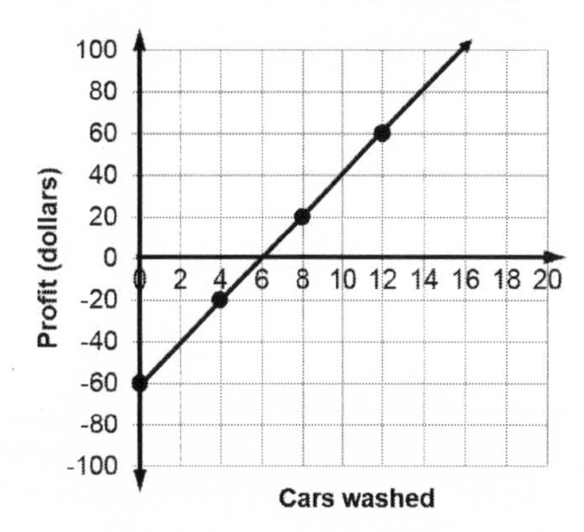

In the same way when the value of *y* is 0, the value of *x* is 6, which makes the *x*-intercept $(6,0)$. The *x*-intercept tells us that after washing 6 cars (*x* is 6) our profit will be zero dollars (*y* is 0). Sometimes this point is called the "break-even" point since, after washing 6 cars, we're able to pay back our original loss of $60.

Before going on I need to mention that even though the *y*-intercept is $(0,-60)$ it's common for people to disregard the 0 coordinate and say, "The y intercept is negative sixty". In the same way, it's common to hear, "The x intercept is six." even though it's actually $(6,0)$.

Let's try another example using a data table instead of a graph. If you look at the data table below, you'll notice that the first row uses the number of minutes since a concert ended to predict the number of cars remaining in a parking lot. As expected, the number of cars decreases after the concert ends. The second row, $(0,630)$ will be the *y*-intercept since the value of *x* is 0. The *y*-intercept implies there were 630 cars in the lot when the concert ended (at 0 minutes). The fifth row, $(90,0)$, is an *x*-intercept since the value of *y* is 0. The *x*-intercept implies all the cars have left the lot an hour and a half (90 minutes) after the concert ends.

Minutes since concert ended	Cars remaining in the parking lot
0	630
30	420
60	210
90	0

Here's some practice finding, and discussing, the intercepts of applied functions.

Practice 3.2.1 Finding and Interpreting Intercepts

Using the given information;

1. *Identify the y-intercept (if there is one) and discuss its meaning or why it's not helpful to know the y-intercept.*
2. *Identify the x-intercept (if there is one) and discuss its meaning or why it's not helpful to know the x-intercept.*

a)

Years since 1974	Movie ticket prices (Dollars)
0	1.98
10	3.28
30	5.88

1. The y-intercept $(0,1.98)$ implies that in 1974 (year 0) the price of a movie ticket was \$1.98. $\Rightarrow$ I started in the left (the x) column and found the pair where the value of x was 0.

2. The x-intercept isn't helpful since movie tickets were never free. $\Rightarrow$ I started in the right (the y) column and tried to find a pair where the value of y was 0. Since y doesn't take on the value 0, there is no known x-intercept. The x-intercept wouldn't be helpful anyway since it would find the years the price of a movie ticket was free (the price dropped to \$0) which isn't going to happen.

b)

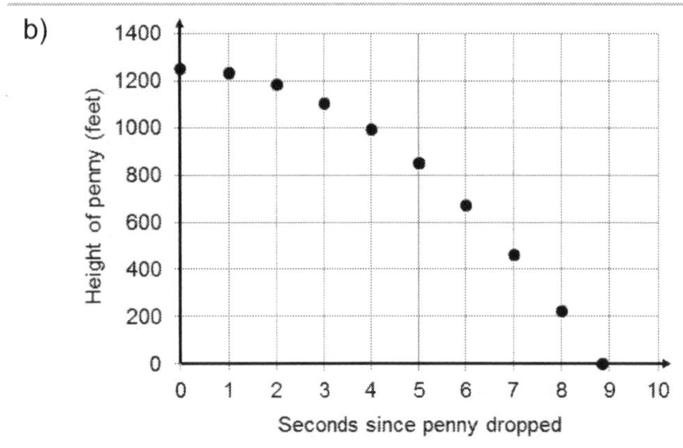

1. The y-intercept $(0,1250)$ implies the penny was dropped from a height of around 1,250 feet. $\Rightarrow$ I started at 0 on the x-axis and moved up the y-axis to the point $(0,1250)$.

2. The x-intercept $(9,0)$ implies that, after being released, it will take the penny around 9 seconds to hit the ground. $\Rightarrow$ I started at 0 on the y-axis (the penny is on the ground) and moved along the x-axis to about $(9,0)$.

Homework 3.2 Using the given information

a) Identify the y-intercept (if there is one) and discuss its meaning or why it's not helpful to know the y-intercept.

b) Identify the x-intercept (if there is one) and discuss its meaning or why it's not helpful to know the x-intercept.

1)

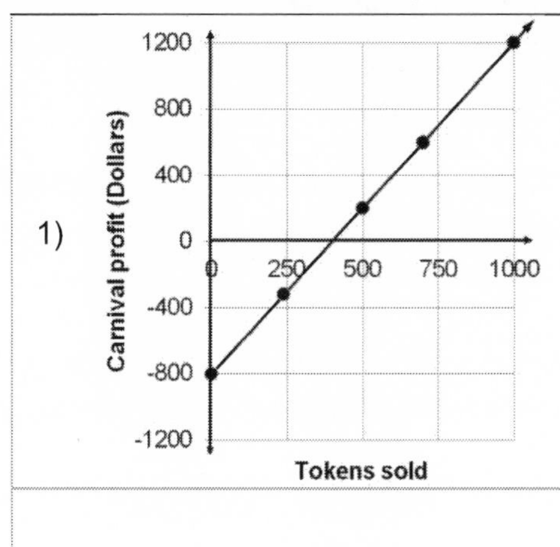

2)

Time traveling in a car (hours)	Distance from home (miles)
0	288
2.25	172
4.5	81
6	0

3)

Months on a diet	Weight (pounds)
0	140
5	129
7	120
10	118

4)

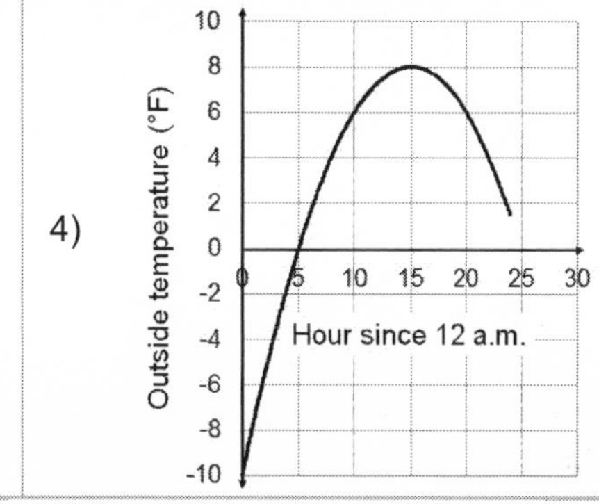

5)

Hours since toxin introduced	Ounces of live bacteria
0	8
5	3
15	0.4
20	0

6)

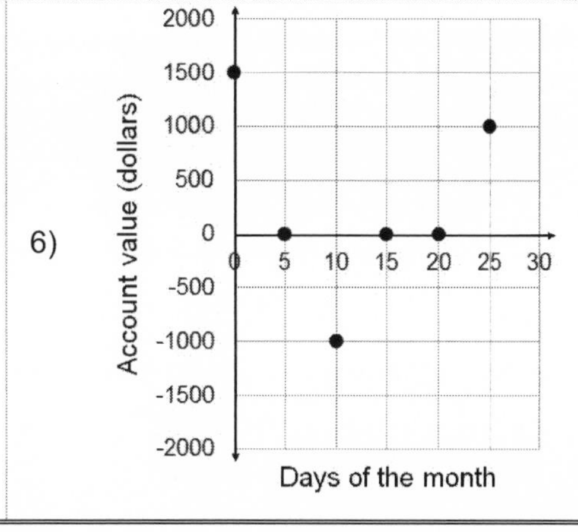

3.2.2 Graphing to Predict the Intercepts

Now let's put together our work about drawing graphs and interpreting intercepts.

Practice 3.2.2 Graphing to Predict the Intercepts

Using the given function

a) Graph the data and extend the pattern to estimate any intercepts.

b) Discuss the meaning of any useful intercept or why you would consider an intercept not to be useful.

Scale the x-axis from 0 to 16 and the y-axis from 0 to 1,000.

Months since the couch was bought	Resale value (Dollars)
4	600
5	550
10	300

b)

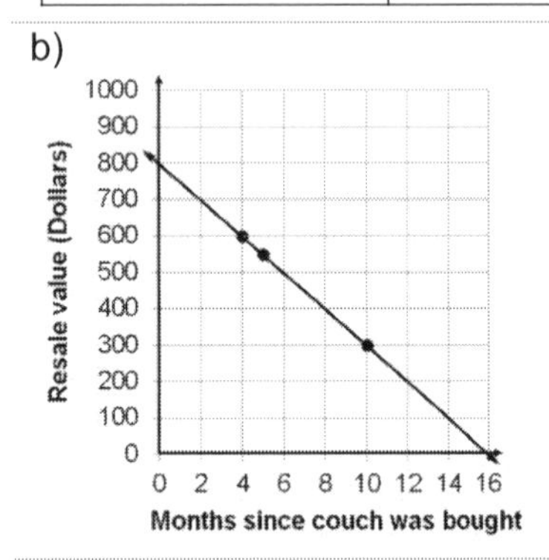

$\Rightarrow$ Plotted the points and connected them with a straight line. Made sure to correctly label both axes.

c) The y-intercept $(0,800)$ says the original value of the coach was about \$800. The x-intercept $(16,0)$ says that after 16 months the resale value of the coach will have dropped to \$0.

$\Rightarrow$ The y-intercept will be the ordered pair when x is 0 and the x-intercept will be the ordered pair when y is 0.

Homework 3.2 Using the given function;

a) Graph the data and extend the pattern to estimate any intercepts.

b) Discuss the meaning of any useful intercept or why you would consider an intercept not to be useful.

7) Scale the x-axis from 0 to 40 and the y-axis from 0 to 12.

Years since 1983	Reported fire rates (per 1,000)
10	8
20	6
30	4

8) Scale the x-axis from 0 to 80 and the y-axis from 0 to 100.

Minutes playing the slot machine	Money left (Dollars)
15	66.25
20	60
45	28.75

9) Scale the x-axis from 0 to 60 and the y-axis from 0 to 300.

Year since 1960	U.S. Population (Millions)
10	202
20	228
30	254

10) Scale the x-axis from 0 to 16 and y-axis from 0 to 30.

Years after 2007	Average hourly earnings ($)
3	22.29
5	23.31
8	24.84

11) Scale the x-axis from 0 to 24 and the y-axis from 0 to 6000.

Months since loan originated	Amount still left (Dollars)
6	4100
12	2600
18	1100

3.2.3 Finding Intercepts Using an Equation

You may have noticed in the last set of exercises that every graph looked close to a straight line. When the graph of our data can be approximated by a straight line we have a linear function. As you'll see latter in this chapter, every linear function has the form $y = mx + b$ where x and y are variables and m and b are constants. In this topic, you'll practice finding intercepts using the algebraic form of the linear function.

When it comes to linear intercepts there's a part that's straightforward and a part that's not. Finding the intercepts is straight forward. To find a y-intercept substitute 0 for x and solve for y. To find any x-intercepts substitute 0 for y and solve for x. The part that isn't straightforward is interpreting the results of your work. To find meaning for your intercepts focus on what the x and y values are counting.

Practice 3.2.3 Finding Intercepts Using an Equation

Identify each intercept and discuss whether its meaning is or is not useful.

a) The function $y = 133x + 1{,}590$ estimates the average yearly cost of tuition and fees for a public U.S. Two-Year College (y), using the number of years it's been since 2001, (x).

The y-intercept is $(0,1590)$. This implies that in 2001 (year 0) the cost was $1,590. $\Rightarrow$

Substituted 0 for x and solved.

$$y = 133x + 1{,}590$$
$$y = 133(0) + 1{,}590.$$
$$y = 1{,}590$$

The x-intercept is approximately $(-12,0)$. This implies that in 1989 (12 years before 2001) U.S. colleges were free (the cost was $0). This is a case where the x-intercept isn't useful. $\Rightarrow$

Substituting 0 for y and solved.

$$y = 133x + 1{,}590$$
$$0 = 133x + 1{,}590.$$
$$-12 \approx x$$

The symbol $\approx$ implies approximately equal.

b) The function $y = 6.5x - 20$ returns the profit (in dollars) made from a car wash, y, given x, the number of cars that have been washed.

The y-intercept is $(0,-20)$ this implies that when no cars have been washed the profit is $-\$20$. (The car wash starts off twenty dollars in debt.)	$\Rightarrow$	Found the y-intercept by substituting 0 for x and solving for y, $y = 6.5x - 20$ $y = 6.5(0) - 20$. $y = -20$
The x-intercept is approximately $(3.1,0)$. This tells us that the group will have to wash 4 cars before their profit will greater than \$0.	$\Rightarrow$	Found the x-intercept by substituting 0 for y and solving for x, $y = 6.5x - 20$ $0 = 6.5x - 20$. $3.1 \approx x$

Homework 3.2 Identify each intercept and discuss whether its meaning is or is not useful.

12) To repay a car loan a student promises to pay her parents \$320 per month until the loan is paid off. If she supplies the number of months she has been paying (x), the function $y = 5{,}200 - 320x$ will return the amount still outstanding on the loan, (y).

13) A neighbor's child is selling cookies to raise money. The mint cookies cost \$3 per box and the coconut cookies cost \$2 per box. When x is 0,2,4 or 6 the function $y = 9 - 1.5x$ shows the number of boxes of coconut cookies you can buy (y), given the number of boxes of mint cookies you buy (x).

14) The function $y = 635x + 23{,}200$ estimates the average yearly cost of tuition and fees for a private U.S. Four-Year College (y), using the number of years it's been since 2001, (x).

15) Depreciation is the loss in value of an asset over time. Straight line depreciation assumes the loss in value remains the same each year for a certain number of years. The linear function $y = 850 - 56.95x$, estimates the current value of a refrigerator where x is the number of years since the refrigerator was purchased and y is the current value of the refrigerator.

16) Although the cost to own and operate an automobile fluctuates year to year (especially due to oil prices), the function $y = 114x + 7{,}434$, where y is the cost in dollars and x is the number of years since 2000, gives a good estimate of this cost.

17) A parking lot closes for the night at 10 p.m. Cars in the lot are allowed to leave, but new cars aren't allowed to enter until 6 a.m. the following morning. The function $y = -45x + 270$ estimates the number of cars remaining in the lot, y, given the number of hours the lot has been closed x.

18) Between 2000 and 2016 the average number of cars per family in the United States could be estimated by the function $y = 1.9$ where y is the number of cars and x is the number of years after 2000.

3.2.4 *Graphing Using the Intercept Method*

I still remember sitting in my first economics course in college as the professor introduced supply and demand curves. Although using the word "curve" was correct, the first examples he graphed on the blackboard (yes it was a while ago) were linear functions so the graphs were straight lines. To graph the function $y = 12 - 2x$, he covered the *x* in the algebraic function with his hand and said, "If *x* is zero then *y* is twelve." and he put a point on the *y*-axis at 12. Then he covered the *y* in the linear function and said, "And if y is zero then x is six." and he put a point on the *x*-axis at 6. Last, he drew a straight line through both points to make the graph of the function. For a few seconds I was confused by what he was doing, then I realized he was using the intercept method.

The intercept method is a fast way to graph a linear function. The idea is to build a small data table by first substituting 0 for *x* and finding the value of *y,* and then substituting 0 for *y* and finding the value of *x*. Since two points define a line, we plot the two points, which will be on the *x* and *y* axes (hence the name intercept method) and draw the line.

Procedure **– Graphing a linear function using the intercept method**
1. Make a data table with three rows. Label the first row using *x* and *y*, set the value of *x* in row two to 0 and the value of *y* in row three to 0.
2. Substitute 0 for *x* in the linear function and find the value of *y*. Complete the ordered pair in row two.
3. Substitute 0 for *y* in the linear function and find the value of *x*. Complete the ordered pair in row three.
4. Plot the points and draw a line through the two points.

Practice 3.2.4 Graphing Using the Intercept Method

Graph the following linear functions using the intercept method.

a) $y = 2x + 4$

x	y
0	
	0

$\Rightarrow$ Built the data table for step 1.

x	y
0	4
	0

$\Rightarrow$ Substituted 0 for *x* and solved to find the value of *y*, $\begin{aligned} y &= 2(0) + 4 \\ y &= 4 \end{aligned}$.

Completed the ordered pair in row 2.

x	y
0	4
−2	0

$\Rightarrow$ Substituted 0 for *y* and solved to find the value of *x*, $\begin{aligned} 0 &= 2x + 4 \\ -2 &= x \end{aligned}$.

Completed the ordered pair in row 3.

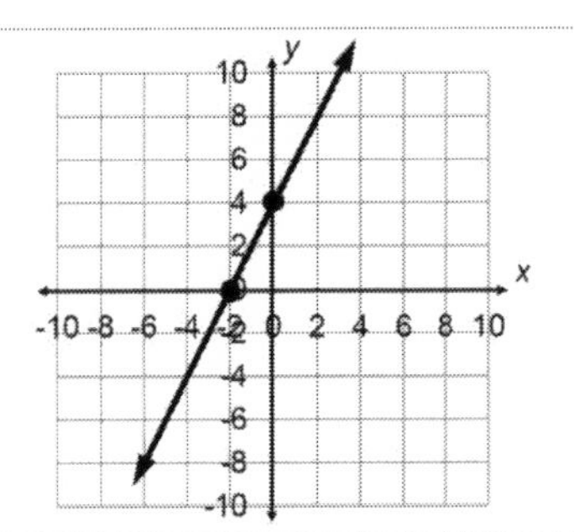

$\Rightarrow$ Graphed the points $(0,4)$, $(-2,0)$ and drew the line.

b) $2x-3y=15$

x	y
0	
	0

$\Rightarrow$ Built the data table.

x	y
0	–5
	0

$\Rightarrow$ Substituted 0 for *x* and solved to find the value of *y*,

$$\begin{aligned} 2(0)-3y&=15 \\ y&=-5 \end{aligned}$$

. Completed the ordered pair in row 2.

x	y
0	–5
7.5	0

$\Rightarrow$ Substituted 0 for *y* and solved to find the value of *x*,

$$\begin{aligned} 2x-3(0)&=15 \\ x&=7.5 \end{aligned}$$

. Completed the ordered pair in row 3.

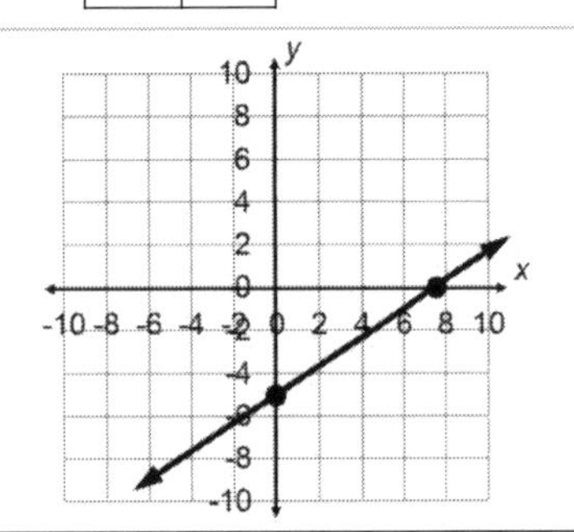

$\Rightarrow$ Graphed the points $(0,-5)$, $(7.5,0)$ and drew the line.

c) $y=-4x$

x	y
0	0
0	0

$\Rightarrow$ Built the data table, substituted 0 for *x* and solved to find the ordered pair is $(0,0)$. The same ordered pair happens if I substitute 0 for *y* and find the value for *x*.

x	y
0	0
0	0
1	–4

$\Rightarrow$ So far, I only have one ordered pair. I'll need two different ordered pairs to draw a line. Substituted 2 for *x* (any non-zero value would work, I picked a value that makes my work easy) and found the value for *y* was -8. I now have two points.

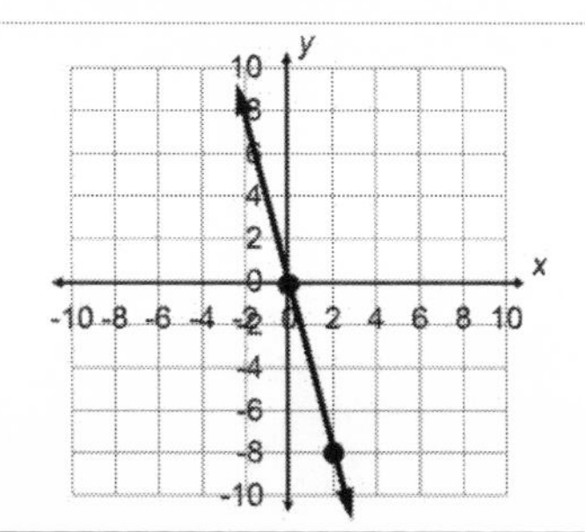

$\Rightarrow$ Graphed the points $(0,0)$, $(1,-4)$ and drew the line.

Homework 3.2 Graph the following linear equations using the intercept method.

19) $y = x+1$	20) $y = -x+4$	21) $y = -3x$	22) $y = \frac{1}{2}x - 2$
23) $-5x+3y-15=0$	24) $\frac{1}{4}x + y - 2 = 0$	25) $2x-3y=18$	26) $x+y=\frac{5}{3}$
27) $10x+5y-42=0$	28) $-\frac{2}{3}x - \frac{1}{2}y = 2$		

Homework 3.2 Answers

1) a) The *y*-intercept, $(0,-800)$, implies that before any tokens are sold, the debt was \$800.
   b) The *x*-intercept, $\approx(375,0)$, implies that we'll break even after selling about 375 tokens.

2) a) The y-intercept, $(0,288)$ tells us the driver started 288 miles from home.
   b) The x-intercept, $(6,0)$ tells us that it took them 6 hours to drive home.

3) a) The y-intercept, $(0,140)$ tells us the person started their diet at 140 pounds.
   b) The *x*-intercept doesn't provide any useful information since the person won't weigh 0 pounds.

4) a) The y-intercept, $(0,-10)$, tells us that that at midnight the temperature was 10 degrees below 0.
   b) the *x*-intercept, $(5,0)$, tells us that the temperature was 0 degrees at 5 a.m..

5) a) The y-intercept, $(0,8)$, shows there were 8 ounces of live bacteria when the toxin was introduced.
   b) The *x*-intercept, $(20,0)$, tells us that all the bacteria were dead after 20 hours.

6) a) The y-intercept $(0,1500)$, shows that at the beginning of the month the value of the account was \$1,500..
   b) The *x*-intercepts, $(5,0)$, $(15,0)$ and $(20,0)$, tells us there was no money in the account on the 5th, 15th and 20th day of the month.

7) a)

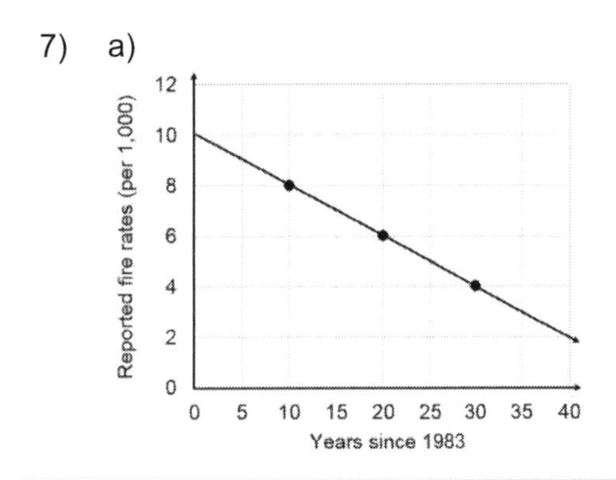

b) The y-intercept, $(0,10)$, implies that in 1983 there were about 10 fires per thousand people. Although the rate of fires has been dropping, and it certainly looks like it may reach 0, The x-intercept will not have any useful information since the rate of fires will probably never drop to 0.

8) a)

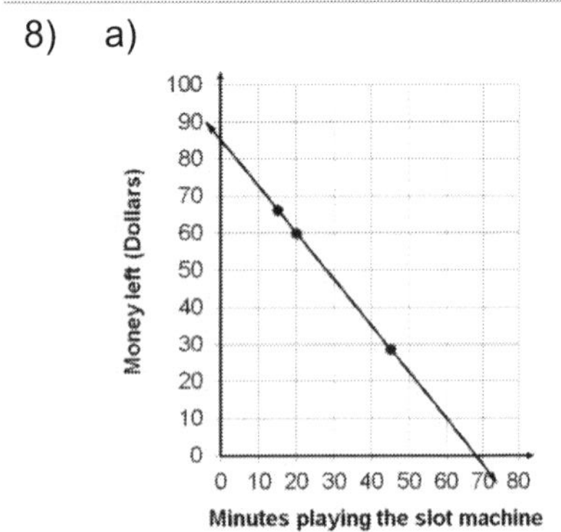

b) The y-intercept, $(0,85)$, implies they had $85 when they started playing. The x-intercept, $(68,0)$. implies that after playing for about an hour and eight minutes they will be out of money.

9) a)

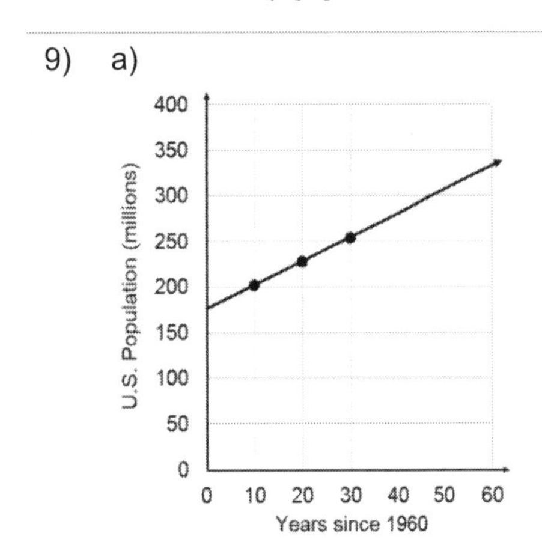

b) The y-intercept, $(0,175)$, estimates that in 1960 there were about 175,000,000 people in the United States. The x-intercept doesn't carry any useful information since the population probably isn't going to drop the 0 through 2020.

10) a)

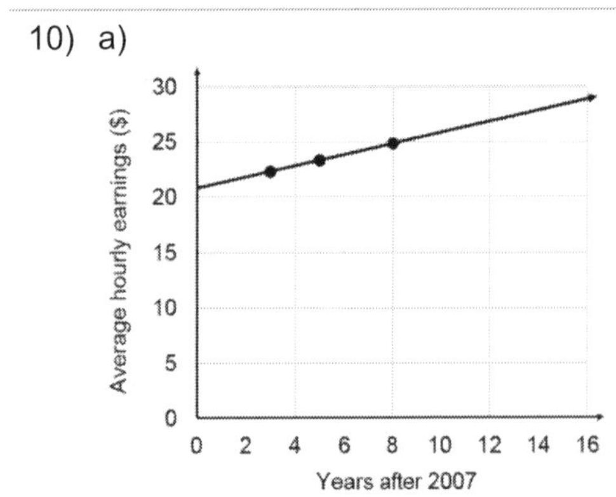

b) The y-intercept, $(0,21)$, estimates that in 2007 average hourly earnings in the United States were around $21. The x-intercept doesn't carry any useful information since earning $0 per hour, on average, is going to happen.

11) a)

Amount still left (dollars)

Months since loan originated

b) The y-intercept, $(0,5600)$, estimates the original amount of the loan was around \$5,600. The x-intercept, $(22,0)$, estimates the loan will be paid off after 23 months.

---

12) The y-intercept, $(0,5200)$, tells us that the original loan amount was \$5,200. The x-intercept, $(16.25,0)$, tells us that the loan will be paid off after 17 months.

---

13) The y-intercept is $(0,9)$ so if you decide not to buy any boxes of mint cookies you can buy 9 boxes of coconut cookies. The x-intercept is $(6,0)$ so if you decide to buy 6 boxes of mint cookies you can't buy any boxes of coconut cookies.

---

14) The y-intercept, $(0,23200)$, tells us that in 2001 it cost \$23,200 for tuition and fees at the private college. The x-intercept, $(-36.5,0)$, tells us that around 1965 you could attend the college for free. This is a case where the x-intercept probably isn't helpful.

---

15) The y-intercept $(0,850)$ tells us the original value of the refrigerator was \$850. The x-intercept $(15,0)$ tells us that after 15 years the value of the refrigerator has dropped to \$0.

---

16) The y-intercept, $(0,7434)$, tells us that it cost \$7,434 to own and operate an automobile during 2000. The x-intercept, $(-65,0)$, tells us that in 1935 the cost to own and operate an automobile was \$0. Here's another case where the x-intercept isn't helpful.

---

17) The y-intercept, $(0,270)$, tells us there were 270 cars in the parking lot when it closed. The x-intercept, $(6,0)$ tells us that all the cars have left the lot by 4 a.m.

---

18) The y-intercept, $(0,1.9)$, tells us that in 2000 the average number of cars per family was 1.9 cars. There won't be an x-intercept since this constant function is a line, starting at the y-intercept, and moving to the right parallel to the x-axis.

---

19)

x	y
0	1
–1	0

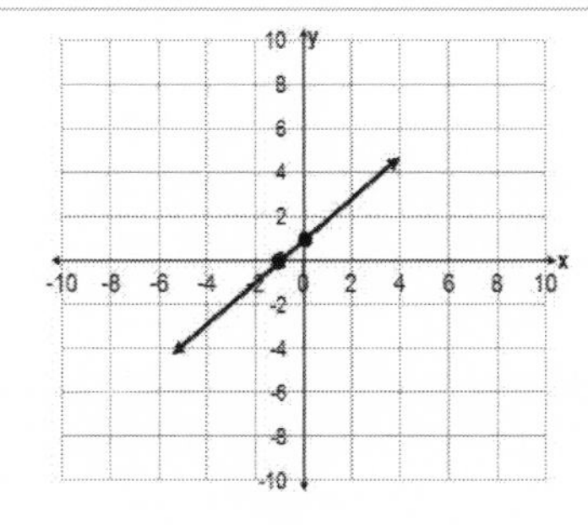

20)

x	y
0	4
4	0

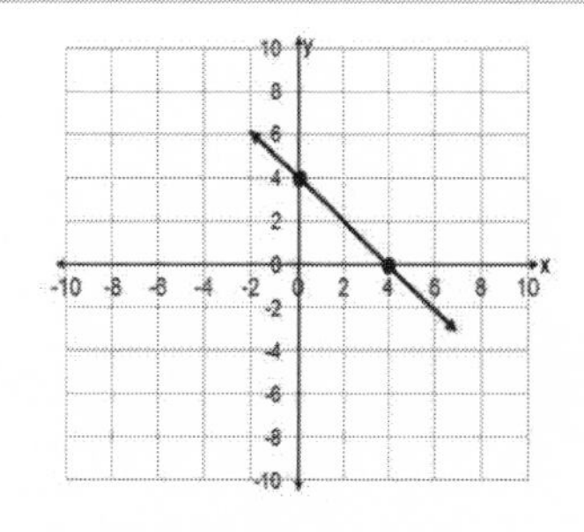

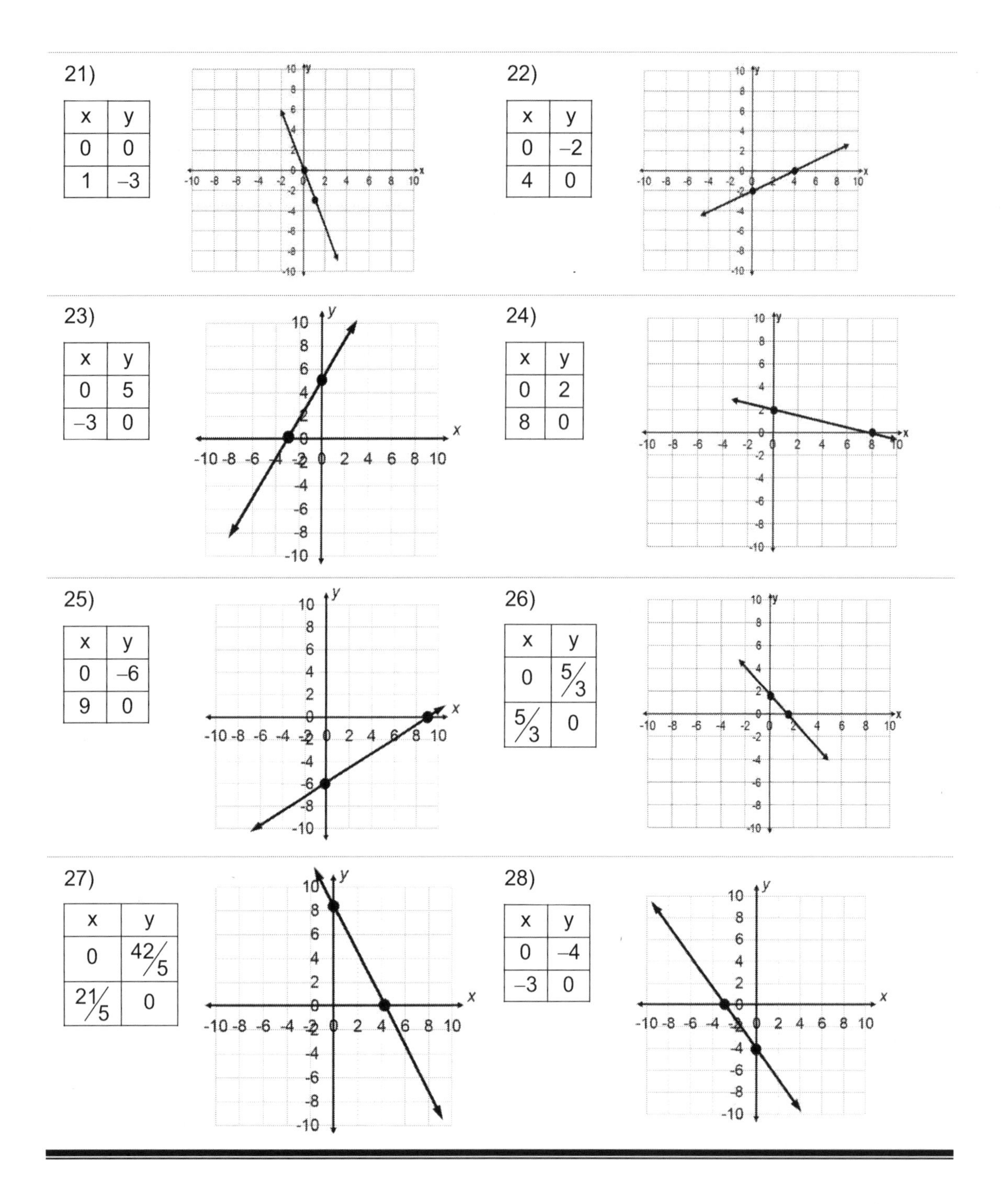

21)

x	y
0	0
1	–3

22)

x	y
0	–2
4	0

23)

x	y
0	5
–3	0

24)

x	y
0	2
8	0

25)

x	y
0	–6
9	0

26)

x	y
0	$\frac{5}{3}$
$\frac{5}{3}$	0

27)

x	y
0	$\frac{42}{5}$
$\frac{21}{5}$	0

28)

x	y
0	–4
–3	0

3.3 The Slope of a Linear Function

A linear function is useful when a straight line would pass close to or through most of the points of our graphed data. When experts look at this straight line they quickly get a feel for how the *y* values are changing as the values of *x* increase from left to right. In general, this idea is known as the rate of change but for linear functions in particular, the idea is called the "slope".

3.3.1 Some Vocabulary for Slope

Numerically, **slope** is a quotient that compares the difference in *y* values to the difference in *x* values for two ordered pairs. There's a connection between our earlier work with increasing, decreasing or constant functions and the sign for the value of the slope.

To describe the slope, we take the point of view that the values of *x* are increasing. If the values of *y* are also increasing we say the linear function has a **positive slope**. The word positive makes sense since the value of the slope for an increasing linear function will be a positive number. The graph to the right shows an increasing linear function with a positive slope.

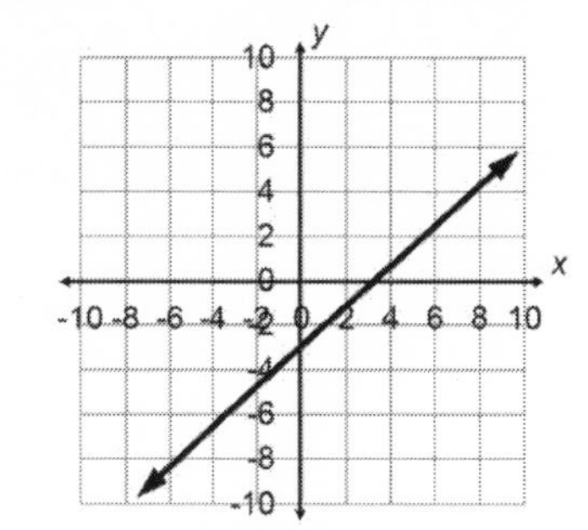

If, as the values of *x* increase, the values of *y* decrease we say the linear function has a **negative slope**. The word negative also makes sense since the value of the slope for a decreasing linear function will be a negative number. The graph to the right shows a decreasing linear function with a negative slope.

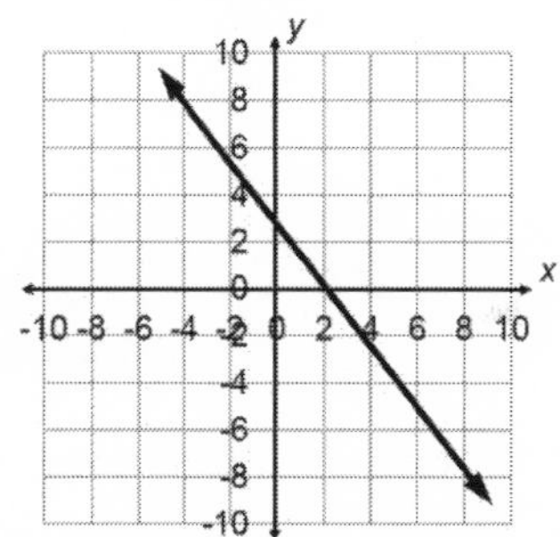

If the values of *y* remain constant as the values of *x* increase, we say our constant linear function has a **zero slope**. The graph to the right shows the constant linear function $y = 5$.

As you might expect the value of the slope for a constant function is 0.

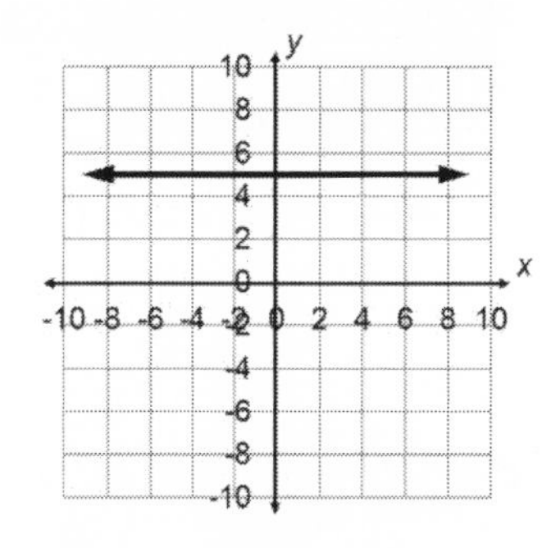

The graph to the right shows an example where the values of *y* change but the value of *x* doesn't. We say a situation like this has an **undefined slope**. We won't spend time on undefined slopes since they don't represent a function. (Notice the situation to the right violates the vertical line test.)

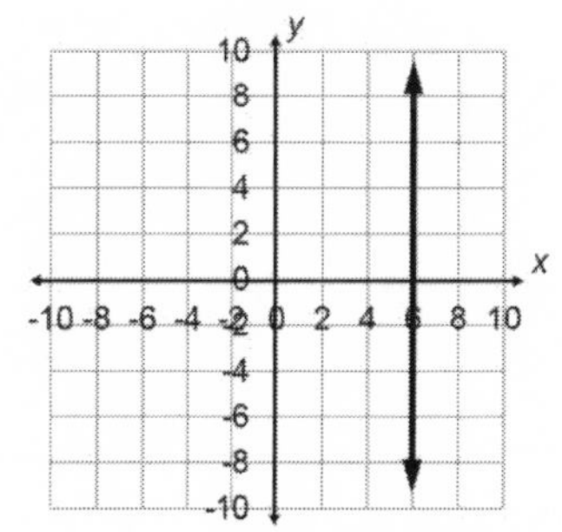

Practice 3.3.1 Some Vocabulary for Slope

Discuss whether the function is increasing, decreasing or constant and whether the slope value should be positive, negative or zero.

a) 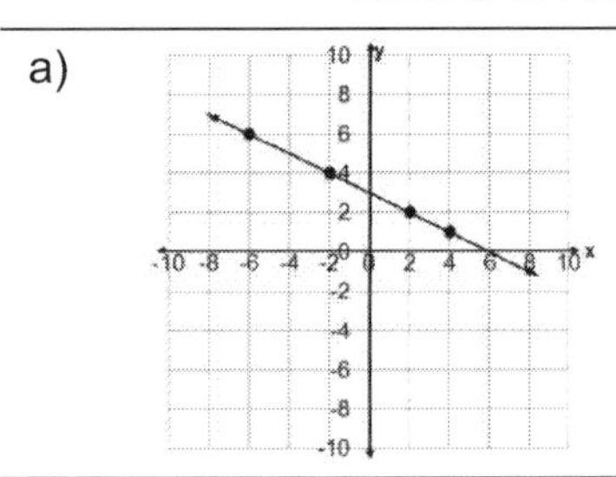 $\Rightarrow$ As the values of x increase the values of y decrease so the function is decreasing and the line has a negative slope.

Homework 3.3 *Discuss whether the function is increasing, decreasing or constant and whether the slope value should be positive, negative or zero.*

1) 2) 3) 4)

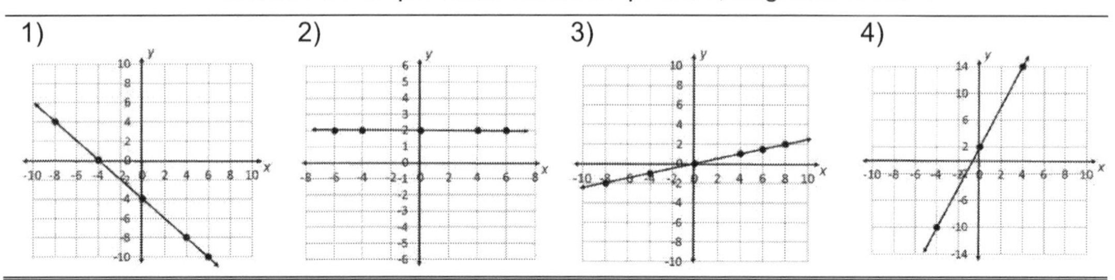

3.3.2 Finding the Value of the Slope from a Graph

Recall that slope is a quotient that compares the difference in y values to the difference in x values for two ordered pairs. Historically, the letter m was chosen to stand in place of the value of the slope so $m = \dfrac{\text{the difference in } y \text{ values}}{\text{the difference in } x \text{ values}}$. For a graph, our ordered pairs will be two points on the line.

To find the value of the slope for the line to the right I've decided to start at the point $(2,0)$ and end at the point $(6,8)$. The difference in my y values will be 8 units (0 to 8 is 8 units) and the difference in my x values will be 4 units (2 to 6 is 4 units). The value of the slope would be $m = \dfrac{\text{difference in } y}{\text{difference in } x} = \dfrac{8}{4} = 2$. Notice the slope is the same if I start at $(6,8)$ and go back to $(2,0)$. In this case y <u>decreases</u> by 8 units (from 8 down to 0) while x <u>decreases</u> by 4 units (from 6 back to 2) so the slope is $\dfrac{-8}{-4}$ which again simplifies to 2. For a linear function the slope is the same regardless of which two points on the line you choose or the order in which you go from one point to the other. In practice though, we usually take the point of view that x values will be increasing (we'll be moving left to right on the x-axis.)

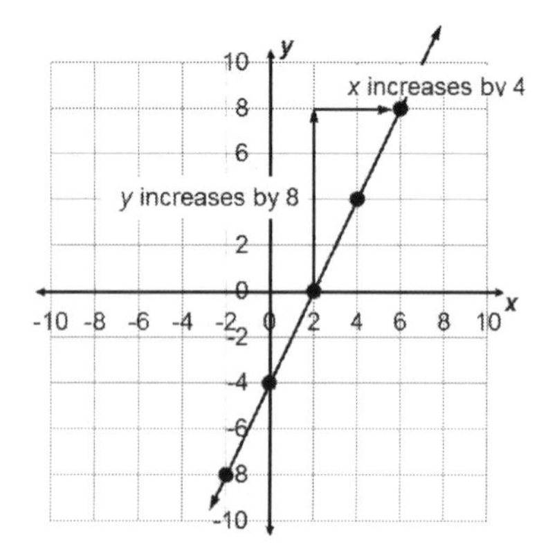

Practice 3.3.2 Finding the Value of the Slope from a Graph

Describe whether you expect the slope to be positive, negative or zero and then find the value of the slope.

a)

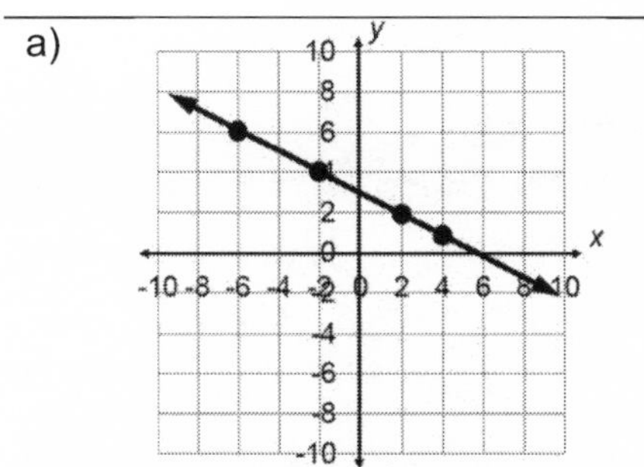

$\Rightarrow$ This is a decreasing function so I expect a negative slope. Two ordered pairs I'm comfortable with are $(-2,4)$ and $(2,2)$. Moving from the point $(-2,4)$ to the point $(2,2)$ the values of *y* <u>decrease</u> by 2 (from 4 down to 2) as the values of *x* increase by 4 (from −2 to 2). The value of the slope is $\frac{-2}{4}$ or $-\frac{1}{2}$.

Homework 3.3 Describe whether you expect the slope to be positive, negative or zero. Then find the value of the slope.

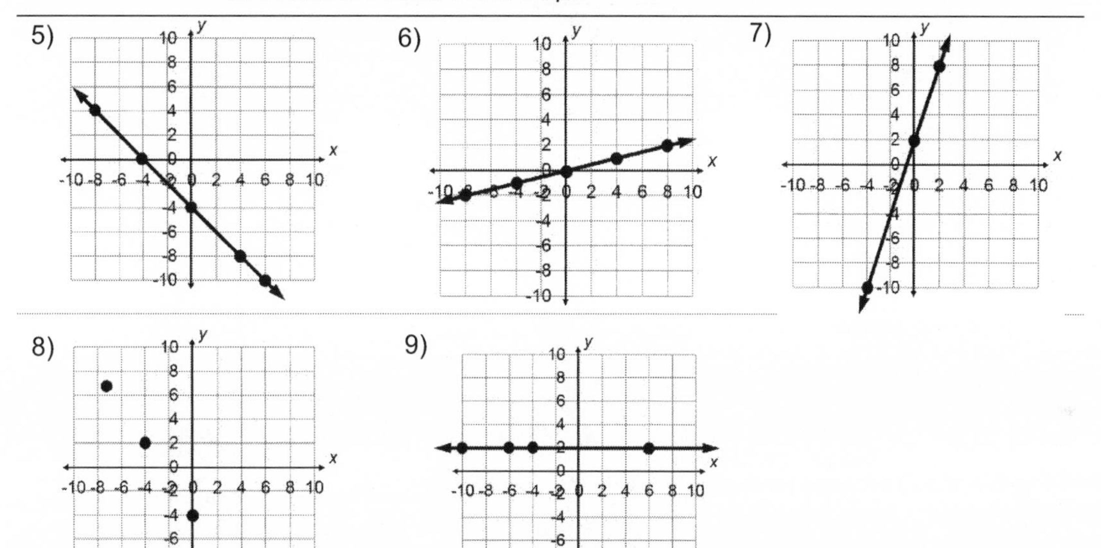

3.3.3 Applying Slope

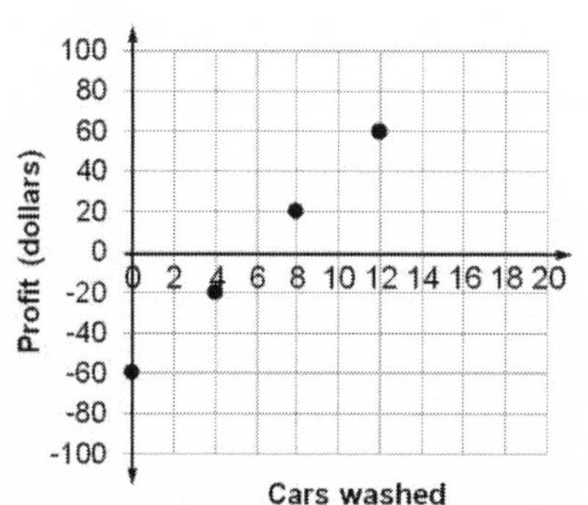

With applied problems, I'll ask you to find the "meaning" of the slope. Meaning comes from combining the value of the slope with the items (labels, units) the *x* and *y* values are counting. For example, to find the meaning of the slope for the function to the right I would first select a couple of ordered pairs, say $(8,20)$ and $(12,60)$. Next, I would realize the increase in *y* of 40 (from 20 to 60) is actually an increase of 40 <u>dollars</u> (from \$20 to \$60). In the same way, the increase in *x* of 4 (from 8 to 12) is actually an increase of 4 <u>cars</u> (from 8 cars to 12 cars). The applied slope would be $\frac{\$40}{4 \text{ cars}} = \frac{\$10}{1 \text{ car}}$ which implies that profit increases by \$10 for every car that's washed.

Notice I said, "Profit <u>increases</u> by \$10", not, "the profit is \$10". "The profit is \$10." implies the profit is constantly \$10 regardless of the number of cars washed. Slope is about how the function is <u>changing</u> as you consider two ordered pairs.

Also, when discussing meaning, it's common to divide and make the value of the denominator a 1 and to move a factor of -1 to the numerator. That way we're discussing the change in y for a 1 unit increase in x. Here's some practice finding the meaning of an applied slope.

Practice 3.3.3 Applying Slope

Find the value, and discuss the meaning of, the slope.

a)

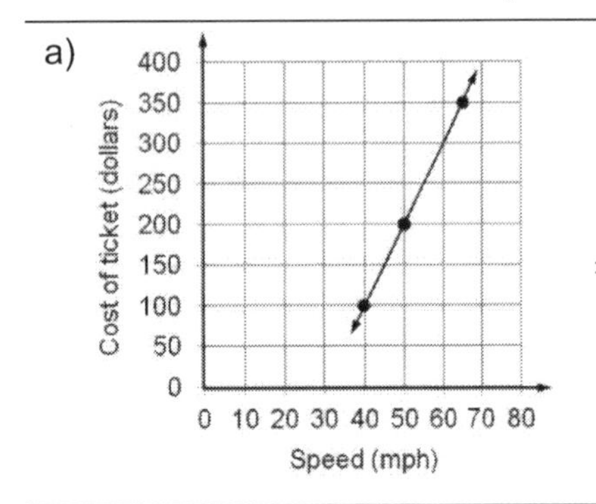

$\Rightarrow$ Using the points $(40,100)$ and $(50,200)$, y increases by \$100 while x increases by 10 mph. The slope would be $\frac{\$100}{10 \text{ mph}} = \frac{\$10}{1 \text{ mph}}$ so the meaning of the slope would be that every mile per hour over the speed limit, increases the cost of the ticket by ten dollars.

b)

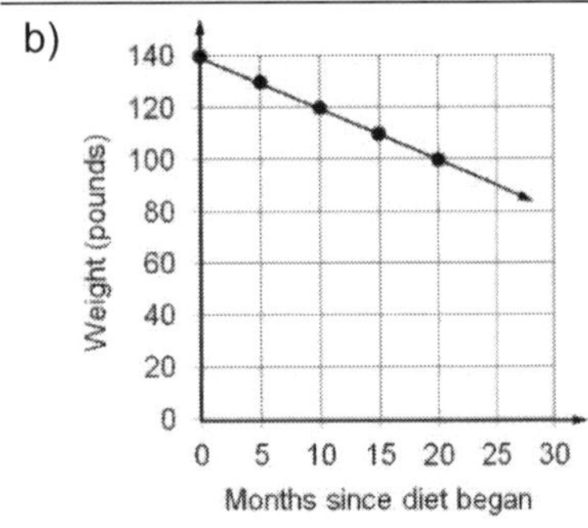

$\Rightarrow$ Using the points $(0,140)$ and $(10,120)$ the slope would be $\frac{-20 \text{ pounds}}{10 \text{ months}} = \frac{-2 \text{ pounds}}{1 \text{ month}}$

(Remembered to move the factor of -1 to the numerator.) The meaning is that the person is losing two pounds per month while on the diet.

Homework 3.3 Find the value, and discuss the meaning of, the slope.

10)

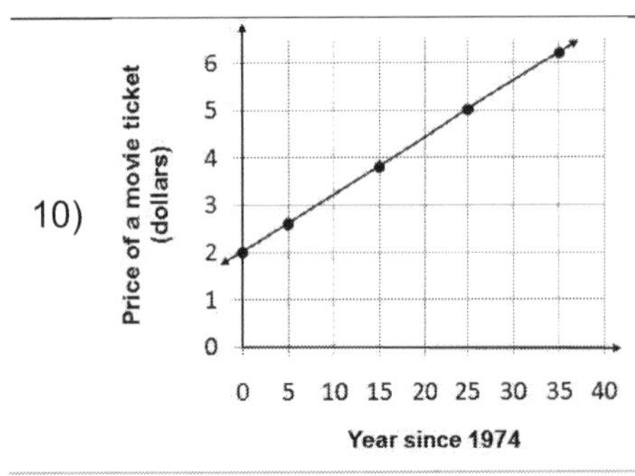

11)

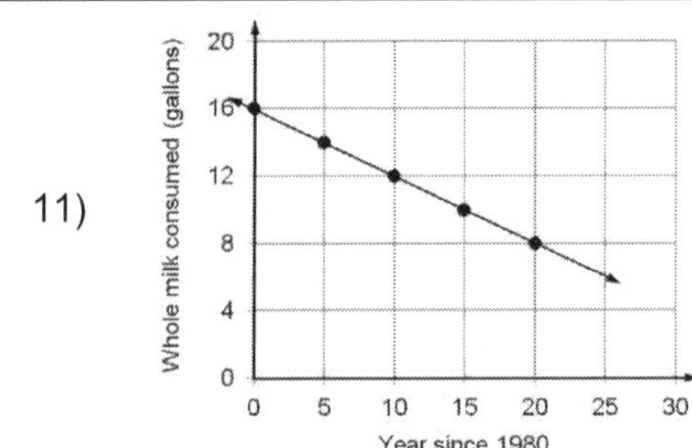

12) 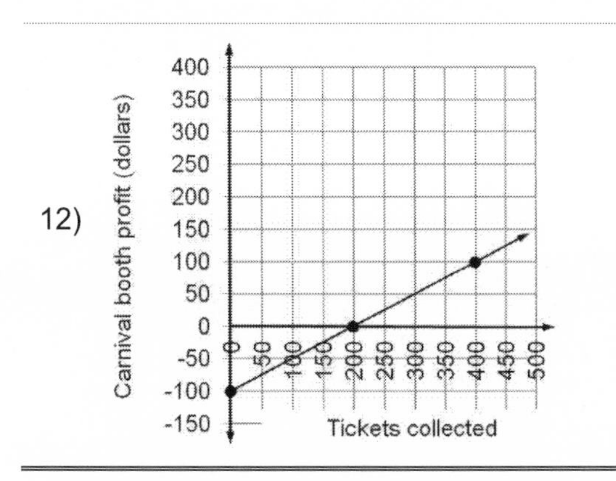

13) 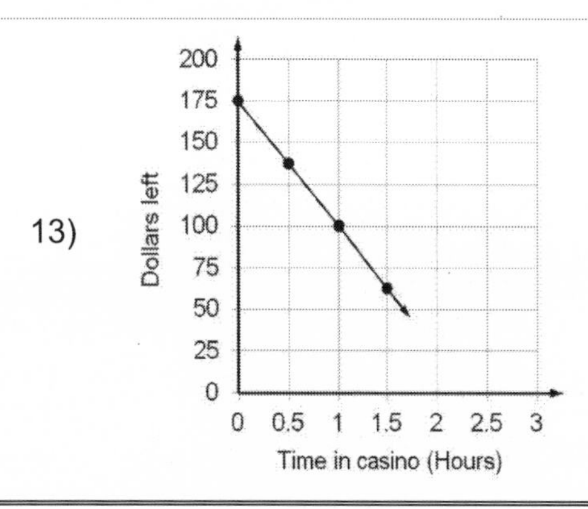

3.3.4 Finding the Slope Algebraically

In practice, people usually find the value of the slope using the **slope formula**, $m=\frac{y_2-y_1}{x_2-x_1}$. The little subscripted numbers, (the 1's and 2's) aren't operators, they're subscripts to help distinguish between two general ordered pairs called (x_1,y_1) and (x_2,y_2).

To use the slope formula, choose two ordered pairs and make one your (x_1,y_1) pair and the other your (x_2,y_2) pair. The value of the slope will be the same regardless of which ordered pair is which. Then substitute the values for *x* and for *y* into the slope formula and simplify. For problems without labels or units it's common to reduce any fraction but not necessarily carry out the final quotient. For instance, if $(-5,-6)$ was our (x_1,y_1) and $(-1,8)$ was our (x_2,y_2) then $m=\frac{y_2-y_1}{x_2-x_1}=\frac{8-(-6)}{-1-(-5)}=\frac{14}{4}$ which we'd reduce to $\frac{7}{2}$ but probably not write as 3.5. Notice how important simplifying signed numbers becomes when we're finding a value for *m*.

Practice 3.3.4 Finding the Slope Algebraically

Find the value of the slope.

a) 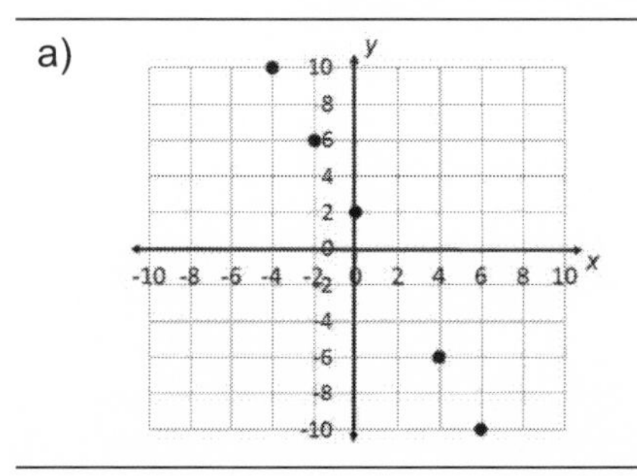

$m=-2$ $\Rightarrow$ Choosing $(-2,6)$ for (x_1,y_1) and $(4,-6)$ for (x_2,y_2) and substituting gives $m=\frac{-6-6}{4-(-2)}=\frac{-12}{6}=-2$.

b)

x	y
−6	−2
−3	0
0	2

$m = \frac{2}{3}$. ⇒ Choosing $(-3,0)$ and $(0,2)$ leads to $m = \frac{2-0}{0-(-3)} = \frac{2}{3}$ so the slope is two-thirds. Any two ordered pairs, in any order, could have been chosen and the slope would have been the same.

Homework 3.3 Find the value of the slope.

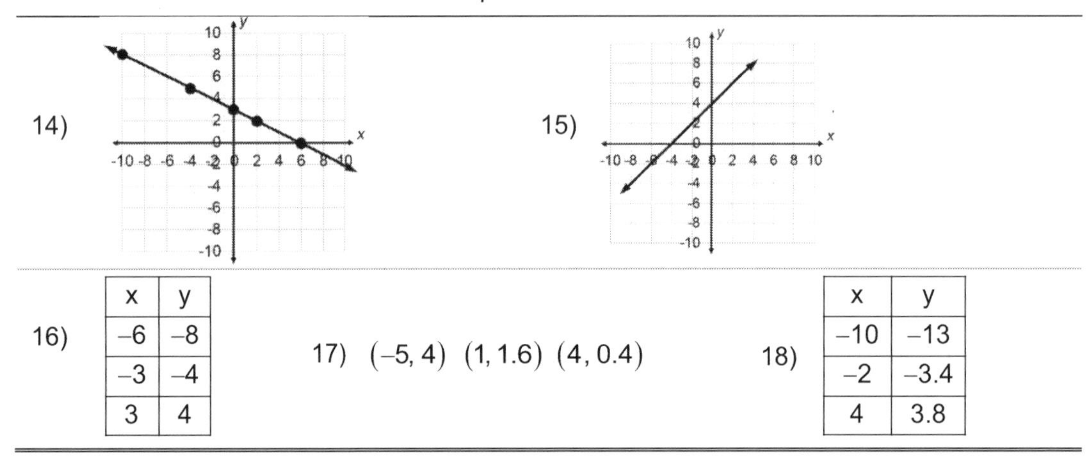

14) 15)

16)

x	y
−6	−8
−3	−4
3	4

17) $(-5, 4)$ $(1, 1.6)$ $(4, 0.4)$

18)

x	y
−10	−13
−2	−3.4
4	3.8

3.3.5 Finding an Applied Slope Algebraically

Now let's use the slope formula to find an applied slope. Here are the average costs for a year at a United States community college from 1980 to 1995. If we pick years 0 and 15, (1980 and 1995) the corresponding costs are $280 and $1,135 and the slope will be $m = \frac{1135 - 280}{15-0} = \frac{\$57}{1 \text{ year}}$. The slope tells us that from 1980 until 1995 the cost of tuition was increasing about $57 per year.

Year since 1980	One year's tuition ($)
0	280
10	850
15	1,135

Practice 3.3.5 Finding an Applied Slope Algebraically

Find, and discuss the meaning of, the slope.

a)

Movies rented	Monthly cost ($)
2	52.90
1	48.95
7	72.65

⇒ Using the points $(2,52.90)$ and $(1,48.95)$ the slope would be $m = \frac{52.90-48.95}{2-1} = \frac{\$3.95}{1 \text{ movie}}$ so each time a movie is rented the monthly cost increases by $3.95.

b)

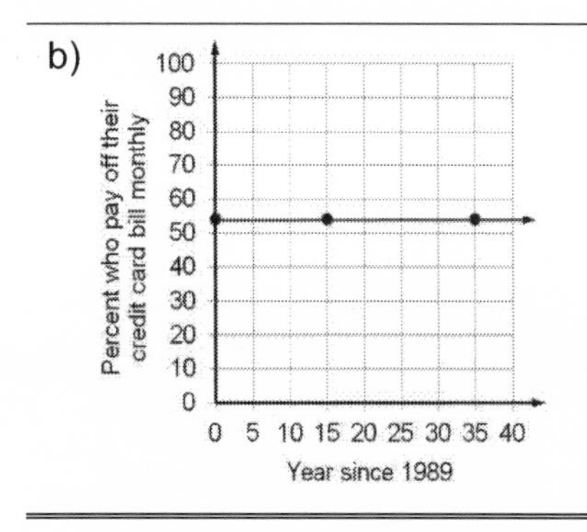

$\Rightarrow$ Using the points $(15,54)$ and $(35,54)$ the slope would be $m = \frac{54-54}{35-15} = \frac{0}{20} = 0$. The slope implies that since 1989 about 54% of all credit card holders consistently pay off their bill every month.

Homework 3.3 Find, and discuss the meaning of, the slope.

19)

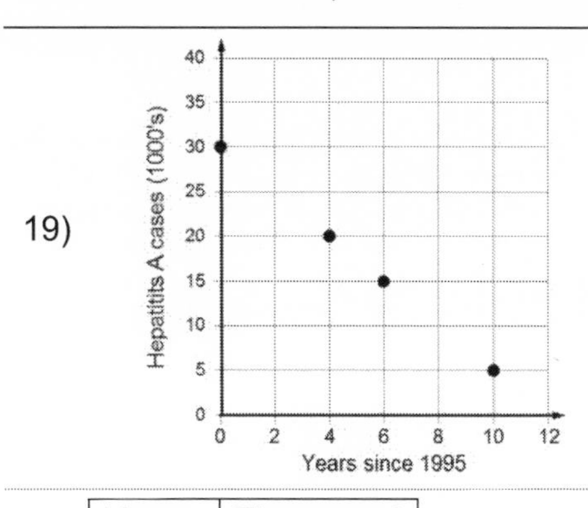

20)

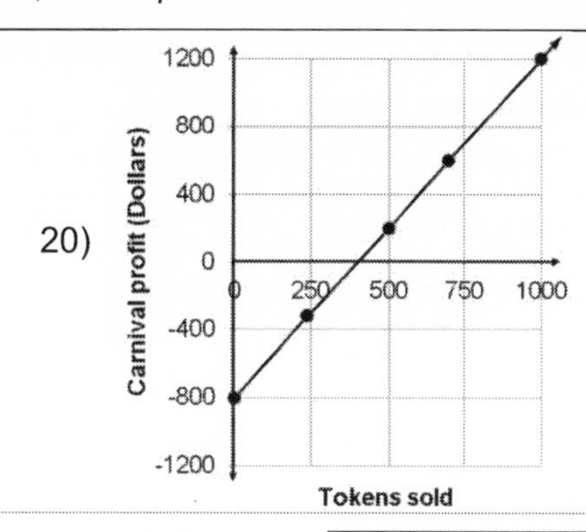

21)

Years since 2005	Time spent sleeping (hours)
0	8.6
3	8.6
4	8.6
10	8.6

22)

Months since diet began	Weight (pounds)
6	195
12	180
18	165
24	150

23)

Degrees Celsius	Degrees Fahrenheit
−18	0
0	32
20	68
100	212

24)

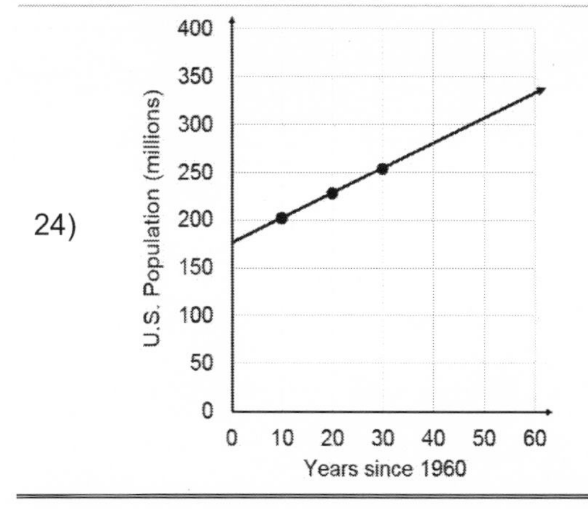

25)

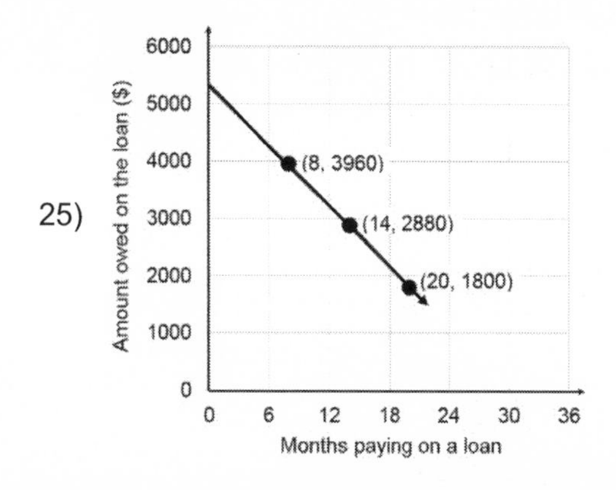

Homework 3.3 Answers

1) The decreasing function has a negative slope.

2) The constant function has a zero slope.

3) The increasing function has a positive slope.

4) The increasing function has a positive slope.

5) A decreasing function has a negative slope. Used the points $(-4,0)$ and $(0,-4)$ to find the value of the slope is -1.

6) An increasing function has a positive slope. Used the points $(-8,-2)$ and $(8,2)$ to find the value of the slope is $\frac{1}{4}$.

7) An increasing function has a positive slope. Used the points $(-4,-10)$ and $(4,14)$ to find the value of the slope is 3.

8) A decreasing function has a negative slope. Used the points $(-4,2)$ and $(0,-4)$ to find the value of the slope is $-\frac{3}{2}$.

9) A constant function has a zero-slope. Used the points $(0,2)$ and $(4,2)$ to find the value of the slope is 0.

10) Using the points $(0,2)$ and $(25,5)$ the slope would be $\frac{\$0.12}{1\text{ year}}$ so every year the price of a movie ticket increases by about twelve cents.

11) Using the points $(10,12)$ and $(20,8)$ the slope would be $\frac{-0.4\text{ gallons}}{1\text{ year}}$ so every year people are consuming about four-tenths less gallons of whole milk.

12) Using the points $(0,-100)$ and $(400,100)$ the slope would be $\frac{\$0.50}{1\text{ ticket}}$ so the profit increases fifty cents per ticket.

13) Using the points $(0,175)$ and $(1,100)$ the slope would be $\frac{-\$75}{1\text{ hour}}$ so the person is losing about \$75 per hour while in the casino.

14) Using $(2,2)$ and $(-10,8)$. $m=\frac{y_2-y_1}{x_2-x_1}=\frac{8-2}{-10-(2)}=\frac{6}{-12}=-\frac{1}{2}$. The slope is $-\frac{1}{2}$.

15) Using $(-4,0)$ and $(0,4)$. $m=\frac{y_2-y_1}{x_2-x_1}=\frac{4-0}{0-(-4)}=\frac{4}{4}=1$. The slope is 1.

16) Using $(3,4)$ and $(-6,-8)$. $m=\frac{y_2-y_1}{x_2-x_1}=\frac{-8-4}{-6-3}=\frac{-12}{-9}=\frac{4}{3}$. The slope is $\frac{4}{3}$.

17) Using $(1,1.6)$ and $(-5,4)$. $m=\frac{y_2-y_1}{x_2-x_1}=\frac{4-1.6}{-5-1}=\frac{2.4}{-6}=\frac{-0.4}{1}$. The slope is -0.4 or $-\frac{4}{10}$.

18) Using $(-10,-13)$ and $(-2,-3.4)$. $m=\frac{y_2-y_1}{x_2-x_1}=\frac{-3.4-(-13)}{-2-(-10)}=\frac{9.6}{8}=\frac{1.2}{1}$. The slope is 1.2.

19) Using the points $(0,30)$ and $(10,5)$ the slope would be $m=\frac{30-5}{0-10}=\frac{-2.5\text{ cases}}{1\text{ year}}$ which implies Hepatitis A cases in the U.S. were decreasing by about 2,500 cases per year.

20) Using the points $(0,-800)$ and $(1000,1200)$ the slope would be $m=\frac{1200--800}{1000-0}=\frac{\$2}{1\text{ token}}$ so the carnival increased their profit by $2 every time they sold a token.

21) Using the points $(4,8.6)$ and $(10,8.6)$ the slope would be $m=\frac{8.6-8.6}{10-0}=\frac{0\text{ Hours}}{1\text{ year}}$ so the average hours spent sleeping has remained constant at around 8.6 hours per day.

22) Using the points $(6,195)$ and $(18,165)$ the slope would be $m=\frac{165-195}{18-6}=\frac{-2.5\text{ pounds}}{1\text{ month}}$. The person's weight decreased by 2.5 pounds per month.

23) Using the points $(100,212)$ and $(0,32)$ the slope would be $m=\frac{212-32}{100-0}=\frac{180}{100}=\frac{1.8°\text{ C}}{1°\text{ F}}$ which says a change of 1.8° Celsius corresponds to a 1° change in degrees Fahrenheit.

24) Using $(10,200)$ and $(30,250)$ the slope would be $m=\frac{250-200}{30-10}=\frac{2.5\text{ million}}{1\text{ year}}$ so the population of the U.S. is increasing by around 2.5 million per year.

25) Using the points $(8,3960)$ and $(20,1800)$ the slope would be $m=\frac{1{,}800-3{,}960}{20-8}=-\frac{\$180}{1\text{ month}}$ so the amount owed is decreasing by $180 per month.

3.4 Linear Functions in Slope-Intercept Form

One reason linear functions are popular is that it's easy to build the linear function for yourself. The most common algebraic form for a linear function is known as slope-intercept form.

3.4.1 Building a Linear Function in Slope-Intercept Form

One common way for describing a linear function is $y = mx + b$ which is known as **slope-intercept form**. Two of the letters, the *x* and the *y*, should be left as variables. The other two letters, the *m* and the *b*, need to be replaced by constants. The letter *m* is replaced with the value for the slope and the letter *b* is replaced with the *y*-coordinate of the *y*-intercept. Let's practice building some linear functions.

Practice 3.4.1 Building a Linear Function in Slope-Intercept Form

Use the information to build the linear function in slope-intercept form.

a)

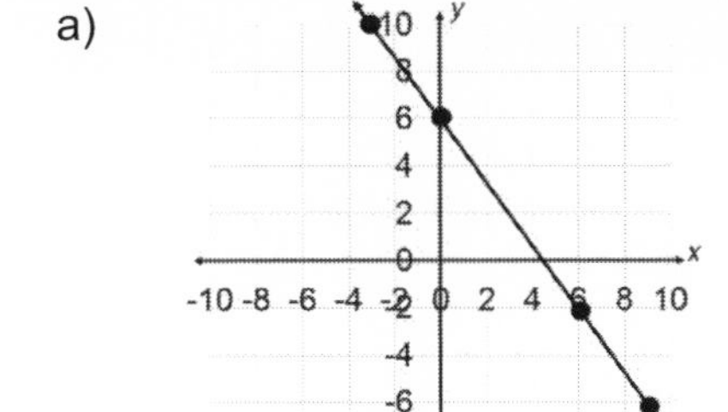

$y = mx + b$	$\Rightarrow$	Began with slope-intercept form.
$y = mx + 6$	$\Rightarrow$	The graph shows the *y*-intercept is $(0,6)$. Substituted the *y*-coordinate of the *y*-intercept for *b*.
$y = \frac{-4}{3}x + 6$	$\Rightarrow$	Used the ordered pairs $(0,6)$ and $(6,-2)$ to find the value of *m* is $-\frac{4}{3}$, $\left(\frac{-2-6}{6-0} = \frac{-8}{6} = \frac{-4}{3}\right)$. Replaced *m* with the value of the slope. I can stop since the linear function has the form $y = mx + b$ with *x* and *y* left variable and *m* and *b* replaced by constants.

b)

x	y
−6	−8
−1	−3
0	−2
4	2

$y = mx + b$	$\Rightarrow$	Began with slope-intercept form.
$y = mx + -2$ $y = mx - 2$	$\Rightarrow$	Noticed that *y* was -2 when *x* was 0 so the *y*-intercept is $(0,-2)$. Replaced *b* with -2 and wrote as a subtraction.

$m = 1$	$\Rightarrow$	Used $(-6,-8)$ and $(-1,-3)$ to find $m = \dfrac{-3-(-8)}{-1-(-6)} = \dfrac{-3+8}{-1+6} = \dfrac{5}{5} = 1$.
$y = x - 2$	$\Rightarrow$	Replaced m with 1. The linear function now has the form $y = mx + b$ with y and x left variable and m and b replaced by constants.

Homework 3.4 Use the information to build the linear function in slope-intercept form.

1)

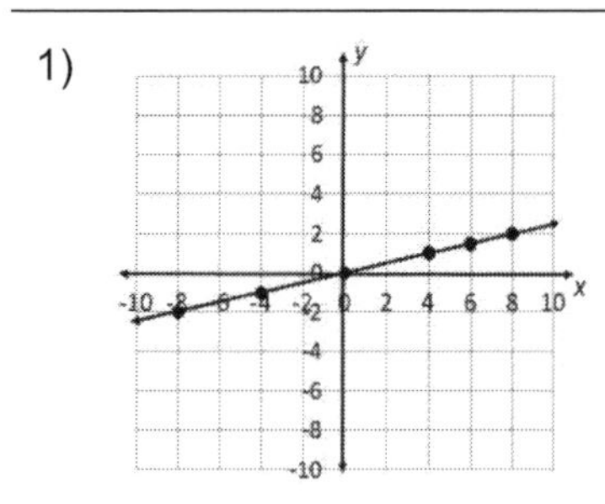

2)

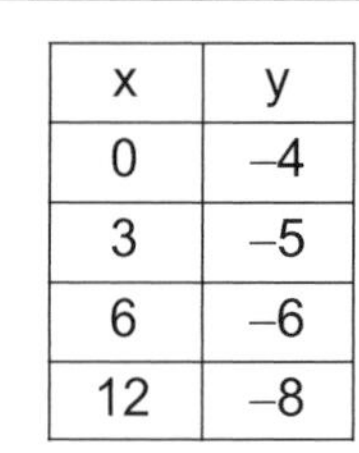

x	y
0	–4
3	–5
6	–6
12	–8

3)

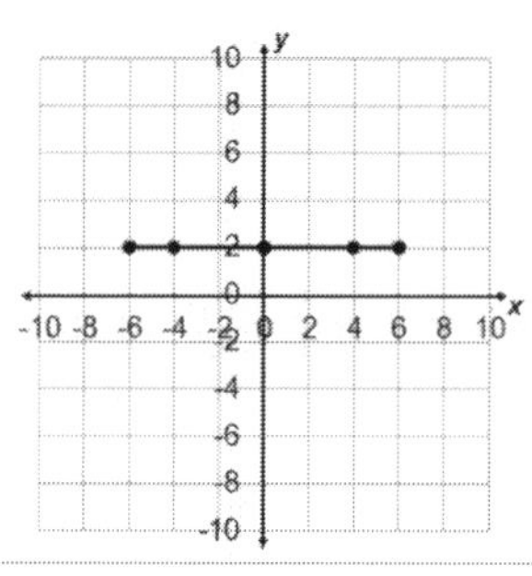

4)

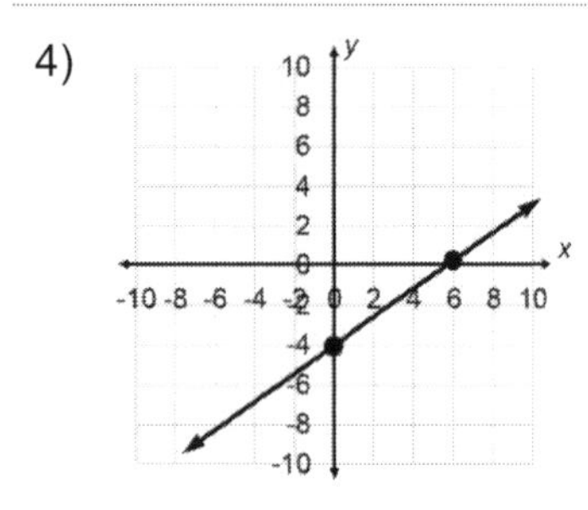

5)

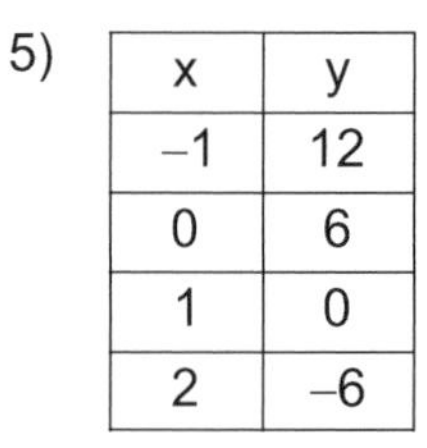

x	y
–1	12
0	6
1	0
2	–6

6)

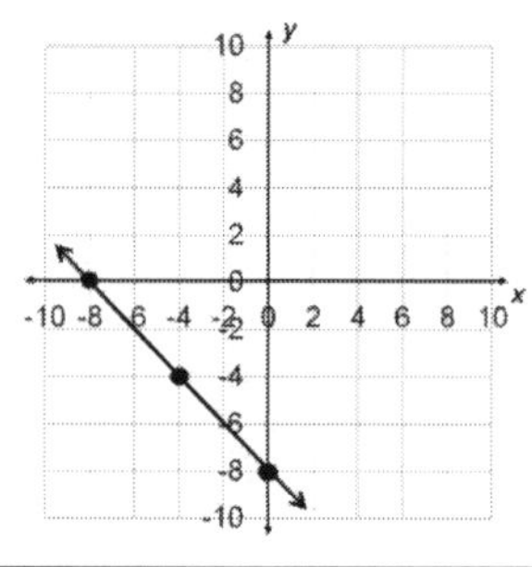

3.4.2 Finding the y-Intercept Algebraically

In the last set of problems you were able to find the *y*-intercept (the value for *b*) by looking for the ordered pair with an *x*-coordinate of 0. In practice, it's common not to have this ordered pair, so we find the value of *b* algebraically. The procedure is to start with the general slope-intercept form, supply values for *y*, *x* and *m* and then solve to find the value for *b*.

Practice 3.4.2 Finding the y-Intercept Algebraically

Use the information to build the linear function in slope-intercept form.

a) The line has a slope of 3 and goes through the point $(-2,6)$.

$y = mx + b$	$\Rightarrow$	Started with the general slope-intercept form.
$y = 3x + b$	$\Rightarrow$	Substituted the supplied value for *m*.
$6 = 3(-2) + b$	$\Rightarrow$	Substituted the values for *y* and *x* from the given point.
$12 = b$	$\Rightarrow$	Solved for *b*.
$y = 3x + 12$	$\Rightarrow$	Made sure the final answer was written in slope-intercept form.

b)

Gallons of gas	Miles driven
15	480
8	256
21	672

$y = mx + b$ ⇒ Started with the general slope-intercept form.

$m = \dfrac{256-480}{8-15} = \dfrac{-224}{-7} = 32$ ⇒ Used the slope formula to find m. Remembered that if the signs are the same then the quotient is positive. The value 32 implies they can drive 32 miles per gallon of gas.

$y = 32x + b$ ⇒ Substituted the value for m.

$672 = 32(21) + b$
$672 = 672 + b$
$0 = b$
⇒ Substituted values for y and x from one of the given ordered pairs. (With a linear function, it doesn't matter which ordered pair is chosen.) The value of b is 0 which implies if their miles are 0 the gallons of gas used is also 0.

$y = 32x$ or $y = 32x + 0$ ⇒ Wrote the linear function in slope-intercept form.

Homework 3.4 Use the information to build the linear function in slope-intercept form.

7) The line has a slope of 4 and goes through the point $(-2,-7)$.

8) The line goes through the points $(7,1)$ and $(10,-5)$.

9) The line has a slope of $-1/3$ and goes through the point $(-3,1)$.

10) The line goes through the points $(-4,-4)$ and $(6,-4)$.

11)

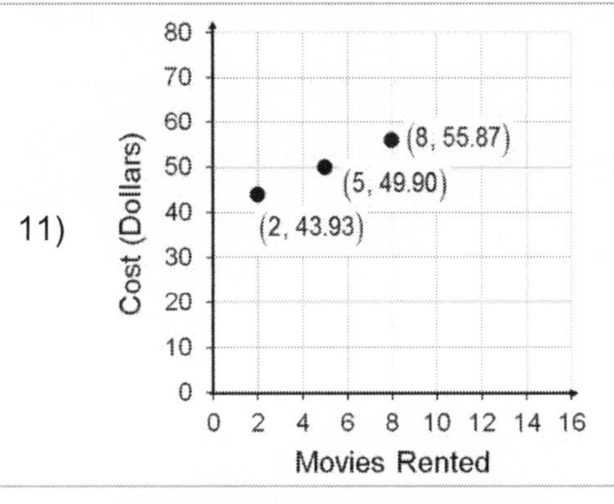

12)

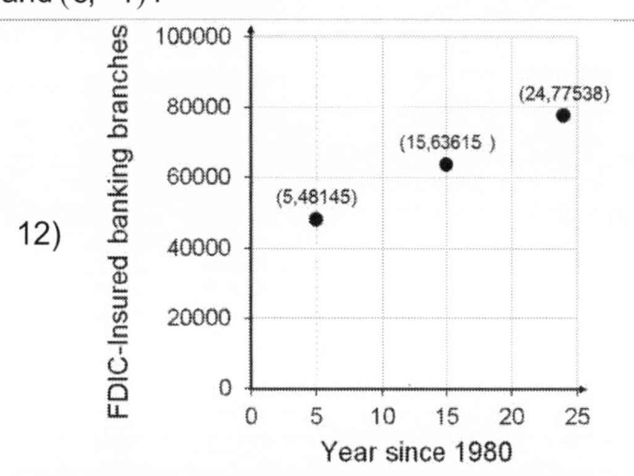

13)

Year since 1984	FDIC-Insured banking institutions
5	12,545
15	9,165
20	7,475
24	6,123

14) From 1990 (year 0) until 2006 (year 16) the number of liver transplants in the United States was increasing. In 1995 there were about 3,809 liver transplants. By 2000 there were about 4,949 transplants. Assume the number of transplants depends on the year since <u>1990</u>. (Let x be the number of years since 1990.)

3.4.3 Applying the Linear Function

Let's practice applying linear functions when both m and b must be found algebraically.

Practice 3.4.3 Applying the Linear Function

Use the given information to answer the following questions.

a)

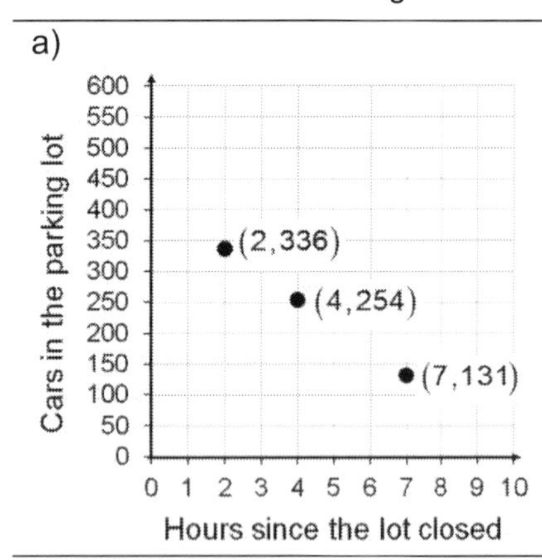

1. Find the slope and discuss its meaning.
2. Find the y-intercept and discuss its meaning.
3. Build the linear function in slope-intercept form.
4. Use the linear function to predict the number of cars left in the lot 6 hours after closing.
5. Use the linear function to predict when 200 cars will be in the lot.
6. Use the linear function to find the x-intercept and discuss its meaning.

1. Every hour there are 41 fewer cars in the lot.	$\Rightarrow$	Used $\frac{254-336}{4-2}=\frac{-41\text{ cars}}{1\text{ hour}}$.
2. There were 418 cars in the lot when it closed.	$\Rightarrow$	Used $336=-41(2)+b \Rightarrow b=418$
3. $y=-41x+418$	$\Rightarrow$	Substituted values for m and b.
4. After 6 hours, I predict there will be around 172 cars in the lot.	$\Rightarrow$	I was given an "x" value and I needed to find the "y" value. Started from the linear function, substituted 6 for and found that $y=172$.
5. There will be 200 cars in the lot around 5.3 hours, or 5 hours 18 minutes after closing.	$\Rightarrow$	Here I was given a "y" value and I needed to find an "x" value. Substituted 200 for y and solved to find the corresponding value for x. Used $\frac{3}{10}=\frac{x}{60}$ to change 0.3 hours to minutes.
6. The x-intercept is very close to $(10.2,0)$. The parking lot will be empty after about 10 hours and 12 minutes.	$\Rightarrow$	Substituted 0 for y and solved to find the corresponding value for x was around 10.2.

Homework 3.4 Use the given information to answer the following questions.

15) A person kept a journal since they began their diet. After 6 months they weighed 195 pounds. After 10 months they weighed 185 pounds. Assume their weight y depends linearly on x, the months since they began the diet.

a) Find the slope and discuss its meaning.
b) Find the y-intercept and discuss its meaning.
c) Build the linear function in slope-intercept form.
d) Predict the person's weight after 1 year of dieting.
e) Predict when they will reach their ideal weight of 150 pounds.

16)

Cars Washed	Profit (dollars)
8	–30
12	–10
20	30

a) Find the slope and discuss its meaning.
b) Find the y-intercept and discuss its meaning.
c) Build the linear function in slope-intercept form.
d) How many cars need to be washed to break even?
e) What will be the profit if 83 cars are washed?
f) How many cars needed to be washed to have $280 more in profit than they had after washing 30 cars?

17)

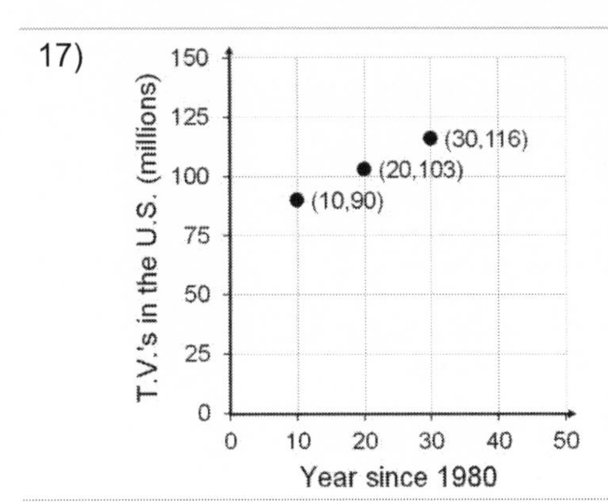

a) Find the slope and discuss its meaning.
b) Find the y-intercept and discuss its meaning.
c) Build the linear function in slope-intercept form.
d) Predict when the number of T.V.'s will first exceed 120,000,000.
e) Predict the number of T.V.s in 2020.
f) When will there first be 10 million more T.V.s then in 1995?

18) The average price of a movie ticket is reflected in the following (x,y) ordered pairs. x is the year since 1995. y is the cost in dollars.

$(5,5.35)$ $(10,6.50)$ $(13,7.19)$

a) Find the slope and discuss its meaning.
b) Find the y-intercept and discuss its meaning.
c) Build the linear function in slope-intercept form.
d) Predict the cost of a movie ticket in 1993.
e) When will the price of a movie ticket first reach $10?
f) Predict the average price of a movie ticket in 2012.

19)

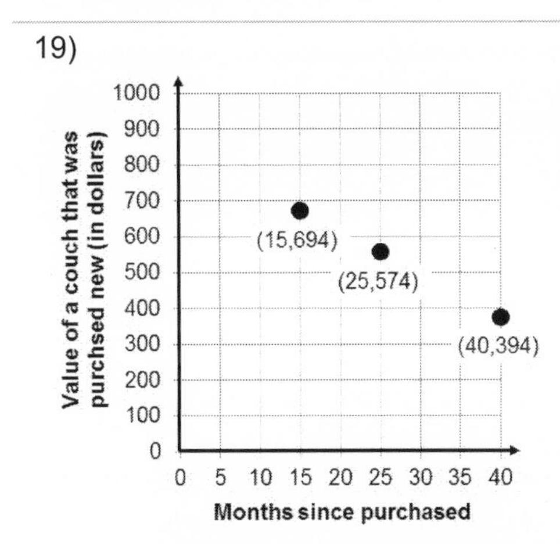

a) Find the slope and discuss its meaning.
b) Find the y-intercept and discuss its meaning.
c) Build the linear function in slope-intercept form.
d) When will the couch be worth half its original price?
e) What's the predicted value of the couch after 1.5 years?
f) Find the x-intercept and discuss its meaning.

20)

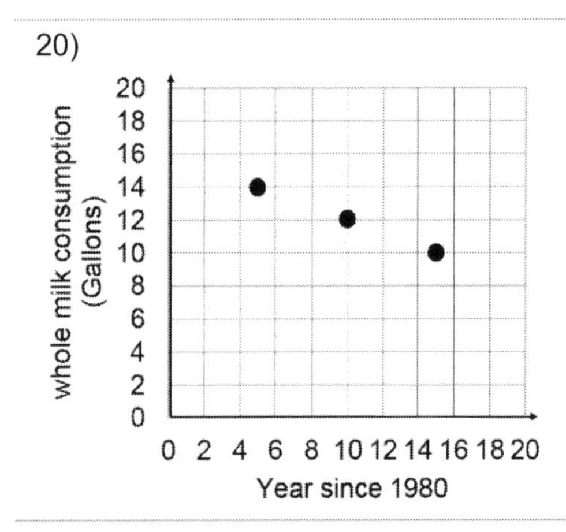

a) Find the slope and discuss its meaning.
b) Find the y-intercept and discuss its meaning.
c) Build the linear function in slope-intercept form.
d) Predict whole milk consumption in 2001.
e) In what year will whole milk consumption first reach 6 gallons?
f) In what year will consumption be half the amount in 1988?

21)

Year since 2000	Per capita consumption of bottled water (gallons)
2	20
7	30
8	32

a) Find the slope and discuss its meaning.
b) Find the y-intercept and discuss its meaning.
c) Build the linear function in slope-intercept form.
d) Predict bottled water consumption in 1999.
e) Find when the per capita consumption of bottled water will be 40 gallons.
f) What will consumption be 10 years after the year consumption was 24 gallons?

22)

Movies purchased during the month	Monthly cost for cable (dollars)
2	52.90
7	72.65
8	76.60

a) Find the slope and discuss its meaning.
b) What is the meaning of the y-intercept?
c) Build the linear function in slope-intercept form.
d) If the monthly bill was $64.75 how many movies were watched?
e) If I don't want my total bill to exceed $90 in any one month, how many movies can I watch?

Homework 3.4 Answers

1) The *y*-intercept is $(0,0)$. The slope is $\frac{1}{4}$. The function is $y = \frac{1}{4}x + 0$ or $y = \frac{1}{4}x$.

2) The *y*-intercept is $(0,-4)$. The slope is $-\frac{1}{3}$. The function is $y = \frac{-1}{3}x - 4$.

3) The *y*-intercept is $(0,2)$. The slope is 0. The function is $y = 0x + 2$ or $y = 2$.

4) The *y*-intercept is $(0,-4)$. The slope is $\frac{2}{3}$. The function is $y = \frac{2}{3}x - 4$.

5) The *y*-intercept is $(0,6)$. The slope is -6. The function is $y = -6x + 6$.

6) The *y*-intercept is $(0,-8)$. The slope is -1. The function is $y = -1x - 8$ or $y = -x - 8$.

7) $y = 4x + 1$ 8) $y = -2x + 15$ 9) $y = -\frac{1}{3}x$ 10) $y = -4$

11) $y = 1.99x + 39.95$ 12) $y = 1547x + 40{,}410$ 13) $y = -338x + 14{,}235$ 14) $y = 228x + 2669$

15) a) Every month the person is losing an additional 2.5 pounds. b) They started their diet at 210 pounds. c) The linear function is $y = -2.5x + 210$. d) After one year of dieting their weight will be 180 pounds. e) After 24 months (2 years).

16) a) The profit increases by \$5 every time they wash a car. b) They started the car wash \$70 in debt. c) The linear function is $y = 5x - 70$. d) 14 cars will cover the initial cost of \$70 e) \$345 f) 86 cars

17) a) Every year since 1980 the number of televisions has increased by 1.3 million. b) In 1980 there were about 77 million T.V.s in the U.S. c) The linear function is $y = 1.3x + 77$ d) Around 2013 e) 129 million f) Around 2002 – 2003.

18) a) Ticket prices are increasing by about 23 cents per year. b) In 1995 the price of a movie ticket was around \$4.20. c) The linear function is $y = 0.23x + 4.20$. d) \$3.74. e) About the year 2020. f) \$8.11

19) a) The value of the couch is decreasing \$12 per month. b) Originally the couch cost \$874. c) The linear function is $y = -12x + 874$. d) After about 3 years. e) \$658. f) The *x*-intercept is $(0, 72.8)$. Theoretically the value of the couch will be \$0 after about 72 months (6 years).

20) a) Every year whole milk consumption is declining an additional 0.4 gallons. b) In 1980 consumption was around 16 gallons. c) The linear function is $y = -0.4x + 16$. d) 7.6 gallons. e) 2005 f) 2004

21) a) Every year consumption is increasing by about 2 gallons. b) In 2000 bottled water consumption was 16 gallons. c) The linear function is $y = 2x + 16$. d) 14 gallons. e) 2012 f) 44 gallons

22) a) The cost increases an additional \$3.95 per month for every movie that's rented. b) The monthly cable bill costs \$45 per month even if no movies are rented. c) The linear function is $y = 3.95x + 45$. d) 5 movies e) 11 movies

3.5 Rewriting Linear Functions into Slope-Intercept Form

It's difficult to look at the function $2x-6y=12$ and quickly "see" the slope is $\frac{1}{3}$ and the y-intercept is -2. In this section you'll practice rewriting $2x-6y=12$ into slope-intercept form $y=\frac{1}{3}x-2$ so it's easy to see the slope and the y-intercept.

The symbolic manipulations necessary to transform a linear function can be confusing at first but it's important to stick with it. As you continue with algebra you'll use these ideas often, not just with linear functions.

3.5.1 Rewriting mx

Recall with $y=mx+b$ that mx is a variable term with coefficient m. Sometimes we need to rewrite the variable term and make the value of m, explicit. For example, $\frac{3x}{2}$, which is not in the form mx, can be rewritten as the product $\left(\frac{3}{2}\right)\left(\frac{x}{1}\right)$ which most people prefer to write as $\frac{3}{2}x$. Now the variable term is in the form mx and it's easy to see the coefficient (the value of m) is $\frac{3}{2}$.

Practice 3.5.1 Rewriting *mx*

Rewrite the following terms into the form mx and identify the coefficient.

a) $\frac{2x}{5}$

$\left(\frac{2}{5}\right)\left(\frac{x}{1}\right)$ $\Rightarrow$ Rewrote the product $\frac{2x}{5}$ as two factors $\left(\frac{2}{5}\right)\left(\frac{x}{1}\right)$.

$\frac{2}{5}x$ The coefficient is $\frac{2}{5}$. $\Rightarrow$ Thought of $\frac{x}{1}$ as x.

b) $\frac{x}{6}$

$\frac{1}{6}x$ The coefficient is $\frac{1}{6}$. $\Rightarrow$ Thought of $\frac{x}{6}$ as $\left(\frac{1}{6}\right)\left(\frac{x}{1}\right)$.

c) $\frac{-5x}{15}$

$\frac{-x}{3}$ $\Rightarrow$ Reduced the common factor of five $\frac{-\cancel{5}x}{3\times\cancel{5}}$.

$-\frac{1}{3}x$ The coefficient is $-\frac{1}{3}$. $\Rightarrow$ Thought of $\frac{-x}{3}$ as $\left(-\frac{1}{3}\right)\left(\frac{x}{1}\right)$.

Homework 3.5 Rewrite the following terms into the form mx and identify the coefficient.

1) $\frac{x}{4}$ 2) $-\frac{-3x}{2}$ 3) $\frac{6x}{8}$ 4) $-\frac{4x}{10}$ 5) $-\frac{x}{7}$ 6) $-\frac{-4x}{2}$

3.5.2 Reordering Terms

If the sum $mx+b$ is written in the form $b+mx$, we'll use the commutative property to reorder the terms. Since the commutative property only works for addition, remember to think of a difference as adding the opposite.

Practice 3.5.2 Reordering Terms

Rewrite the following terms into the form $mx+b$.

a) $4-2x$

$4+-2x$	$\Rightarrow$	Wrote the subtraction as adding the opposite.
$-2x+4$	$\Rightarrow$	Used the commutative property of addition to reorder the terms.

b) $-6+x$

$x+-6$	$\Rightarrow$	Used the commutative property to reorder the terms.
$x-6$	$\Rightarrow$	Adding an opposite is usually rewritten as subtraction.

c) $3-\frac{6x}{5}$

$-\frac{6x}{5}+3$	$\Rightarrow$	Thought of subtraction as adding the opposite and reordered the terms.
$-\frac{6}{5}x+3$	$\Rightarrow$	Wrote the first term in the form mx.

Homework 3.5 Rewrite the following terms into the form $mx+b$.

7) $-6+4x$ 8) $-3-x$ 9) $\frac{1}{2}+\frac{4x}{5}$ 10) $\frac{2}{9}-\frac{3x}{9}$ 11) $\frac{3}{4}-\frac{-12x}{4}$ 12) $-\frac{1}{7}+\frac{-x}{3}$

3.5.3 Rewriting Quotients

A common mistake with a quotient like $\frac{3x+1}{3}$ is "reducing" the 3's, $\frac{\cancel{3}x+1}{\cancel{3}}$ and ending with $x+1$. This isn't right since the dividend is a sum, not a product and only common factors can be reduced. The quotient can be rewritten as two fractions though, each with the original denominator, $\frac{3x}{3}+\frac{1}{3}$. Now we can reduce $\frac{\cancel{3}x}{\cancel{3}}+\frac{1}{3}=x+\frac{1}{3}$. Here's some practice.

Practice 3.5.3 Rewriting *Quotients*

Rewrite the following quotients into the form $mx+b$.

a) $\frac{6x-5}{3}$

$\frac{6x}{3}-\frac{5}{3}$	$\Rightarrow$	Wrote the quotient as two fractions each with the common denominator.
$2x-\frac{5}{3}$	$\Rightarrow$	Reduced the first term. The expression is now in the form $mx+b$.

b) $\frac{4-2x}{4}$

$\frac{4}{4}-\frac{2x}{4}$	$\Rightarrow$	Wrote the single term as two terms with a common denominator.
$1-\frac{x}{2}$	$\Rightarrow$	Reduced common factors.
$-\frac{x}{2}+1$	$\Rightarrow$	Thought of subtraction as adding the opposite and reordered the terms.
$-\frac{1}{2}x+1$	$\Rightarrow$	Rewrote the first term in the form mx.

Homework 3.5 Rewrite the following quotients into the form $mx+b$.

13) $\frac{3x+4}{5}$ 14) $\frac{x+9}{3}$ 15) $\frac{-12+8x}{4}$ 16) $\frac{0.5+1.1x}{0.1}$ 17) $\frac{-2-9x}{2}$ 18) $\frac{18-6x}{4}$

3.5.4 A Factor of -1

When the denominator of our fraction has a factor of -1, you'll need to pay particular attention to the signs of the terms as you simplify and reorder.

Practice 3.5.4 A Factor of -1

Rewrite the following terms into the form $mx+b$.

a) $\frac{\frac{3}{8}x+5}{-1}$

$\frac{\frac{3}{8}x}{-1}+\frac{5}{-1}$	$\Rightarrow$	Wrote the fraction as two fractions with the common denominator.
$-\frac{3}{8}x+-5$	$\Rightarrow$	Simplified the first term and then the second term. When dividing, if the signs are different, the result is negative.
$-\frac{3}{8}x-5$	$\Rightarrow$	Simplified the second term. Adding the opposite is usually written as subtraction.

b) $\frac{6-8x}{-2}$

$\frac{6}{-2}-\frac{8x}{-2}$	$\Rightarrow$	Wrote the fraction as two fractions with the common denominator.
$-3+4x$	$\Rightarrow$	Simplified
$4x-3$	$\Rightarrow$	Reordered the terms and wrote adding an opposite as subtraction.

Homework 3.5 Rewrite the following terms into the form $mx+b$.

19) $\frac{x-4}{-1}$ 20) $\frac{4+2x}{-2}$ 21) $\frac{-1.5x-4.2}{-0.6}$ 22) $\frac{3-4x}{-3}$ 23) $\frac{-14+4x}{-6}$

3.5.5 Rewriting Linear Functions into Slope-Intercept Form

Now that you've practiced some of the individual skills, let's put those skills together and practice rewriting a linear function into slope-intercept form.

Practice 3.5.5 Rewriting Linear Functions into Slope-Intercept Form

Rewrite the function into slope-intercept form and identify m and b.

a) $4x+2y+6=0$

$2y=-4x-6$	$\Rightarrow$	Subtracted 6 and 4x from both sides to isolate the y term.
$y=\frac{-4x-6}{2}$	$\Rightarrow$	Divided both sides by 2.
$y=\frac{-4x}{2}-\frac{6}{2}$	$\Rightarrow$	Wrote the single fraction as two fractions with a common denominator.
$y=-2x-3$	$\Rightarrow$	Reduced both fractions. m is -2 and b is -3. (Remembered to think of $-2x-3$ as $-2x+-3$.)

b) $\frac{1}{3}x+\frac{1}{4}y=1$

$12\left(\frac{1}{3}x+\frac{1}{4}y\right)=(12)(1)$	$\Rightarrow$	Multiplied both expressions by the LCD of all the terms.
$\frac{12}{3}x+\frac{12}{4}y=(12)(1)$ $4x+3y=12$	$\Rightarrow$	Distributed and reduced on the left and simplified on the right.
$3y=-4x+12$ $y=\frac{-4x}{3}+\frac{12}{3}$ $y=\frac{-4}{3}x+4$	$\Rightarrow$	Subtracted $4x$ from both expressions, divided both expressions by 3, rewrote the first term and simplified the second. $m=-\frac{4}{3}$ and $b=4$.

c) $5-y=\frac{1}{2}x$

$5-y-5=\frac{1}{2}x-5$ $-y=\frac{1}{2}x-5$	$\Rightarrow$	Subtracted 5 from both expressions to isolate the y term.
$\frac{-y}{-1}=\frac{\frac{1}{2}x-5}{-1}$	$\Rightarrow$	Divided both expressions by -1.
$y=\frac{\frac{1}{2}x}{-1}-\frac{5}{-1}$	$\Rightarrow$	Simplified the left expression and thought of the right expression as two fractions with a common denominator of -1.
$y=-\frac{1}{2}x+5$	$\Rightarrow$	Simplified. m is $-\frac{1}{2}$ and b is 5.

Homework 3.5 Rewrite the function into slope-intercept form and identify m and b.

24) $y+x=8$	25) $y-2x=-6$	26) $-y=-x+4$	27) $x+y+1=0$
28) $-8x+2y=2$	29) $3y-9=x$	30) $5x-2y-3=0$	31) $-8x-2y=12$
32) $2x-6y+10=0$	33) $20x+40y+80=0$	34) $-x+5=-15y$	
35) $y+\frac{3}{4}=\frac{2x}{3}$	36) $-\frac{x}{2}+\frac{1}{4}y=0$	37) $-\frac{1}{6}y-x=-1$	
38) $8y-7=4x$	39) $\frac{2}{3}y+\frac{x}{3}=1$	40) $-4x-2y-\frac{1}{2}=0$	

Homework 3.5 Answers

1) $\frac{1}{4}x$ The coefficient is $\frac{1}{4}$. 2) $\frac{3}{2}x$ The coefficient is $\frac{3}{2}$. 3) $\frac{3}{4}x$ The coefficient is $\frac{3}{4}$.

4) $-\frac{2}{5}x$ The coefficient is $-\frac{2}{5}$. 5) $-\frac{1}{7}x$ The coefficient is $-\frac{1}{7}$. 6) $2x$ The coefficient is 2.

7) $4x-6$ 8) $-x-3$ 9) $\frac{4}{5}x+\frac{1}{2}$ 10) $-\frac{1}{3}x+\frac{2}{9}$ 11) $3x+\frac{3}{4}$ 12) $-\frac{1}{3}x-\frac{1}{7}$

13) $\frac{3}{5}x+\frac{4}{5}$ 14) $\frac{1}{3}x+3$ 15) $2x-3$ 16) $11x+5$ 17) $-\frac{9}{2}x-1$ 18) $-\frac{3}{2}x+\frac{9}{2}$

19) $-x+4$ 20) $-x-2$ 21) $2.5x+7$ 22) $\frac{4}{3}x-1$ 23) $-\frac{2}{3}x+\frac{7}{3}$ 24) $y=-x+8$, $m=-1, b=8$

25) $y=2x-6$, $m=2, b=-6$

26) $y=x-4$, $m=1, b=-4$

27) $y=-x-1$, $m=-1, b=-1$

28) $y=4x+1$, $m=4, b=1$

29) $y=\frac{1}{3}x+3$, $m=\frac{1}{3}, b=3$

30) $y=\frac{5}{2}x-\frac{3}{2}$, $m=\frac{5}{2}, b=\frac{-3}{2}$

31) $y=-4x-6$, $m=-4, b=-6$

32) $y=\frac{1}{3}x+\frac{5}{3}$, $m=\frac{1}{3}, b=\frac{5}{3}$

33) $y=-\frac{1}{2}x-2$, $m=-\frac{1}{2}, b=-2$

34) $y=\frac{1}{15}x-\frac{1}{3}$, $m=\frac{1}{15}, b=-\frac{1}{3}$

35) $y=\frac{2}{3}x-\frac{3}{4}$, $m=\frac{2}{3}, b=-\frac{3}{4}$

36) $y=2x$, $m=2, b=0$

37) $y=-6x+6$, $m=-6, b=6$

38) $y=\frac{1}{2}x+\frac{7}{8}$, $m=\frac{1}{2}, b=\frac{7}{8}$

39) $y=-\frac{1}{2}x+\frac{3}{2}$, $m=-\frac{1}{2}, b=\frac{3}{2}$

40) $y=-2x-\frac{1}{4}$, $m=-2, b=-\frac{1}{4}$

Section 3.6 Other Linear Function Topics

Here are some other topics that involve linear functions.

3.6.1 Point-Slope Form for the Equation of a Line

Although slope-intercept form is a common way to write linear functions there's a second common form for the equation of a line called point-slope form. Point-slope form can be built from the slope formula if the (x_2, y_2) pair is left variable. Here's the idea.

$\dfrac{y-y_1}{x-x_1}=m$	Started with the slope formula. Noticed that the (x_2, y_2) ordered pair has been left variable.
$(x-x_1)\dfrac{y-y_1}{x-x_1}=m(x-x_1)$	Multiplied both sides by $x-x_1$ and reduced on the left side.
$y-y_1=m(x-x_1)$	This is point-slope form.

Here's some practice with point-slope form.

Practice 3.6.1 Point-Slope Form for the Equation of a Line.

Find the equation of the line. Begin with point-slope form and end with slope-intercept form.

a) The line goes through the points $(3,1)$ and has a slope of -2.

$y-y_1=m(x-x_1)$	$\Rightarrow$	Started with the general point-slope form.
$y-y_1=-2(x-x_1)$	$\Rightarrow$	Substituted the supplied value for m.
$y-1=-2(x-3)$	$\Rightarrow$	Substituted values for y_1 and x_1.
$y-1=-2x+6$ $y=-2x+7$	$\Rightarrow$	Solved for y. Finished by rewriting in slope-intercept form.

b)

Percent discount	Customers per hour
5	28
7	32
15	48
20	58

$y-y_1=m(x-x_1)$	$\Rightarrow$	Started with the general point-slope form.
$m=\dfrac{32-28}{7-5}=\dfrac{4}{2}=2$	$\Rightarrow$	Used the slope formula to find m.
$y-28=2(x-5)$	$\Rightarrow$	Substituted values for m, y_1 and x_1. Any ordered pair from the data table will work.
$y-28=2x-10$ $y=2x+18$	$\Rightarrow$	Solved for y. The function is now in slope-intercept form.

Homework 3.6 Find the equation of the line. Begin with point-slope form and end with slope-intercept form.

1) The line has a slope of $\frac{3}{4}$ and goes through the point $(-8,1)$.

2) The line goes through the points $(-6,-3)$ and $(12,-6)$.

3) The line has a slope of $-\frac{1}{3}$ and goes through the point $(-3,1)$.

4) The line goes through $(5,6)$ and the *x* coordinate of the *x*-intercept is 11.

5) The line has the same *y*-intercept as the line $-4x+6y=12$ and has the same slope as the line that goes through $(15,-12)$ and $(11,-8)$.

6)

Year since 1984	FDIC-Insured banks
5	12,545
15	9,165
24	6,123

7) A student borrows money for college and agrees to pay a set amount each month until the loan is paid back. After 6 months, they owe \$3800. After 12 months, they owe \$2600. Assume the outstanding loan amount is dependent on the number of months since the loan began.

8)

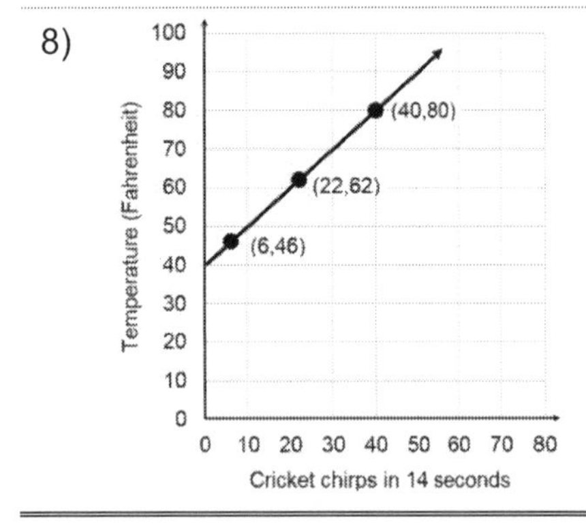

9)

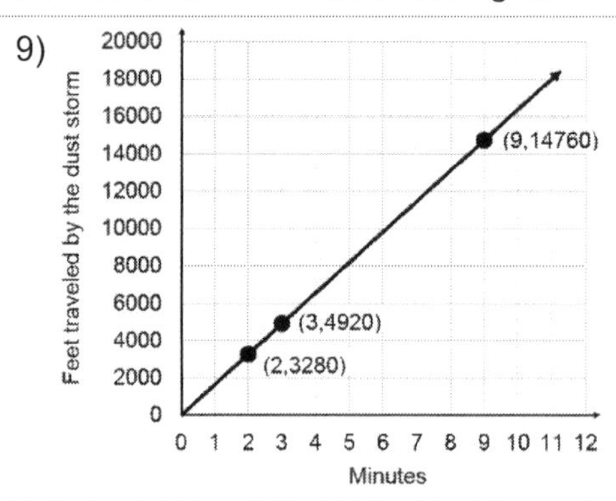

3.6.2 Parallel and Perpendicular Lines

Two lines are parallel if they have the same slope. For example, the lines $y=-2x+6$ and $y=-2x-8$, which are graphed to the right, are parallel since they both have slope -2. Parallel lines with different *y*-intercepts never meet while those with the same *y*-intercept are considered the same line.

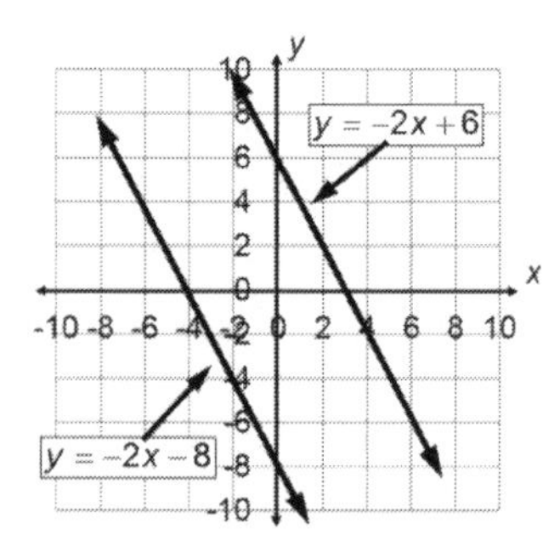

Lines that are perpendicular meet in a "right" (90 degree) angle. The slopes of perpendicular lines are opposite reciprocals (as long as neither has a divisor of 0.) The graph to the right shows two lines that are perpendicular. Notice one line has a slope of $\frac{1}{2}$ and the other has a slope of -2. Since the reciprocal of one-half is two, and the slopes have different signs, the lines are perpendicular.

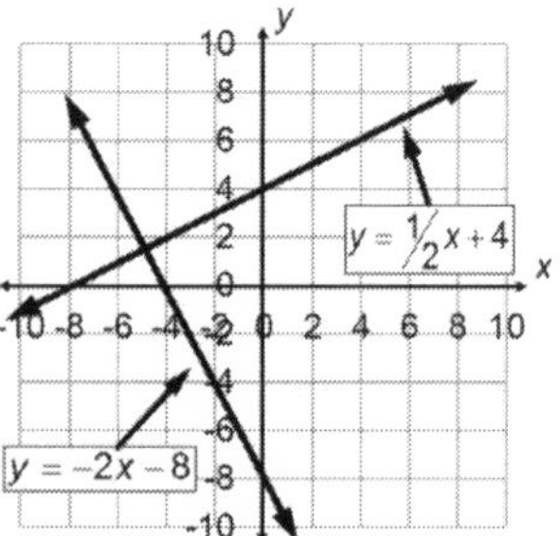

Practice 3.6.2 Parallel and Perpendicular Lines

Use point-slope form and/or slope-intercept form to answer the question.

a) Find the equation of the line that's parallel to $3x+y-8=0$ and goes through $(1,4)$.

$y=-3x+8$	$\Rightarrow$	Solved the first line for y and found the slope is -3. Since the lines are parallel, the slope of the second line must also be -3.
$y-4=-3(x-1)$	$\Rightarrow$	Started with point-slope form and substitute the supplied value for the slope and the values for the point on the second line.
$y-4=-3x+3$ $y=-3x+7$	$\Rightarrow$	Distributed and solved for y to find the equation of the line in slope-intercept form.

b) Decide if the line that goes through the points $(-7,12)$ and $(-3,6)$ is perpendicular to the line $2x-3y=1$.

$m=\frac{12-6}{-7-(-3)}=\frac{6}{-4}=-\frac{3}{2}$	$\Rightarrow$	Found the slope of the first line.
$y=\frac{2}{3}x-\frac{1}{3}$	$\Rightarrow$	Solved the second line for y and found the slope is $\frac{2}{3}$.
$m=-\frac{3}{2}$ and $m=\frac{2}{3}$	$\Rightarrow$	The lines are perpendicular since one slope is the opposite reciprocal of the other.

Homework 3.6 Use point-slope form and/or slope-intercept form to answer the question.

10) Show the line that goes through the points $(8,-4)$ and $(6,-1)$ is parallel to the line that goes through the points $(1,1)$ and $(-3,7)$.

11) Find the equation of the line that goes through the point $(-4,9)$ and is perpendicular to the line that goes through the points $(4,7)$ and $(3,3)$.

12) Find the equation of the line that goes through $(-6,-6)$ and is perpendicular to the line $\frac{1}{2}y-x=-2$.

13) Find the equation of the line if the x coordinate of the x-intercept is -5 and the line is parallel to the line $-4x-y-1=0$.

14) Assuming we're using the word "intercept" to discuss the non-zero coordinate, show the line with an x-intercept of 7 and a y-intercept of -5 is perpendicular to the line with an x-intercept of 5 and a y-intercept of 7.

15) Find the equation of the line that goes through $(-1,9)$ that is parallel to the line that goes through $(4,5),(7,5)$.

16) Two perpendicular lines both go through the point $(2,2)$. If one of the lines also goes through the point $(4,6)$ find the equation of the <u>other</u> line in slope-intercept form.

17) Find the equation of the line that is perpendicular to the line that goes through the points $(-9,14)$, $(30,14)$ and has the same x-intercept as the line $9x-y=9$.

18) Decide if the line through $(1,-1)$, $(2,2)$ ever touches the line through $(-1,-2)$, $(2,7)$.

Homework 3.6 Answers

1) $y=\frac{3}{4}x+7$ 2) $y=-\frac{1}{6}x-4$ 3) $y=-\frac{1}{3}x$ 4) $y=-x+11$ 5) $y=-x+2$

6) $y=-338x+14{,}235$ 7) $y=-200x+5000$ 8) $y=x+40$ 9) $y=1640x$

10) Both lines have the slope $-3/2$ so they are parallel.

11) The slope of the line through $(4,7)$ and $(3,3)$ is 4, the opposite reciprocal is $-\frac{1}{4}$. The equation of the other line is $y=-\frac{1}{4}x+8$.

12) The slope of the line $\frac{1}{2}y-x=-2$ is 2. The perpendicular line will have slope $-\frac{1}{2}$. The equation of the line will be $y=-\frac{1}{2}x-9$.

13) Remember the *x*-intercept is the point $(-5,0)$. The equation of the line is $y=-4x-20$.

14) The equation of the first line is $y=\frac{5}{7}x-5$. The equation of the second line is $y=-\frac{7}{5}x+7$.

15) The equation of the line is $y=9$. Remember a line with a slope of 0 is parallel to the *x*-axis.

16) The slope for the first line is 2 so the slope for the perpendicular line is $-\frac{1}{2}$. Using point-slope form for the second line gives $y-2=-\frac{1}{2}(x-2)$ which in slope-intercept form would be $y=-\frac{1}{2}x+3$.

17) The line that goes through the points $(-9,14)$ and $(30,14)$ is $y=14$. The slope for this line is 0. The reciprocal of 0 is undefined. The *x*-intercept for the line $9x-y=9$ would be $(1,0)$. The line $x=1$ is perpendicular to $y=14$ and has the *x*-intercept $(1,0)$.

18) The slope of the first line is 3. The slope of the second line is also 3 so the lines are parallel. The *y*-intercept of the first line is $(0,-4)$. The *y*-intercept of the second line is $(0,1)$. The lines are parallel but have different *y*-intercepts. The lines don't share any points in common.

Chapter 4

Exponentiation

4.1 An Introduction to Exponentiation

In this section, we'll move beyond the four basic operations of adding, subtracting, multiplying and dividing and begin working with exponentiation. Exponentiation is in line 1 of the order of operations.

4.1.1 Exponentiation and the Idea of Power

Exponentiation is an operation that requires a base and an exponent and is written in **exponential form** with the exponent written as a superscript to the right of the base, $\text{Base}^{\text{exponent}}$. For instance, 3^2 is written in exponential form with 3 as the base and 2 as the exponent. If the exponent is a 2 it's common to say the base is, "squared" and if the exponent is a 3, to say the base is, "cubed". It's also common not to write exponents of 1. For example, 3^1 is usually written as 3.

If we restrict the value of our exponent to the positive integers, we have a **power**. To simplify a power, we use the exponent to decide how many factors of the base we should multiply together. For example, the second power of three, 3^2, is nine, $3^2 = 3\times 3 = 9$.

Before continuing, I need to mention that it's common for people to use the word power and exponent interchangeably, even though they're different. I suspect this comes from thinking that "Two to the fourth power." is describing an exponent of 4 instead of the power 16. To lesson this confusion, let's agree, from this point on, to refer to an expression like 3^2 as, "The second power of three." instead of, "Three to the second power." Here's some practice with the vocabulary for powers.

Practice 4.1.1 Exponentiation and the Idea of Power

Fill in the blank using term, sum, factor, product, minuend, subtrahend, difference, dividend, divisor, quotient, base, exponent or power.

---

a) Before simplifying $4^2 + 3^2$,

4 is a(n) ___, 3 is a(n) ___, 2 is a(n) ___, 4^2 is both a(n) ___ and a(n) ___, 3^2 is both a(n) ___ and a(n) ___ and $4^2 + 3^2$ is a(n) ___.

base, base, exponent, power, term, power, term, sum

---

b) Before simplifying $5(6-4^2)^3$,

6 is the___, 4 is a(n)___, 2 is a(n)___, 4^2 is both a(n) __ and the __, $6-4^2$ is a(n) __,

$(6-4^2)$ is a(n) __, 3 is a(n) ___, $(6-4^2)^3$ is both a(n) ___ and a(n) ___, 5 is a(n) __,

and $5(6-4^2)^3$ is a(n) ___.

minuend, base, exponent, power, subtrahend, difference, base, exponent, power, factor, factor, product

Homework 4.1 Fill in the blank using term, sum, factor, product, minuend, subtrahend, difference, dividend, divisor, quotient, base, exponent or power.

1) Before simplifying $(5-3)^2$,

5 is the ___, 3 is the ___, $5-3$ is a(n) ___, $(5-3)$ is a(n) ___, 2 is a(n) ___, and

$(5-3)^2$ is a(n) ___.

2) Before simplifying $(-7)^2-4(-2)(-3)$,

(-7) is a(n) ___, 2 is a(n) ___, $(-7)^2$ is both a(n) ___ and the___, 4 is a(n) ___, (-2)

is a(n) ___ and (-3) is a(n) ___, $4(-2)(-3)$ is both a(n) ___ and the___ and

$(-7)^2-4(-2)(-3)$ is a(n) ___.

3) Before simplifying $(-3)^{-2(-4)}$,

-2 is a(n) ___, (-4) is a(n) ___, $-2(-4)$ is both a(n) ___ and a(n) ___, (-3) is a(n)

___, and $(-3)^{-2(-4)}$ is a(n) ___.

4) Before simplifying $\left(\frac{4}{5^2}\right)^3$,

4 is the ___, 5 is a(n) ___, 2 is a(n)___, 5^2 is both a(n) ___ and the___, $\frac{4}{5^2}$ is a(n)

___, $\left(\frac{4}{5^2}\right)$ is a(n) ___, 3 is a(n) ___ and $\left(\frac{4}{5^2}\right)^3$ is a(n) ___.

5) Before simplifying $7{,}000(4)^{(0.05)(20)}$,

7,000 is a(n) ___, 0.05 is a(n) __, (20) is a(n) __, $(0.05)(20)$ is both a(n) ___ and a(n)

___, (4) is a(n) __, $(4)^{(0.05)(20)}$ is both a(n) ___ and a(n) ___, and $7{,}000(4)^{(0.05)(20)}$ is

a(n) ___.

4.1.2 Beginning to Simplify Powers

It's useful to memorize the first few perfect squares and perfect cubes so you're able to sometimes simplify powers without using a calculator. Here are the first few perfect squares.

$$0^2 = 0 \quad 1^2 = 1 \quad 2^2 = 4 \quad 3^2 = 9 \quad 4^2 = 16 \quad 5^2 = 25$$
$$6^2 = 36 \quad 7^2 = 49 \quad 8^2 = 64 \quad 9^2 = 81 \quad 10^2 = 100$$

And the first few perfect cubes.

$$0^3 = 0 \quad 1^3 = 1 \quad 2^3 = 8 \quad 3^3 = 27 \quad 4^3 = 64 \quad 5^3 = 125$$
$$(-1)^3 = -1 \quad (-2)^3 = -8 \quad (-3)^3 = -27 \quad (-4)^3 = -64 \quad (-5)^3 = -125$$

A scientific calculator will help when you can't simplify an exponential expression mentally. Here are some of the most common exponent keys although there might be others,

$$\boxed{x^y} \text{ or } \boxed{y^x} \text{ or } \boxed{\wedge} \text{ or } \boxed{x^{\square}}\ .$$

Usually you'll enter the base, push the exponent key, enter the exponent and then push the equal key. Here's some practice simplifying powers.

Practice 4.1.2 Beginning to Simplify Powers

Simplify. Only use a calculator when necessary.

a) 5^2

$25 \Rightarrow$ Multiplied two factors of the base 5×5 .

b) $(-4)^3$

$-64 \Rightarrow$ Multiplied $(-4)(-4)(-4)$. The parentheses tell us the base is -4

c) -2^2

$-4 \Rightarrow$ This one is subtle; the base is 2 not -2. It's often useful to think of the dash symbol as a factor of negative one, $-1 \times 2^2 = -1 \times 2 \times 2$

d) $\left(\frac{2}{3}\right)^3$

$\frac{8}{27} \Rightarrow$ Multiplied three factors of the base $\left(\frac{2}{3}\right)\left(\frac{2}{3}\right)\left(\frac{2}{3}\right)$.

Homework 4.1 Simplify. Only use a calculator when necessary.

6) -3^2	7) 1^{14}	8) 4^3	9) $(-3)^4$	10) 0^3	11) $-(-8)^2$
12) $(-1)^{716}$	13) $(-4)^3$	14) $\left(\frac{1}{2}\right)^5$	15) -19^4	16) $\left(\frac{7}{5}\right)^1$	17) $\left(-\frac{5}{9}\right)^3$

4.1.3 Powers and the Order of Operations

As I mentioned earlier, powers are simplified before multiplying or dividing.

Practice 4.1.3 Powers and the Order of Operations

Count the number of operations, name the operations using the correct order and then simplify the expression.

a) $2^4 \times (-3)^3$

There are three operations. By the order of operations powers are simplified first left to right and multiplication is simplified last.

16×-27	$\Rightarrow$	Simplified the powers left to right.
-432	$\Rightarrow$	Multiplied.

b) $9^2 - 4(2)(-3)$

There are four operations. Power is first, then multiplication and finally subtraction.

$81 - 4(2)(-3)$	$\Rightarrow$	Squared nine.
$81 - (-24)$	$\Rightarrow$	Multiplied left to right.
105	$\Rightarrow$	Thought of the subtraction of an opposite as addition.

Homework 4.1 Count the number of operations, name the operations using the correct order and then simplify the expression.

18) $-5^2 - 3^2$	19) $(5)^2 - 4(3)(2)$	20) $(-4)^4 - 4^3$	21) $6 \div 2 \times 3 - 2^4$
22) $(-7)^2 - 4(-2)(-3)$	23) $3 - 3^5 + 5^3$	24) $-6^4 \div 6^3$	25) $(-1)^2 - 5(-2)(9)$
26) $2^3 \times 2^2 - 2^5$	27) $-4^2 + (-4)^2$		

4.1.4 Powers and Grouping

With power you'll need to consider both explicit and implicit grouping. Explicit grouping with parentheses, brackets or braces remains the same. Operations in the exponent are a type of implicit grouping and must be finished before simplifying the power. Here's some practice.

Practice 4.1.4 Powers and Grouping

Count the number of operations, name the operations using the correct order and then simplify the expression.

a) $(6+3)^2 - (6^2 + 3^2)$

There are six operations. First, find the sum in the first set of parentheses and then simplify powers in the second set of parentheses. Next, find the power in the minuend and the sum in the subtrahend. Last, find the difference.

$(9)^2 - (36+9)$	$\Rightarrow$	Added inside the first set of parentheses and found powers in the second set of parentheses.
$81 - (45)$ 36	$\Rightarrow$	Squared nine, found the sum inside the parentheses and then subtracted.

b) $3^{4-1}+4^{3-1}$		
There are five operations. Operations in the exponents are first followed by powers and then addition.		
3^3+4^2	$\Rightarrow$	Simplified the implicit grouping first.
$27+16$	$\Rightarrow$	Found the third power of three and the second power of four.
43	$\Rightarrow$	Added.

c) $(7-12)^{2-(-1)}-(10-6)^{2+3}$		
There are seven operations. Operations inside parentheses and operations in exponents must be completed first, powers are then simplified left to right and subtraction is last.		
$(-5)^{2-(-1)}-(4)^{2+3}$	$\Rightarrow$	Explicit grouping was simplified left to right. (The order in which you do any specific implicit or explicit grouping isn't going to matter. Just make sure all grouping is completed before you move on to powers.)
$(-5)^3-(4)^5$	$\Rightarrow$	Implicit grouping was simplified left to right.
$-125-1024$	$\Rightarrow$	Simplified powers left to right.
$-1{,}149$	$\Rightarrow$	Subtracted.

Homework: 4.1 Count the number of operations, name the operations using the correct order and then simplify the expression.

28) $(2^3-3^3)(2-3)^3$ 29) $\dfrac{2^2-4^2}{(2-4)^2}$ 30) $-3^{-2(-4)}$ 31) $4(6-4^2)^3$

32) $3^{3+3}-2^{3\times3}$ 33) $-16^3+(5--3)^4$ 34) $(-4-(-2))^2+(6-11)^2$

35) $(7-1)^2-(7^2-1^2)$ 36) $-2^{-6}--3^2$ 37) $2^{-6+10}-4^{4-1}$

Homework 4.1 Answers

1) minuend, subtrahend, difference, base, exponent, power

2) base, exponent, power, minuend, factor, factor, factor, product, subtrahend, difference

3) factor, factor, product, exponent, base, power

4) dividend, base, exponent, power, divisor, quotient, base, exponent, power

5) factor, factor, factor, product, exponent, base, power, factor, product

6) -9	7) 1	8) 64	9) 81	10) 0	11) -64
12) 1	13) -64	14) $1/32$	15) $-130{,}321$	16) $7/5$	17) $-125/729$

18) There are three operations. Powers are first left to right and subtraction is last. –34

19) There are four operations. Squaring is first, then multiplication left to right, and subtraction is last. 1

20) There are three operations. Powers are simplified first left to right and subtraction is last. 192

21) There are four operations. Squaring is first, then division and multiplication left to right and subtraction is last. –7

22) There are four operations. Power is first, then multiplication left to right and subtraction is last. 25

23) There are four operations. Powers are simplified first left to right and then subtraction and addition is simplified left to right. -115

24) There are three operations. Powers are simplified first left to right, and division is last. –6

25) There are four operations. Power is first, then multiplication left to right and subtraction is last. 91

26) There are five operations. Powers are simplified first, then multiplication and subtraction is last. 0

27) There are three operations. Powers are simplified first, and addition is last. 0

28) There are six operations. Powers and subtraction inside the first set of parentheses is first. Subtraction in the second set of parentheses and cubing are next. Multiplication is last. 19

29) There are six operations. Powers and subtraction in the dividend are first. Subtraction and squaring in the divisor are next. Dividing is last. –3

30) There are two operations. The exponent multiplication is first and power is last. –6,561

31) There are four operations. Squaring and subtraction inside the parentheses is first, cubing is next and multiplication is last. –4,000

32) There are five operations. Exponent operations are first, powers are next and subtraction is last. 217

33) There are four operations. Subtraction is first, powers are next left to right and addition is last. 0

34) There are five operations. Subtraction left to right is first, powers are next and addition is last. 29

35) There are six operations. Subtraction in the first parentheses and squaring in the second parentheses are first. Squaring the power in the minuend is next. Then subtraction inside the second set of parentheses and finally subtraction. –12

36) Begin in the exponent with squaring and subtraction then find the power. –8

37) There are five operations. Operations in exponents are first left to right, powers are next left to right, and subtraction is last. –48

4.2 Applying Powers

In this section, we'll practice evaluating powers. Recall the direction, "evaluate" implies we'll substitute a value in place of our variable(s) and simplify.

4.2.1 Checking Answers for a Quadratic Equation

In chapter two I mentioned that an equation like $3x-5=2(x+8)$ is called linear because the exponent on the variable base was a 1, $3x^1-5=2(x^1+8)$. (Recall it's common not to write exponents of 1.) As you saw in chapter two a linear equation often has a single solution.

An equation like $2x^2+4x-1=0$, where 2 is the largest exponent on a variable, is known as a quadratic equation. With a quadratic equation, it's common to have two solutions. For example, the solution set for $x^2=4$ is $\{-2,2\}$. You'll solve quadratic equations later in the course. For now, we'll practice checking answers for a quadratic equation.

Practice 4.2.1 Checking Answers for a Quadratic Equation

Decide whether each supplied answer is or is not a solution.

a) $x^2-8x-20=0;\ x=-2$ and $x=10$

Work		Explanation
$(-2)^2-8(-2)-20 \quad 0$	⇒	Began by substituting -2 for every instance of x. It's always best to substitute into parentheses.
$4+16-20 \quad 0$ 0	⇒	Simplified the left side to find that both expressions have the same value. -2 is probably a solution.
$(10)^2-8(10)-20 \quad 0$	⇒	This time substituted 10 for the variable.
$100-80-20 \quad 0$ 0	⇒	Simplified the left expression. Since the expressions are the same, 10 is probably a solution also.

b) $2(y^2+2)=-y^2-7y;\ y=-1$ and $y=1$

Work		Explanation
$2((-1)^2+2) \quad -(-1)^2-7(-1)$	⇒	Began by substituting -1 for every instance of y.
$2(1+2) \quad -1+7$ $2(3) \quad 6$ 6	⇒	Simplified and found the expressions are the same value. -1 is a solution.
$2((1)^2+2) \quad -(1)^2-7(1)$	⇒	Began by substituting 1 for every instance of y.
$2(1+2) \quad -1-7$ $2(3) \quad -8$ 6	⇒	Simplified and found the expressions are not the same value. 1 is not a solution.

Homework 4.2 Decide whether each supplied answer is or is not a solution.

1) $m^2 - 10m + 24 = 0$; $m = 4$ and $m = 6$ 2) $k^2 - 7k = -10$; $k = -2$ and $k = 5$

3) $2p^2 + 13p + 6 =$; $p = -6$ and $p = 3$ 4) $3m^2 - 2m - 1 = 0$; $m = -1$ and $m = 1$

5) $3k^2 + 15k = -9(k+5)$; $k = -5$ and $k = -3$

6) $x^2 - 2.25x + 1.125 = 0$; $x = 0.75$ and $x = 1.5$

7) $2(y+6)^2 - 32 = 0$; $y = -10$ and $y = -2$ 8) $-4(x-3)^2 + 1 = -35$; $t = 5$ and $t = 0$

4.2.2 Evaluating the Discriminant

Later in this course we'll solve quadratic equations using the quadratic formula.

The Quadratic Formula

The solutions to the quadratic equation $ax^2 + bx + c = 0$ are given by the quadratic formula $x = \dfrac{-b \pm \sqrt{b^2 - 4ac}}{2a}$.

As you can see the quadratic formula uses square-roots which we'll cover in chapter 6. Today we'll practice simplifying the **discriminant**, which is the expression $b^2 - 4ac$ that is "under" the square root symbol.

To evaluate the discriminant, you'll substitute values for the letters a, b and c and simplify. Here's some practice.

Practice 4.2.2 Evaluating the Discriminant

Evaluate the discriminant, $b^2 - 4ac$, for the given values.

a) $a = 1$, $b = -2$ and $c = 2$

$b^2 - 4ac$ $(-2)^2 - 4(1)(2)$	$\Rightarrow$	Substituted the supplied values. Made sure to substitute into parentheses.
$4 - 4(1)(2)$ $4 - 8$ -4	$\Rightarrow$	Followed the order of operations and squared first, found the product second and subtracted last.

Homework 4.2 Evaluate the discriminant, $b^2 - 4ac$, for the given values.

9) $a = -2$, $b = 1$ and $c = -1$ 10) $a = 2$, $b = -6$ and $c = 1$ 11) $a = -3$, $b = -3$ and $c = -9$

12) $a = 1$, $b = 4$ and $c = 4$ 13) $a = 3$, $b = -7$ and $c = -5$ 14) $a = 3$, $b = -2$ and $c = -1$

4.2.3 Evaluating Quadratic Functions

We'll end this section by using quadratic functions to answer some questions.

Practice 4.2.3 Evaluating Quadratic Functions

Use the function to answer the following questions.

a) Between 1980 and 2010 the number of cable subscribers in the U.S. (in millions) could be estimated by the function $y = -0.1x^2 + 4.5x + 16.7$ where x is the years since 1980.

1. Find the number of subscribers in 1980
2. Find the number of subscribers in 2005

1. In 1980 there were 16,700,000 subscribers.	$\Rightarrow$	Evaluated the right side using 0. $C = -0.1(0)^2 + 4.5(0) + 16.7 = 16.7$
2. In 2005 there were 66,700,000 subscribers.	$\Rightarrow$	Evaluated the right side using 25. $C = -0.1(25)^2 + 4.5(25) + 16.7 = 66.7$

Homework 4.2 Use the function to answer the following questions.

15) Between 1982 and 1997 the function $y = -0.9x^2 + 7.5x + 168.8$ modeled the number of cases of Chickenpox (in thousands) in the United States. x stands for the number of years after 1982.

a) Find the number of cases of Chickenpox in 1982 (year 0).
b) Find the number of cases of Chickenpox in 1986.
c) Find the number of cases of Chickenpox in 1994.

16) The number of farms (in thousands) in the United States between 1975 and 2009 can be modeled using the function $y = 0.6x^2 - 32x + 2{,}544$ where x is the number of years after 1975.

a) Find the number of farms in 1975.
b) Find the number of farms in 1998.
c) Find the number of farms in 2009.

17) If an object is thrown into the air with an initial speed of 88 feet per second the function $y = -16x^2 + 88x$ describes its height, in feet, as a function of x, the time in seconds. (Disregard the height of the source that is throwing the object.)

a) Find the height after 0 seconds.
b) The object will reach its maximum height after 2.75 seconds. Find the maximum height.
c) What is happening after 5.5 seconds?

18) A person who makes "vintage" bicycles finds if they charge $\$x$ per bicycle then the function $y = -0.2x^2 + 242x - 42{,}000$ estimates their profit (or loss) for the year.

a) What happens if the person sets their price at $0?
b) The price $605 will maximize the profit for this business. What will the maximum profit be?
c) Find the profit if the person sets their price at $800 for a bicycle.

19) The function $y = 0.03x^2 - 0.7x + 5.6$ predicts the cases of Rocky Mountain Spotted Fever per one-million persons in the United States, y, using the number of years after 1980 (x).

a) A minimum number of cases occurred around 1992. What was the minimum number of cases per one-million persons?

b) Estimate the number of cases per one-million persons that will occur in 2017.

20) The percent of 12th grade boys who have smoked in the last 30 days can be modeled using the function $y = -0.037x^2 + 0.981x + 25.54$ where x is the number of years after 1980.

a) The highest percent occurred around 1993. Find the percent of boys smoking that year.

b) Use the function to estimate the percent of 12th grade boys smoking in 2018.

21) Since 1970 the function $y = 5.5x^2 - 46.4x + 692$ has given a good approximation of the cost for one year of tuition and fees (in dollars) at a United States four-year public college. Assume x is the number of years after 1970.

a) Find the cost for a year of college in 1970.

b) Find the cost for a year of college in 2018.

c) If the trend continues estimate how much a year of college will be in 2025.

Homework 4.2 Answers

1) Both values are a solution. 2) Only 5 is a solution. 3) Only -6 is a solution.

4) Only 1 is a solution. 5) Both values are a solution. 6) Both values are a solution.

7) Both values are a solution. 8) Only 0 is a solution. 9) -7 10) 28

11) -99 12) 0 13) 109 14) 16

15) a) 168,800 cases, b) 184,400 cases, c)129,200 cases

16) a) 2,544,000 farms, b) 2,125,400 farms, c) 2,149,600 farms

17) a) After 0 seconds the object is at height 0 feet. b) The maximum height will be 121 feet. c) Since the height is again 0 feet the ball must have returned to its original height.

18) a) If they give their bicycles away (set the price at \$0) the profit is $-\$42,000$. b) \$31,205 c) \$23,600

19) a) About 1.5 cases per one-million persons b) Around 21 cases per one-million persons.

20) a) Around 32% of 12th grade boys. b) Around 9.4%

21) a) About \$692, b) \$11,137 c) \$14,778

4.3 An Introduction to Exponent Rules

The exponent rules allow us to quickly multiply, divide or find powers of expressions written in exponential form. You can memorize exponent rules but it's stronger if you understand how they're based on the commutative property of multiplication (reordering the factors), the associative property of multiplication (regrouping the factors), the multiplicative identity (reducing factors of 1) and writing the final product or quotient in exponential form. Please notice that the exponent rules are for products and quotients only, they're not for sums and differences.

The Rules for Exponents (For real number variables *x* and *y* and constants *m* and *n*.)			
The Exponent Rules for Products			
Product Rule	$x^m x^n = x^{m+n}$	$3^4 \times 3^2 = 3^{4+2} = 3^6$	$k^7 \times k = k^{7+1} = k^8$
Power Rule	$(x^m)^n = x^{m \times n}$	$(3^4)^2 = 3^{4\times2} = 3^8$	$(y^4)^2 = y^{4\times2} = y^8$
Power of a Product Rule	$(xy)^m = x^m y^m$	$(2\times10)^3 = 2^3 \times 10^3$	$(2n)^3 = 2^3 n^3$
The Exponent Rules for Quotients ($x \neq 0$) ($y \neq 0$)			
Quotient Rule	$\frac{x^m}{x^n} = x^{m-n}$ $\frac{x^m}{x^n} = \frac{1}{x^{n-m}}$	$\frac{2^5}{2^2} = 2^{5-2} = 2^3$ $\frac{2^5}{2^2} = \frac{1}{2^{2-5}} = \frac{1}{2^{-3}}$	$\frac{y^5}{y^2} = y^{5-2} = y^3$ $\frac{y^5}{y^2} = \frac{1}{y^{2-5}} = \frac{1}{y^{-3}}$
Power of a Quotient Rule	$\left(\frac{x}{y}\right)^m = \frac{x^m}{y^m}$	$\left(\frac{2^2}{3}\right)^3 = \frac{(2^2)^3}{3^3}$	$\left(\frac{5}{y^3}\right)^3 = \frac{5^3}{(y^3)^3}$
Rules for Integer Exponents Less Than 1 ($x \neq 0$) ($y \neq 0$)			
Zero-Exponent Rule	$x^0 = 1$	$4^0 = 1$	$(2x)^0 = 1$
Negative Exponent Rule	$x^{-m} = \frac{1}{x^m}$ $\frac{1}{x^{-m}} = x^m$	$3^{-2} = \frac{1}{3^2}$ $\frac{1}{3^{-2}} = 3^2$	$y^{-2} = \frac{1}{y^2}$ $\frac{1}{y^{-2}} = y^2$

4.3.1 The Product Rule

The product rule groups similar factors to quickly multiply powers with the same base. For instance, the product $3^4 \times 3^2$, which has six factors of 3, can be replaced by 3^6.

$$\underbrace{(3\times3\times3\times3)}_{3^4} \times \underbrace{(3\times3)}_{3^2}$$
$$(3\times3\times3\times3\times3\times3)$$
$$3^6 = 729$$

Careful!	Don't change the value of the base
	Don't change the value of the base when you use the exponent rules. For instance, $3^4 \times 3^2$ simplifies to a power of three, 3^6, not a power of 9.

Practice 4.3.1 The Product Rule
Simplify.

a) $4 \times 4^3 \times 4^4$

4^{1+3+4}	$\Rightarrow$	Used the product rule. Remembered $4 = 4^1$.
4^8	$\Rightarrow$	Added the exponents and kept the common base.
$65{,}536$	$\Rightarrow$	Calculated the power using a calculator.

b) $k^7 \times k^{14}$

k^{7+14}	$\Rightarrow$	Used the product rule. Remembered to keep the common base.
k^{21}	$\Rightarrow$	Added the exponents.

Homework 4.3 Simplify.

1) $(-3)^3(-3)^2$ 2) $c^3 \times c \times c^2$ 3) $(-1)^{18}(-1)^5(-1)^7$ 4) $-y^4 \times y^4$

5) $2k^7(6k^4)$ 6) $(-4x^2)(5x^4)$

4.3.2 The Power Rule

The power rule helps when the base is itself a power. To the right you can see an example of why the rule is true. By seeing the original power, a^4, as the base for the power of 2, we can use the definition of exponentiation, and the product rule, to find the final power of *a*.

$$(a^4)^2$$
$$a^4 \times a^4$$
$$a^{4+4}$$
$$a^8$$

Practice 4.3.2 The Power Rule
Simplify.

a) $(4^2)^3$

$4^{2\times 3} = 4^6$	$\Rightarrow$	Used the power rule.
$4{,}096$	$\Rightarrow$	Used a calculator to find the sixth power of 4.

Homework 4.3 Simplify.

7) $(5^4)^2$ 8) $(m^7)^3$ 9) $-(k^2)^5$ 10) $\left[(-3)^3\right]^3$ 11) $\left[c^8\right]^8$

4.3.3 *The Power of a Product Rule*

This rule is helpful when your base has more than one type of factor. It's also an important tool when you're trying to make decisions about the final sign of a power.

To the right, you can see how this rule uses the definition of exponentiation, the commutative property (to reorder the factors) and the associative property (to regroup the factors) to rewrite $(2k)^3$ as 2^3k^3.

$$(2k)^3$$
$$(2k)(2k)(2k)$$
$$2\times2\times2\times k\times k\times k$$
$$(2\times2\times2)\times(k\times k\times k)$$
$$2^3k^3$$

Practice 4.3.3 The Power of a Product Rule

Simplify.

a) $(-7)^4$

$(-1\times7)^4$	$\Rightarrow$	Thought of the dash symbol as a factor of negative one.
$(-1)^4\times7^4$	$\Rightarrow$	Used the power of a product rule.
2,401	$\Rightarrow$	Simplified both factors.

b) $(5n)^3$

5^3n^3	$\Rightarrow$	Used the power of a product rule.
$125n^3$	$\Rightarrow$	Calculated the third power of 5.

Homework 4.3 Simplify.

12) $(-y)^5$ 13) $(ab)^2$ 14) $-(3h)^3$ 15) $(-2x)^4$ 16) $(-xy)^4$

4.3.4 *The Quotient Rule*

To simplify $\frac{2^5}{2^2}$ we could reduce factors of 1 and write the result in exponential form like I have to the right. Another way, since the bases are the same, is to find the difference in the exponents, $\frac{2^5}{2^2}=2^{5-2}=2^3$ to find the factors that remain after reducing. Using this difference in the exponents is the idea behind the quotient rule.

$$\frac{2^5}{2^2}$$
$$\frac{2\times2\times2\times2\times2}{2\times2}$$
$$\frac{\cancel{2}\times\cancel{2}\times2\times2\times2}{\cancel{2}\times\cancel{2}}$$
$$2^3=2^{5-2}$$

Practice 4.3.4 The Quotient Rule

Simplify.

a) $\dfrac{10^{24}}{10^{22}}$

$10^{24-22} = 10^2$	$\Rightarrow$ Since the bases were the same I used the quotient rule.
100	$\Rightarrow$ Simplified.

b) $\dfrac{x^5y^2}{x^3y}$

$x^{5-3}y^{2-1}$	$\Rightarrow$ There are two bases, x and y. Subtracted the exponents using the quotient rule for the base of x and then for the base of y. Remembered to think of the factor of y in the divisor as y^1.
x^2y	$\Rightarrow$ Found the difference in the exponents.

Homework 4.3 Simplify.

17) $8^4/8^2$ 18) $\dfrac{(-4)^{17}}{(-4)^{12}}$ 19) $\dfrac{-q^{14}}{q^8}$ 20) $\dfrac{2^{15}y^8}{2^{12}y^6}$ 21) $\dfrac{6^{14}n^9}{6^{12}n^6}$

4.3.5 The Power of a Quotient Rule

Like the power of a product rule, the power of a quotient rule is often used when we have different bases in the dividend and divisor of a quotient. To the right, you can see how this rule uses the definition of power, the commutative property and the associative property to rewrite $\left(\dfrac{2}{y}\right)^3$ as $\dfrac{2^3}{y^3}$.

$$\left(\frac{2}{y}\right)^3$$

$$\frac{2}{y}\times\frac{2}{y}\times\frac{2}{y}$$

$$\frac{2\times2\times2}{y\times y\times y}$$

$$\frac{2^3}{y^3}=\frac{8}{y^3}$$

Practice 4.3.5 The Power of a Quotient Rule

Simplify.

a) $\left(\dfrac{7}{w}\right)^4$

$\dfrac{7^4}{(w)^4}$	$\Rightarrow$ Used the power of a quotient rule.
$\dfrac{2{,}401}{w^4}$	$\Rightarrow$ Calculated the fourth power of seven using a calculator.

b) $\left(\frac{-x}{y}\right)^3$

$\frac{(-x)^3}{y^3}$	$\Rightarrow$ Used the power of a quotient rule.
$\frac{-x^3}{y^3}$	$\Rightarrow$ Simplified the dividend.

Homework 4.3 Simplify.

22) $\left(\frac{3}{5}\right)^2$	23) $\left(\frac{-a}{8}\right)^2$	24) $\left(-\frac{x}{y}\right)^4$	25) $\left(\frac{-6}{h}\right)^3$	26) $\left(-\frac{m}{n}\right)^5$

4.3.6 Integer Exponents Less Than 1

Since multiplication (and addition) is commutative, we didn't have to worry about the order of the factors when we worked with the product rules for exponents. For example, when adding exponents, both y^5y^3 and y^3y^5 simplify to y^8. Since subtraction isn't commutative, this isn't the case with the quotient rule. For example, if we start with $\frac{y^5}{y^3}$, and use the quotient rule to subtract to the dividend, $\frac{y^{5-3}}{1}$, our new exponent is a positive two, $y^{5-3} = y^2$. If we subtract to the divisor instead, $\frac{1}{y^{3-5}}$, we get an exponent of negative two, $\frac{1}{y^{-2}}$.

To use the full power of the quotient rule we'll have to move beyond natural number exponents to integer exponents. First, I'll discuss exponents of 0 and then exponents that are negative integers.

4.3.7 The Zero–Exponent Rule

If you ask people to simplify 5^0 almost no one, unless they've had some recent practice with algebra, thinks the answer is 1. Yet 1 follows naturally from the quotient rule. Here's an argument using a base of 5, but any nonzero base could be used. $1 = \frac{5^2}{5^2} = 5^{2-2} = 5^0$.

Practice 4.3.7 The Zero–Exponent Rule
Simplify.

a) -4^0

-1 $\Rightarrow$ Used the zero-exponent rule. Only 4 is the base.

b) $(-3)^0$

$1 \Rightarrow$ Used the zero-exponent rule. This time the base is -3.

c) $(-5k)^0$

$1 \Rightarrow$ The base is $-5k$.

Homework 4.3 Simplify.

27) x^0 28) $-b^0$ 29) 200^0 30) $(-v)^0$ 31) $-\left(\frac{m^{15}}{n^2}\right)^0$

4.3.8 The Negative Exponent Rule

If you ask people to simplify 2^{-1} most people don't realize the correct answer is $\frac{1}{2}$. Yet $\frac{1}{2}$ follows naturally from the quotient rule as I'll show using a base of 2 and our previous idea of a zero exponent. $2^{-1} = 2^{0-1} = \frac{2^0}{2^1} = \frac{1}{2}$. In a similar way, you can show that $\frac{1}{2^{-1}} = 2$.

Probably the most common way to simplify negative exponents is to take the reciprocal and make the exponent positive. For instance, to simplify h^{-4}, we'd take the reciprocal, $\frac{1}{h^{-4}}$, and make the exponent positive $\frac{1}{h^4}$. To simplify $\frac{1}{m^{-3}}$ we'd take the reciprocal and make the exponent positive, $\frac{1}{m^{-3}} = m^3$.

Here's some practice with negative exponents.

Practice 4.3.8 The Negative Exponent Rule

Simplify. All final exponents should be positive.

a) 3^{-2}

$\frac{1}{3^2} = \frac{1}{9} \Rightarrow$ Used the negative exponent rule and simplified.

b) $2k^{-4}$

$\frac{2}{k^4} \Rightarrow$ Used the negative exponent rule. Only k is the base.

c) 0^{-1}

$\frac{1}{0^1} = \frac{1}{0}$ Undefined $\Rightarrow$ Took the reciprocal of the base and made the exponent positive but a divisor of 0 is undefined. If you look back at the original rules you'll notice the base can't be 0.

d) $\dfrac{1}{y^{-5}}$

y^5 $\Rightarrow$	Used the negative exponent rule.

e) $\dfrac{n^3}{4^{-3}}$

4^3n^3 $\Rightarrow$	Used the negative exponent rule. The base of n already has a positive exponent.
$64n^3$ $\Rightarrow$	Found the third power of four.

Homework 4.3 Simplify. All final exponents should be positive.

32) 3^{-4}	33) $\dfrac{7}{x^{-7}}$	34) $(-2)^{-2}$	35) $2y^{-3}$	36) $\dfrac{1}{-8^{-2}}$
37) $\dfrac{n^{-4}}{m}$	38) $\dfrac{-1}{2k^{-4}}$	39) $\dfrac{n^{-7}}{m^{-4}}$	40) $\dfrac{a^3}{5^{-2}}$	41) $\dfrac{4^{-1}h^3}{k^{-2}}$

Homework 4.3 Answers

1) $(-3)^5 = -243$ 2) $c^{3+1+2} = c^6$ 3) $(-1)^{30} = 1$ 4) $-1 \times y^4 \times y^4 = -y^8$ 5) $12k^{11}$

6) $-20x^6$ 7) $5^8 = 390{,}625$ 8) m^{21} 9) $-k^{10}$ 10) $(-3)^9 = -19{,}683$

11) c^{64} 12) $-y^5$ 13) a^2b^2 14) $-27h^3$ 15) $16x^4$

16) x^4y^4 17) $8^2 = 64$ 18) $(-4)^5 = -1{,}024$ 19) $-q^6$ 20) $2^3y^2 = 8y^2$

21) $6^2n^3 = 36n^3$ 22) $\frac{3^2}{5^2} = \frac{9}{25}$ 23) $\frac{(-a)^2}{8^2} = \frac{a^2}{64}$ 24) $(-1)^4 \frac{x^4}{y^4} = \frac{x^4}{y^4}$

25) $\frac{(-6)^3}{h^3} = \frac{-216}{h^3}$ 26) $-\frac{m^5}{n^5}$ 27) 1 28) -1 29) 1

30) 1 31) -1 32) $\frac{1}{3^4} = \frac{1}{81}$ 33) $7x^7$ 34) $\frac{1}{(-2)^2} = \frac{1}{4}$ 35) $\frac{2}{y^3}$

36) $\frac{8^2}{-1} = -64$ 37) $\frac{1}{mn^4}$ 38) $\frac{-k^4}{2}$ 39) $\frac{m^4}{n^7}$ 40) $25a^3$ 41) $\frac{k^2h^3}{4}$

4.4 Continuing with Exponent Rules

In this section we'll continue working with the exponent rules. You'll be able to do this first set of problems using a single rule. The problems that follow will require more than one rule.

4.4.1 Choosing the Right Rule

All these problems can be done in one step, if you choose the right rule.

Practice 4.4.1 Choosing the Right Rule

Simplify. All final exponents should be positive. Calculate small numerical powers.

a) $(4y)^3$

$4^3 y^3$	$\Rightarrow$	Used the power of a product rule.
$64y^3$	$\Rightarrow$	Simplified 4^3.

b) $\dfrac{10^{-11}}{10^{-23}}$

$10^{-11-(-23)}$	$\Rightarrow$	Used the quotient rule. Decided to subtract to the numerator so the final exponent would be positive.
$10^{-11+23} = 10^{12}$	$\Rightarrow$	Simplified the exponent. Decided not to calculate the power of ten.

c) $-3k^{-7}$

$\dfrac{-3}{k^7}$	$\Rightarrow$	Used the negative exponent rule. Remembered that only k was the base.

d) $p^{-8}p^{10}$

$p^{-8+10} = p^2$	$\Rightarrow$	Used the product rule.

Homework 4.4 Simplify. All final exponents should be positive. Calculate small numerical powers.

1) $7^{-5} \times 7^8$ 2) $\left(y^{-1}\right)^{-2}$ 3) $10^7 \times 10^{23}$ 4) $\left(\dfrac{x}{6}\right)^2$ 5) $k^{11}k^7k^{-15}$

6) $\dfrac{4^{-1}m^5}{5}$ 7) $\dfrac{a^{-8}}{a^2}$ 8) $\left(-\dfrac{8}{3}\right)^2$ 9) $\dfrac{-2}{3x^{-3}}$ 10) $-\left(m^{-7}\right)^{-5}$

11) $-(5k)^2$ 12) $\dfrac{6r^0}{-2}$ 13) $((-2)k)^6$ 14) $\dfrac{-2}{(3x)^3}$ 15) $(-3n^6)(-7n^3)$

16) $\dfrac{(-4)^{17}}{(-4)^{15}}$ 17) $-b^0-(-b)^0$ 18) $\left(\dfrac{y^{-1}}{5^{-2}}\right)^{-2}$ 19) $\dfrac{x^{-2}}{x^{-1}}$ 20) $\left(\dfrac{-3}{p^8}\right)\left(-\dfrac{4}{p^{-6}}\right)$

4.4.2 *Multistep Exponent Problems*

Now let's practice simplifying expressions where you'll need to use more than one rule. In this context simplifying implies your final exponents are positive and you've calculated any small numerical powers.

Please realize there are often multiple "right" ways to simplify each expression. Just because I work a problem a certain way doesn't mean it's the only way or even the "best" way. What's important is that you're able to justify each of <u>your</u> steps using one of the exponent rules.

Practice 4.4.1 Multistep Exponent Problems

Simplify.

a) $(4y)^3 y^{-7}$

Step		Explanation
$4^3 y^3 y^{-7}$	$\Rightarrow$	Used the power of a product rule.
$64y^{3+-7} = 64y^{-4}$	$\Rightarrow$	Simplified 4^3 and used the power rule.
$64/y^4$	$\Rightarrow$	Used the rule for a negative exponent. Only *y* is the base.

b) $\dfrac{10^{-11}}{10^{-23}} \times \dfrac{10^{15}}{10^{27}}$

Step		Explanation
$10^{-11-(-23)} \times 10^{15-27}$ $10^{12} \times 10^{-12}$	$\Rightarrow$	Subtracted to the numerator using the quotient rule
$10^{12-12} = 10^0$	$\Rightarrow$	Added the exponents using the product rule.
1	$\Rightarrow$	Used the zero-exponent rule.

c) $(xy^4)^2 (x^3y^2)^3$

Step		Explanation
$x^2 (y^4)^2 (x^3)^3 (y^2)^3$	$\Rightarrow$	Used the power of a product rule.
$x^2y^8x^9y^6$	$\Rightarrow$	Used the power rule. Often the above step and this step are done as a single step.
$x^{11}y^{14}$	$\Rightarrow$	Used the power rule.

d) $\left(\dfrac{m^{-1}n^2}{n^{-2}m^2}\right)^5$

Step		Explanation
$\left(\dfrac{n^4}{m^3}\right)^5$	$\Rightarrow$	Used the quotient rule inside the parentheses. $\left(\dfrac{n^{2-(-2)}}{m^{2-(-1)}}\right)^5$
$\dfrac{n^{20}}{m^{15}}$	$\Rightarrow$	Often the power of a quotient rule and the power rule are written as a single step. $\dfrac{n^{4\times5}}{m^{3\times5}} = \dfrac{n^{20}}{m^{15}}$

Homework 4.4 Simplify

21) $7^5 \times 7^{-8}$

22) $(k^3k^2)^2$

23) $\left(\dfrac{c^{14}}{c^{11}}\right)^4$

24) $\dfrac{y^{15}y^8}{y^{20}}$

25) $\dfrac{6r^3r^{-3}}{-2}$

26) $\dfrac{(8^5)^3}{8^{13}}$

27) $(a^{-5}a^{-2})^{-3}$

28) $\dfrac{(n^5)^2}{n^5n^6}$

29) $(-2^{-1}x^7)^3$

30) $\dfrac{(p^{-4})^3}{(p^6)^{-2}}$

31) $\dfrac{m^5}{m^2} \times \dfrac{m^{-1}}{m^3}$

32) $\left(\dfrac{4}{x^{-2}}\right)^{-2}$

33) $\dfrac{4x^{-1}}{(2x)^{-2}}$

34) $\left(\dfrac{y^{-3}}{y^2}\right)^{-4}$

35) $(xy^4)^2(x^3y^2)^3$

36) $\dfrac{10^{-18} \times 10^{-5}}{10^{10}}$

37) $\left(\dfrac{-x}{y^{-2}}\right)^4$

38) $(-x^2)^2(-3x^2)^3$

39) $\dfrac{(-6h^{-2})^2}{-6h}$

40) $\dfrac{1}{a^{-8}(a^2)^3}$

41) $(y^{-1})^{-2}(y^3)^{-1}$

42) $\dfrac{w^7}{w^{-3}} \times \dfrac{w^{-9}}{w^{-2}}$

43) $\dfrac{3^{-2}y^{-2}}{3^{-3}y^5}$

44) $\left(\dfrac{p^{-4}}{p}\right)^2\left(\dfrac{p^4}{p^{-1}}\right)^{-1}$

45) $(x^4x^2)^3(x^8x)^{-2}$

46) $\dfrac{2x^{-2}}{(2x^{-1})^{-2}}$

47) $\dfrac{(ab^{-2})^{-1}}{(a^{-4}b^2)^2}$

48) $\dfrac{(-8^3)^5 \times 8^{-12}}{(8^{-2})^3 \times -8^6}$

Homework 4.4 Answers

1) 343	2) y^2	3) 10^{30}	4) $\frac{x^2}{36}$	5) k^3	6) $\frac{m^5}{20}$	7) $\frac{1}{a^{10}}$
8) $\frac{64}{9}$	9) $-\frac{2}{3}x^3$	10) $-m^{35}$	11) $-25k^2$	12) -3	13) $64k^6$	14) $-\frac{2}{27x^3}$
15) $21n^9$	16) 16	17) -2	18) $\frac{y^2}{625}$	19) $\frac{1}{x}$	20) $\frac{12}{p^2}$	21) $\frac{1}{343}$
22) k^{10}	23) c^{12}	24) y^3	25) -3	26) 64	27) a^{21}	28) $\frac{1}{n}$
29) $-\frac{x^{21}}{8}$	30) 1	31) $\frac{1}{m}$	32) $\frac{1}{16x^4}$	33) $16x$	34) y^{20}	35) $x^{11}y^{14}$
36) $\frac{1}{10^{33}}$	37) x^4y^8	38) $-27x^{10}$	39) $\frac{-6}{h^5}$	40) a^2	41) $\frac{1}{y}$	42) w^3
43) $\frac{3}{y^7}$	44) $\frac{1}{p^{15}}$	45) 1	46) $\frac{8}{x^4}$	47) $\frac{a^7}{b^2}$	48) 512	

4.5 Scientific Notation

Scientific notation is used in fields like chemistry, astronomy, physics or geology to quickly multiply or divide numbers that are very large or very small. A number written in scientific notation is the product of two factors. The first factor has an absolute value that's greater than or equal to 1 but less than 10 and the second factor has a base of 10 and an integer exponent. An example would be 2×10^3 which is 2,000 written in scientific notation. Notice we're not simplifying, just rewriting two thousand from decimal notation 2,000, to scientific notation 2×10^3.

4.5.1 Converting from Decimal to Scientific Notation

To convert from decimal notation to scientific notation, people often talk about "moving" the decimal point. If the absolute value of the number is 10 or greater, the decimal will move to the left until the absolute value of the number is greater than or equal to 1 but less than 10. Since each place value to the left is a factor of 10, each move will increase the exponent by 1.

For instance, to convert 2,000 to scientific notation I realize that even though I don't see a decimal point, it's to the right of the last 0. Next, I move the decimal to the left once and add 1 to the exponent of my base of 10. Since my first factor, 200, isn't yet greater than or equal to 1 but less than 10, I move the decimal to the left again to get 20×10^2. Since my first factor, 20, is still not greater than or equal to 1 but less than 10, I'll need to move the decimal yet again to the left to get 2.000×10^3. Since my first factor, 2, is greater than or equal to 1 but less than 10, I stop moving the decimal, disregard the 0's to the right of the decimal point and write my final value as 2×10^3.

$$
\begin{gathered}
2{,}000. \\
200.0 \times 10^1 \\
20.00 \times 10^2 \\
2.000 \times 10^3 \\
2 \times 10^3
\end{gathered}
$$

Practice 4.5.1 Converting from Decimal to Scientific Notation

Convert from decimal to scientific notation.

a) $-17{,}000$

-1.7000 4 places	$\Rightarrow$	Moved the decimal four places to the left. Stopped because the absolute value of -1.7 is greater than or equal to 1 but less than 10. The exponent of 10 increased by 4.
-1.7×10^4	$\Rightarrow$	The final answer.

b) 1,590,000,000

$\underbrace{1{,}590{,}000{,}000}_{\text{9 places}}$	$\Rightarrow$	Moved the decimal nine places to the left. Stopped because 1.59 is greater than or equal to 1 and less than 10. The exponent of 10 increased by 9.
1.59×10^9	$\Rightarrow$	The final answer.

Homework 4.5 Convert from decimal to scientific notation.

1) 78,000,000	2) 102,400	3) 8	4) -100.117

4.5.2 Converting from Decimal to Scientific Notation – Part 2

If the absolute value of your original number is more than 0 but less than 1, the decimal moves to the right until the absolute value of the number is greater than or equal to 1 but less than 10. Each place value to the right of the decimal is a factor of $\frac{1}{10}$ so each move will <u>decrease</u> the exponent by 1. Remember $\frac{1}{10}=\frac{1}{10^1}=10^{-1}$, $\frac{1}{100}=\frac{1}{10^2}=10^{-2}$ and so on.

For example, to write 0.0037 in scientific notation move the decimal three places to the right until you get the number 3.7 (which is less than or equal to 1 but less than 10). Since the decimal moved three places to the right, the exponent would be -3 so 0.0037 can be rewritten as 3.7×10^{-3}. If you look to the right I've carried out the process step by step.

$$0.0037$$
$$00.037\times10^{-1}$$
$$000.37\times10^{-2}$$
$$0003.7\times10^{-3}$$
$$3.7\times10^{-3}$$

What about numbers whose absolute value is greater than or equal to 1 but less than 10? You can write these numbers in scientific notation using an exponent of 0. So -7 for example, can be thought of as -7×1 and then as -7×10^0 since 10^0 is 1.

Practice 4.5.2 Converting from Decimal to Scientific Notation – Part 2

Convert from decimal to scientific notation.

a) -0.00024

-0.00024 4 places	$\Rightarrow$	Moved the decimal four places to the right. Stopped because the absolute value of -2.4 is greater than or equal to 1 but less than 10. The exponent of 10 decreases by 4.
-2.4×10^{-4}	$\Rightarrow$	The final answer.

Homework 4.5 Convert from decimal to scientific notation.

5) 0.0000112	6) -0.00091	7) 0.0772	8) -0.4

4.5.3 Converting from Scientific to Decimal Notation

A number written in scientific notation can also be written in decimal notation by "moving" the decimal point. If the exponent of 10 is <u>positive</u> move the decimal to the <u>right</u> as many place values as the value of the exponent. If the exponent of 10 is <u>negative</u> move the decimal to the <u>left</u> as many place values as the absolute value of the exponent. Sometimes your beginning product won't have a visible decimal point. In that case the decimal is to the right of the first factor and is usually followed by a 0. For instance, 2×10^3 is thought of as 2.0×10^3.

To see why moving the decimal works realize that 2×10^3 is $2\times10\times10\times10$ and each factor of 10 is another place value. So multiplying $2\times10\times10\times10$ left to right we would get twenty, then two hundred and finally two thousand. On the other hand, 2×10^{-3} is $2\times\frac{1}{10^3}$ or $2\times\frac{1}{10}\times\frac{1}{10}\times\frac{1}{10}$. Multiplying left to right would give us two-tenths, then two-hundredths and finally two-thousandths.

Practice 4.5.3 Converting from Scientific to Decimal Notation
Convert from scientific to decimal notation.

a) 1.2×10^4

$12{,}000$	$\Rightarrow$	Since the exponent is positive four moved the decimal four places to the right $1.\underline{2000}$.

b) -6×10^{-12}

-6.0×10^{-12}	$\Rightarrow$	Since no decimal is visible thought of -6 as -6.0 .
-0.000000000006	$\Rightarrow$	The exponent is negative so moved the decimal 12 places to the left. There is no longer a need for a 0 to the right of 6.

Homework 4.5 Convert from scientific to decimal notation.

9) 7×10^4	10) -5.002×10^5	11) 3.33×10^{-2}	12) 9.09×10^{-1}
13) 1.5×10^1	14) -8×10^{-4}	15) -2.0001×10^4	

4.5.4 Adding and Subtracting Like Terms

As usual, we can add and subtract like terms written in scientific notation using the distributive property. Terms in scientific notation are like if they count the same power of 10. For instance, 3×10^2 and 5×10^2 are like since they're both counting second powers of 10. To add $3\times10^2+5\times10^2$ we use the distributive property to factor out 10^2, $(3+5)\times10^2$ and then add to get 8×10^2. We can't find the sum of $3\times10^4+5\times10^2$ in the same way since the terms don't directly share a common power of ten as a factor. In practice we usually don't write out all the steps with the distributive property. Instead, we add or subtract the first factors and keep a single common power of ten factor.

Practice 4.5.4 Adding and Subtracting Like Terms
Simplify. Write your answer in scientific notation.

a) $3\times10^{-4}+4\times10^{-4}-12\times10^{-4}$

$7\times10^{-4}-12\times10^{-4}$	$\Rightarrow$	Since the first two terms were like, (were the same power of 10) I added the first factors and kept a single factor of the power of ten.
-5×10^{-4}	$\Rightarrow$	Continued by subtracting the first factors of the like terms.

b) $7\times10^{-3}+5\times10^{-3}$

12×10^{-3}	$\Rightarrow$	Added the first factors of the like terms.
$1.2\times10^{1}\times10^{-3}$	$\Rightarrow$	Noticed the first factor, 12, was not greater than or equal to 1 but less than 10. Rewrote the first factor in scientific notation.
$1.2\times10^{1+-3}$ 1.2×10^{-2}	$\Rightarrow$	Used the product rule to simplify the powers of 10. I can stop since the absolute value of 1.2 is greater than or equal to 1 but less than 10.

c) $1.1\times10^{9}-1.8\times10^{9}$

$(1.1-1.8)\times10^{9}$ -0.7×10^{9}	$\Rightarrow$	Subtracted the first factors of the like terms. Noticed the absolute value of the first factor, -0.7, is not greater than or equal to 1 but less than 10.
$-7.0\times10^{-1}\times10^{9}$ -7×10^{8}	$\Rightarrow$	Rewrote the first factor into scientific notation and used the product property for exponents.

Homework 4.5 Simplify. Write your answer in scientific notation.

16) $3\times10^{7}+5\times10^{7}$ 17) $-7\times10^{-11}-2\times10^{-11}$ 18) $-8\times10^{5}-3\times10^{5}+4\times10^{5}$

19) $8\times10^{15}+8\times10^{15}$ 20) $2\times10^{3}-2.6\times10^{3}$ 21) $2\times10^{-1}+8\times10^{-1}$

22) $-3\times10^{-5}-4\times10^{-5}-5\times10^{-5}$ 23) $8\times10^{-31}+9\times10^{-31}+7\times10^{-31}$

24) $8.75\times10^{19}+9.44\times10^{19}+5.8\times10^{19}$ 25) $-9.9\times10^{-21}-7.5\times10^{-21}+1.7\times10^{-21}$

4.5.5 Multiplying and Dividing Using Scientific Notation

Scientific notation give us a way to quickly multiply or divide using the exponent rules. For instance, to quickly multiply $(3\times10^{5})(2\times10^{4})$ I'd reorder the factors using the commutative property, $3\times2\times10^{5}\times10^{4}$, regroup the factors using the associative property, $(3\times2)(10^{5}\times10^{4})$, and then simplify the products 6×10^{9}.

Practice 4.5.5 Multiplying and Dividing Using Scientific Notation
Multiply or divide. Write your answer in scientific notation.

a) $(5\times10^{12})(-3\times10^{-8})$

$(5\times-3)(10^{12}\times10^{-8})$ -15×10^{4}	$\Rightarrow$	Reordered and regrouped the factors. Then simplified inside the parentheses.
$-1.5\times10^{1}\times10^{4}$ -1.5×10^{5}	$\Rightarrow$	Rewrote -15 in scientific notation and used the product rule to simplify the powers of 10.

b) $\dfrac{4.5\times10^{-8}}{9\times10^{-12}}$

$\left(\dfrac{4.5}{9}\right)\left(\dfrac{10^{-8}}{10^{-12}}\right)$	$\Rightarrow$	Reordered and regrouped the factors.
$0.5\times10^{-8--12}$ $0.5\times10^{-8+12}$ 0.5×10^{4}	$\Rightarrow$	Divided. Used the quotient rule to subtract exponents in the second factor. Noticed the absolute value of the first factor, 0.5, is not greater than or equal to 1 but less than 10.
$5\times10^{-1}\times10^{4}$ 5×10^{3}	$\Rightarrow$	Rewrote 0.5 into scientific notation and then simplified the powers of 10 using the product rule.

Homework 4.5 Multiply or divide. Write your answer in scientific notation

26) $(3\times10^{2})(3\times10^{3})$ 27) $\dfrac{18\times10^{-22}}{-6\times10^{-25}}$ 28) $(5\times10^{8})(2\times10^{8})$

29) $\dfrac{1.2\times10^{9}}{4\times10^{7}}$ 30) $(9.8\times10^{-7})(-8.5\times10^{-8})$ 31) $-\dfrac{2\times10^{5}}{8\times10^{13}}$

32) $(8.8\times10^{-22})(6\times10^{-8})$ 33) $(1.002\times10^{15})(0.46\times10^{5})(0.005\times10^{13})$

34) $\dfrac{(200\times10^{-11})(8\times10^{-18})}{0.1\times10^{11}}$ 35) $\dfrac{1.7\times10^{7}}{2\times10^{2}}\times\dfrac{4\times10^{-3}}{8\times10^{2}}$ 36) $9\times10^{17}\times\dfrac{12\times10^{-12}}{0.8\times10^{-18}}$

37) $\dfrac{6\times10^{-24}}{(1.6\times10^{2})(1.875\times10^{-26})}$ 38) $\dfrac{1.6\times10^{7}}{2\times10^{2}}\times\dfrac{4\times10^{-3}}{0.8\times10^{2}}\div\dfrac{6\times10^{-3}}{30\times10^{2}}$

4.5.6 Multiplying and Dividing with Decimal Notation

Sometimes scientific notation helps us quickly multiply or divide numbers written in decimal notation. For example, to multiply $40{,}000\times0.02$ change both factors to scientific notation $(4\times10^{4})(2\times10^{-2})$ and use our previous ideas, 8×10^{2}. The final result would likely be written in decimal notation, 800, since we began with numbers written in decimal notation.

Practice 4.5.6 Multiplying and Dividing with Decimal Notation
Simplify using scientific notation. Write your answer in decimal notation.

a) 2500×2000

$(2.5\times10^{3})(2\times10^{3})$	$\Rightarrow$	Wrote in scientific notation
5×10^{6}	$\Rightarrow$	Multiplied and used the product property of exponents.
5,000,000	$\Rightarrow$	Rewrote in decimal notation

b) $\dfrac{600}{0.003}$

$\dfrac{6\times10^{2}}{3\times10^{-3}}$	$\Rightarrow$	Changed to scientific notation.
2×10^{5}	$\Rightarrow$	Divided and used the quotient property of exponents.
$200{,}000$	$\Rightarrow$	Rewrote in decimal notation.

Homework 4.5 Simplify using scientific notation. Write your answer in decimal notation.

39) $\dfrac{-24{,}000}{1{,}200}$ 40) 0.00006×0.001 41) $0.024\times2{,}000{,}000$ 42) $\dfrac{0.0006}{0.02}$

43) $\dfrac{4{,}000}{0.002}$ 44) $\dfrac{3000\times-0.04}{0.006}$ 45) 0.00008×0.0005 46) $-\dfrac{0.00012}{0.00003\times4000}$

Homework 4.5 Answers

1) 7.8×10^{7} 2) 1.024×10^{5} 3) 8×10^{0} 4) -1.00117×10^{2}

5) 1.12×10^{-5} 6) -9.1×10^{-4} 7) 7.72×10^{-2} 8) -4×10^{-1}

9) 70,000 10) $-500{,}200$ 11) 0.0333 12) 0.909 13) 15

14) -0.0008 15) $-20{,}001$ 16) 8×10^{7} 17) -9×10^{-11} 18) -7×10^{5}

19) $16\times10^{15}=1.6\times10^{16}$ 20) $-0.6\times10^{3}=-6\times10^{2}$ 21) $10\times10^{-1}=1\times10^{0}$

22) $-12\times10^{-5}=-1.2\times10^{-4}$ 23) $24\times10^{-31}=2.4\times10^{-30}$

24) $23.99\times10^{19}=2.399\times10^{20}$ 25) $-15.7\times10^{-21}=-1.57\times10^{-20}$

26) 9×10^{5} 27) -3×10^{3} 28) 1×10^{17} 29) 3×10^{1}

30) -8.33×10^{-14} 31) -2.5×10^{-9} 32) 5.28×10^{-29} 33) 2.3046×10^{30}

34) 1.6×10^{-36} 35) 4.25×10^{-1} 36) 1.35×10^{25} 37) 2 or 2×10^{0}

38) 2×10^{6} 39) $-2\times10^{1}=-20$ 40) $6\times10^{-8}=0.00000006$

41) $4.8\times10^{4}=48{,}000$ 42) $3\times10^{-2}=0.03$ 43) $2\times10^{6}=2{,}000{,}000$

44) $-2\times10^{4}=-20{,}000$ 45) $4\times10^{-8}=0.00000004$ 46) $-1\times10^{-3}=-0.001$

Chapter 5

Polynomials

5.1 An Introduction to Polynomials

In this chapter, we'll discuss and practice operating on, polynomials.

5.1.1 Defining a Polynomial

Let's begin by defining, "polynomial".

***Definition* – Polynomial**
A polynomial in x is a single term, or a sum of terms, where each term is a variable term or a constant. Every variable term has a coefficient, the variable x, and an exponent on x that is a whole number. Example: $2x^3 - 3x + 5$

I hope you noticed that the definition relies on a polynomial being a sum but that $2x^3 - 3x + 5$ also has a difference. As you may suspect, with polynomials we take the point of view that a subtraction is adding an opposite. That is, we see $2x^3 - 3x + 5$ as the sum $2x^3 + -3x + 5$.

5.1.2 The Terms of a Polynomial

Polynomial terms are either constant or variable. A term like 5 is a constant term since its value is consistently 5 regardless of the value taken on by the variable. A term like $3x$ is a variable term since the product will vary depending on the value x takes on. For example, $3x$ has the value 6 if x is 2 or the value -12 if x is -4. Each variable term is the product of a coefficient and a variable factor with a degree. The **coefficient** is the factor that is multiplied to the variable factor. The **degree** is the exponent on the variable base. For example, the term $2x^3$ is a third-degree term (because the exponent is a 3) with a coefficient of 2. With a coefficient or exponent of 1 or a coefficient of -1 we often don't "see" the value. Some examples would be, $-3x$ which is thought of as $-3x^1$, the term x^2 which is thought of as $1x^2$ and $-x^5$ which would be thought of as $-1x^5$. Let's practice naming some polynomial terms.

Practice 5.1.2 The Terms of a Polynomial
Name the degree and coefficient of each term.

a) $4y^3$ ⇒ This is a third-degree term with a coefficient of four.

Homework 5.1 Name the degree and coefficient of each term.

1) $-6m^3$	2) $-x$	3) h^2	4) -12

5.1.3 Naming a Polynomial

To name the entire polynomial, we mention the degree of the polynomial and the number of terms the polynomial has. The degree of the polynomial matches the degree of the largest degree term. To describe the number of terms we can either count the number of terms or, when the polynomial has one, two or three terms, we can use the special names **monomial** (for one term), **binomial** (for two terms) or **trinomial** (for three terms). So $2x^3 - 3x + 5$ is a third-degree (because $2x^3$ is third-degree) trinomial (because the expression has three terms) while $1 + 4p^2$ is a second-degree (because $4p^2$ is second-degree) binomial (because the expression has two terms). Although the names we just used are probably the most common, there would be nothing wrong with referring to $2x^3 - 3x + 5$ as a third-degree three term polynomial and $1 + 4p^2$ as a second-degree two term polynomial.

Again, always keep in mind that a polynomial is a sum so $y^3 - y^2 - 3$ should be thought of as $1y^3 + -1y^2 + -3$ so the coefficient of the third-degree term is 1, the coefficient of the second-degree term is -1 and the constant is -3. Here's some practice naming polynomials.

Practice 5.1.3 Naming a Polynomial

Name the entire polynomial and the degree and coefficient of each term.

a) $4y^3 - 2y^2 - 3$

a) This is a third-degree trinomial. The third-degree term has a coefficient of four, the second-degree term has a coefficient of negative two, and the constant is negative three.	$\Rightarrow$	The greatest exponent on any term is 3 so that's the degree of the three-term polynomial (trinomial). Thought of the polynomial as the sum, $4y^3 + -2y^2 + -3$ which makes the coefficient of the second-degree term -2 and the constant -3.

Homework 5.1 Name the entire polynomial and the degree and coefficient of each term.

5) $3x^2 - 4x - 4$ 6) $x - 1$ 7) $-h^3 - h$ 8) $\frac{1}{2}w^3 + \frac{5}{8}w^2 + w + 2$

5.1.4 Standard Form

Once the polynomial is a sum, the commutative property allows us to reorder the terms into standard form. A polynomial is in s**tandard form** when the terms of the polynomial are written in descending order of degree from left to right. For instance, $4x - 3 + 2x^3 - x^2$ is not in standard form but can be reordered as $2x^3 - x^2 + 4x - 3$ which is in standard form. To "reorder" the subtractions, I thought of the polynomial as a sum, $4x + -3 + 2x^3 + -x^2$, used the commutative property to reorder the terms, $2x^3 + -x^2 + 4x + -3$ and then rewrote each addition of a negative back to subtraction $2x^3 - x^2 + 4x - 3$.

Practice 5.1.4 Standard Form

Write the polynomial in standard form.

a) $-5y + 12y^3 + 2 - 8y^2$

$-5y + 12y^3 + 2 + -8y^2$	$\Rightarrow$	Thought of subtractions as adding an inverse. Only addition is commutative, subtraction is not.
$12y^3 + -8y^2 + -5y + 2$	$\Rightarrow$	Reordered the terms using the commutative property.
$12y^3 - 8y^2 - 5y + 2$	$\Rightarrow$	Rewrote adding a negative as subtraction. In time people usually change a subtraction to adding an inverse in their head, not in their written work.

Homework 5.1 Write the polynomial in standard form.

9) $1 - y - y^2$ 10) $2 + 5w^2 + 4w^3 - w$ 11) $-y - \frac{11}{4} + \frac{4}{3}y^2$ 12) $-21x + 5x^2 + 75 - 2x^3$

5.1.5 Like Polynomial Terms

In the next topic, I'll show you how adding and subtracting polynomials is based on the distributive property. Since the distributive property uses like terms, we'll need to identify when polynomial terms are "like".

Polynomials terms are like when they are the same power of the same letter(s). For example, $\frac{1}{2}x^3$, $-7x^3$, and x^3 are like terms since they're all third powers of x. On the other hand, $5x^3$ and $5x^2$ are not like terms because one's a third power of x and the other's a second power of x. $5x^3$ and $5y^3$ are also not like because, even though they're both third powers, one's a third power of x and the other's a third power of y. Notice that coefficients have nothing to do with whether the terms are like. Constants are also considered like terms.

Practice 5.1.5 Like Polynomial Terms

Identify the like terms.

a) $-7w^2$, $5y^2$, $3w$, $-y^2$, w^3

$5y^2, -y^2$ $\Rightarrow$ Only these terms are the same power of the same letter.

Homework 5.1 Identity the like terms.

13) $2x^2$, $2x$, 3, $-x$, -1 14) $k/2$, $2k$, $-0.2k$, $12k/19$ 15) $5m^2$, $-2m$, $3n$, m, -1

16) $-4x$, x, $-2x^2$, $-2x$, $3x^2$ 17) $\frac{1}{2}y^3$, $\frac{2}{3}w^3$, $\frac{2}{3}w$, $-y^2$

5.1.6 Adding and Subtracting Polynomial Terms

We use the commutative property of addition, the associative property of addition and the distributive property to add and subtract polynomial terms. Below I'll explicitly show the properties I'm using as I add and subtract some polynomial terms.

First, since subtraction is not commutative or associative, I need to change all subtractions to adding the opposite. Next, I'd use the commutative property to reorder the terms. To insure the final sum is in standard form I'd put the second-degree terms to the left, the first-degree terms next and the constants last. Now, I'd regroup like variable terms using the associative property and then, using the distributive property, I'd factor out the common variable factor. Last, I'd simplify and write all adding a negative as a subtraction.

$$2x^2+x+5-8+3x^2-x-6x^2$$
$$2x^2+x+5+-8+3x^2+-x+-6x^2$$
$$2x^2+3x^2+-6x^2+x+-x+5+-8$$
$$\left(2x^2+3x^2+-6x^2\right)+(x+-x)+5+-8$$
$$(2+3+-6)x^2+(1+-1)x+5+-8$$
$$-1x^2+0x+-3$$
$$-x^2-3$$

In practice, people usually don't explicitly carry out these steps. Instead, starting with the highest degree terms, they simplify the coefficients for each group of like variable terms "in their head" and then simplify any constants.

Practice 5.1.6 Adding and Subtracting Polynomial Terms

Simplify. Write you answer in standard form.

a) $-3x+4x^3-x+1-6x^3+9$

$-2x^3$	$\Rightarrow$	Began with the third-degree terms, thought of subtraction as adding the inverse and added the coefficients, $4+-6=-2$.
$-2x^3-4x$	$\Rightarrow$	Moved on to the first-degree terms and again thought of subtractions as adding the inverse, $-3+-1=-4$.
$-2x^3-4x+10$	$\Rightarrow$	Added the constants. The problem is finished since the remaining terms aren't like and the trinomial is in standard form.

b) $4x^2-3x^4-4x^4+8x+x^2+11x^4$

$4x^4$	$\Rightarrow$	Began by adding the fourth-degree terms, $-3+-4+11=4$.
$4x^4+5x^2$	$\Rightarrow$	Moved on to the second-degree terms, $4+1=5$.
$4x^4+5x^2+8x$	$\Rightarrow$	Finished with the first-degree term.

Homework 5.1 Simplify. Write your answer in standard form.

18) $w^3 - 2w^3 + 3w^2 - w^3$

19) $c^2 - c^3 + 4c^2 - 5c^2 - 3c^3$

20) $p^2 - 12 - 3p + p^2 + 16 - 6p$

21) $x^3 - 4x^2 - 2x^3 + 7 - x^2$

22) $-4x + x - 2x^2 - 2x + 3x^2$

23) $-14 - m^5 + 2m^2 - 2m + 3m^5 + 6$

24) $5k^7 + 14k^4 - 7k^7 - 3 + 2k^4 - k + 12 + 9k - 15k^4$

25) $-t + 15 + t^2 - 16t - 3 + 22t^2 - 17 - 23t^2$

26) $6 - 8k^2 - 7 - 3k + k + 12k^2 + 14 + 2k$

27) $5m^3 + 4m^3 - 6 - 9m^3 - m^3 - 2 + 8$

Homework 5.1 Answers

1) This third-degree variable term has a coefficient of negative six.

2) This first-degree variable term has a coefficient of negative one.

3) This second-degree variable term has a coefficient of one.

4) This is a constant.

5) This is a second-degree trinomial. The second-degree term has a coefficient of three. The first-degree term has a coefficient of negative four and the constant is negative four.

6) This is a first-degree binomial. The first-degree term has a coefficient of one. The constant is negative one.

7) This is a third-degree binomial. Both the third-degree term and the first-degree term have a coefficient of negative one.

8) This is a third-degree four term polynomial. The third-degree term has a coefficient of one-half, the second-degree term has a coefficient of five-eighths, the first-degree term has a coefficient of one and two is a constant.

9) $-y^2 - y + 1$ 10) $4w^3 + 5w^2 - w + 2$ 11) $\frac{4}{3}y^2 - y - \frac{11}{4}$ 12) $-2x^3 + 5x^2 - 21x + 75$

13) $2x$ and $-x$ are like as are 3 and -1. 14) All the terms are like.

15) $-2m$ and m are like. 16) $-2x^2$ and $3x^2$ are like as are $-4x, x$ and $-2x$

17) None of the terms are like.

18) $-2w^3 + 3w^2$ 19) $-4c^3$ 20) $2p^2 - 9p + 4$ 21) $-x^3 - 5x^2 + 7$

22) $x^2 - 5x$ 23) $2m^5 + 2m^2 - 2m - 8$ 24) $-2k^7 + k^4 + 8k + 9$

25) $-17t - 5$ 26) $4k^2 + 13$ 27) $-m^3$

5.2 Beginning Polynomial Multiplication

Now that you've had some practice adding and subtracting polynomials, we'll begin multiplying polynomials.

5.2.1 *Some Vocabulary for Products*

Let's begin by practicing some of the vocabulary you'll need.

Practice 5.2.1 Some Vocabulary for Products
Fill in the blanks using the words term, sum, factor, product, minuend, subtrahend, difference, dividend, divisor, quotient, base, exponent or power.

---

a) Before simplifying $(x-5)(x+8)$,

$x-5$ is a ____, $x+8$ is a ____, $(x-5)$ is a ____, $(x+8)$ is a ____, and $(x-5)(x+8)$ is a ____. In the first factor x is the ____ and 5 is the ____. In the second factor x is a ____, and 8 is a ____.

Difference, sum, factor, factor, product, minuend, subtrahend, term, term	$\Rightarrow$	In the first factor the subtrahend 5 is being subtracted from the minuend x so $x-5$ is a difference. In the second factor both x and 8 are terms so $x+8$ is a sum. The entire expression is a product.

Homework 5.2 Fill in the blanks using the words term, sum, factor, product, minuend, subtrahend, difference, dividend, divisor, quotient, base, exponent or power.

---

1) Given $(4h^2)^3 + 5h^6$,

h is a ______, h^2 is a ______ and a ______, 4 is a______, $4h^2$ is a ______ ,

$(4h^2)$ is a ______, 3 is an ______, $(4h^2)^3$ is a ______ and a______, 5 is a ______,

h^6 is a ______ and a______, $5h^6$ is a ______ and a______, and the entire expression is a ______.

2) Given $3y(2y^2-4y)$,

$3y$ can be thought of as both a ______ and a______, y^2 is both a ______ and a______,

$2y^2$ is a ______ and the ______, $4y$ is a ______ and the______, $2y^2-4y$ is a ______,

$(2y^2-4y)$ is a ______ and $3y(2y^2-4y)$ is a ______.

---

3) Given $(3k+7)(k-5)^2$;

$3k$ is both a______ and a______, $3k+7$ is a ______, $(3k+7)$ is a ______, $k-5$ is a ______, $(k-5)$ is a ______, $(k-5)^2$ is a ______ and a______ and $(3k+7)(k-5)$ is a ______.

5.2.2 *Multiplying Monomials*

Earlier you practiced simplifying $(2y)(3y^2)$ to $6y^3$ using the commutative property, $2\times 3\times y\times y^2$, the associative property, $(2\times 3)(y^2y)$ and the product rule, $6y^3$. In practice, you'll probably choose not to write out each of these steps as you simplify polynomials, but try to remain aware that each step needs to be supported by a property, rule or definition.

Practice 5.2.2 Multiplying Monomials
Simplify.

a) $(-4y^2)(3y^3)$

$(-4\times 3)(y^2y^3)$	$\Rightarrow$	Reordered and regrouped the factors. This step is often done mentally.
$-12y^5$	$\Rightarrow$	Multiplied the coefficients and used the product rule.

b) $3y^2(y^3)-(6yy^4)$

$3y^5-6y^5$	$\Rightarrow$	Simplified inside the parentheses and then used the product rule with the minuend.
$-3y^5$	$\Rightarrow$	Found the difference.

Homework 5.2 Simplify.

4) $(x^3x^4)(-3x^3)$ 5) $-2(3n)(-5n^2)$ 6) $5y^3(-y^2)$

7) $2a(a^2)+3a^3$ 8) $-2x^2(-x)-4x^3$ 9) $(2a^3)(3a^2)+(4a^6)$

10) $(y^4y^2)+3y(y^5)-2y^3(y^3)$ 11) $-3h^7-3h^4(-2h^2)-(6h^2)(-2h^5)$

5.2.3 *Distributing a Monomial*

As you work through this next set of problems notice how the distributive property, the product rule and the idea of like terms work together.

Practice 5.2.3 Distributing a Monomial
Simplify.

a) $-2x(x^2+3x)$

$-2x(x^2)+-2x(3x)$	$\Rightarrow$	Distributed the factor of $-2x$ to both terms inside the parentheses.
$-2x^3+-2x(3x)$	$\Rightarrow$	Simplified the first term using the product rule.
$-2x^3-6x^2$	$\Rightarrow$	Simplified the second term using the product rule and rewrote adding a negative as subtraction. I can stop since the polynomial is in standard form and the terms aren't like.

b) $3y(2y^2 - 4y + 7)$	
$3y(2y^2) - 3y(4y) + 3y(7)$ $\Rightarrow$	Distributed $3y$ to the terms inside the parentheses.
$6y^3 - 12y^2 + 21y$ $\Rightarrow$	Found the products.

c) $2k^2 - k(k^2 - 8k)$	
$2k^2 + -k(k^2 + -8k)$ $\Rightarrow$	Decided to think of subtractions as adding an opposite.
$2k^2 + -k(k^2) + k(8k)$ $\Rightarrow$	Distributed to the terms inside the parentheses. Paid particular attention to the sign of the second product.
$2k^2 - k^3 + 8k^2$ $\Rightarrow$	Multiplied.
$-k^3 + 10k^2$ $\Rightarrow$	Added the second-degree terms and wrote the final answer in standard form.

Homework 5.2 Simplify.

12) $x^3(x^2 + x)$	13) $-3m(-2m^2 + m)$	14) $t^3(-1 - t^2 + 2t)$
15) $4p^2(3 + p^2 + 2p)$	16) $-2w(-2w - 2 - w)$	17) $y(y^2 + 4y) - y(y^2 - 8y)$
18) $4x(-x^2 + 12x^3) + x^2(x^2 + 6x - 5)$	19) $2(y + y^3 - y^2) - 2(y - y^2 - 4y^3)$	

5.2.4 Multiplying Binomial Factors Using FOIL

When multiplying two binomial factors most students, at first, prefer using FOIL. FOIL is an acronym that uses the first letter from First, Outside, Inside, Last to help make sure your product includes all four terms. Here's some practice.

Practice 5.2.4 Multiplying Binomial Factors Using FOIL
Simplify.

a) $(3x - 1)(x + 4)$	
$(\boxed{3x} - 1)(\boxed{x} + 4)$ $3x^2$ $\Rightarrow$	Multiplied the "first" terms in each binomial.
$(\boxed{3x} - 1)(x + \boxed{4})$ $3x^2 + 12x$ $\Rightarrow$	Multiplied the "outside" terms.
$(3x\boxed{-1})(\boxed{x} + 4)$ $3x^2 + 12x - x$ $\Rightarrow$	Multiplied the "inside" terms. Remembered to think of $(3x - 1)$ as $(3x + -1)$ so the product was -1 times x.
$(3x\boxed{-1})(x + \boxed{4})$ $3x^2 + 12x - x - 4$ $\Rightarrow$	Multiplied the "last" terms in each binomial.
$3x^2 + 11x - 4$ $\Rightarrow$	Collected like terms to finish the problem.

Homework 5.2 *Simplify.*		
20) $(x+4)(x+3)$	21) $(y+1)(y-1)$	22) $(a-4)(a+2)$
23) $(m-3)(m-4)$	24) $(2y+1)(2y-1)$	25) $(4-p)(2p+7)$
26) $(2v+6)(3v+4)$	27) $(3z-2)(-z-3)$	28) $(7-3k)(3k+7)$

5.2.5 FOIL and Nonlinear Factors

Sometimes we'll need to FOIL with a variable term that has an exponent greater than 1.

Practice 5.2.5 FOIL and Nonlinear Factors
Simplify.

a) $(x^3-1)(x^2+4)$

$x^5+4x^3-x^2-4$	$\Rightarrow$	Multiplied each term in the first factor to every term in the second factor and simplified using the product rule.

b) $(-2x^2+9)(x^2+4)$

$-2x^4-8x^2+9x^2+36$	$\Rightarrow$	Multiplied.
$-2x^4+x^2+36$	$\Rightarrow$	Combined like terms.

Homework 5.2 Simplify.

29) $(t-5)(t^2-5)$ 30) $(a^3+a^2)(2a^2+a)$ 31) $(6-q^2)(6+q^2)$ 32) $(6-k^2)(k^2-6)$

33) $(z^4-z^2)(-5z^2-z^4)$ 34) $(8x^3+3x^8)(2x^2+3x^3)$

Homework 5.2 Answers

1) base, power, factor, factor, product, base, exponent, power, term, factor, power, factor, product, term, sum

2) product, factor, power, factor, product, minuend, product, subtrahend, difference, factor, product

3) product, term, sum, factor, difference, base, power, factor, product

4) $-3x^{10}$ 5) $30n^3$ 6) $-5y^5$ 7) $5a^3$ 8) $-2x^3$ 9) $4a^6+6a^5$ 10) $2y^6$

11) $9h^7+6h^6$ 12) x^5+x^4 13) $6m^3-3m^2$ 14) $-t^5+2t^4-t^3$ 15) $4p^4+8p^3+12p^2$

16) $6w^2+4w$ 17) $12y^2$ 18) $49x^4+2x^3-5x^2$ 19) $10y^3$ 20) $x^2+7x+12$

21) y^2-1 22) a^2-2a-8 23) $m^2-7m+12$ 24) $4y^2-1$ 25) $-2p^2+p+28$

26) $6v^2+26v+24$ 27) $-3z^2-7z+6$ 28) $-9k^2+49$ 29) $t^3-5t^2-5t+25$

30) $2a^5+3a^4+a^3$ 31) $36-q^4$ 32) $-k^4+12k^2-36$ 33) $-z^8-4z^6+5z^4$

34) $9x^{11}+6x^{10}+24x^6+16x^5$

5.3 Operations with Multivariate Polynomials

Multivariate polynomials have more than one kind of variable. For instance, $7xy$ is a multivariate term with the two variables x and y. In this section, we'll practice operating on multivariate polynomials.

5.3.1 Some Vocabulary for Multivariate Polynomials

For the most part, the vocabulary for single variable polynomials carries over to multivariate polynomials. The exception happens when we're discussing the degree of a term. The degree of a mixed term like $-7x^3y$ is found by adding the exponents. To find the degree of $-7x^3y$ we'd add 3, the exponent of x, and 1 the exponent of y, to get 4. So $-7x^3y$ is a fourth-degree term.

When it comes to standard form with multivariate polynomials, things get a little arbitrary. We still write the terms left to right from highest to lowest degree. Sometimes though, two unlike terms will have the same degree. For instance, the polynomial $2ab+b^2+3a^3b$ has two second degree terms and a fourth-degree term. We'd certainly want to write the fourth-degree term furthest to the left, but then which second-degree term should go next? That is, should it be $3a^3b+2ab+b^2$ or $3a^3b+b^2+2ab$? The answer is that either are fine.

5.3.2 Like Multivariate Terms

As usual, multivariate terms are considered like if they have the same powers of the same letters. Keep in mind, that by the commutative property of multiplication, $4a^2b$ and $-5ba^2$ would be like terms, since the order of variable factors a^2 and b won't affect the product. Although both $4a^2b$ and $-5ba^2$ are mathematically equivalent, it's more common for a final answer to be in alphabetical order with a^2 to the left of b.

Practice 5.3.2 Like Multivariate Terms

Identify the like terms.

a) $-7yx^2$, $5yx$, $3xy$, $-xy^2$, $-x^2y$

$-7yx^2$ and $-x^2y$ are like. $5yx$ and $3xy$ are also like.	$\Rightarrow$	By the commutative property, the order of the factors doesn't affect the product.

Homework 5.3 Identify the like terms.

1) $3mn, -m^2, 2nm, 2m$	2) $x^2y, -2yx^2, xy^2, 2x^2y$	3) $-5+7h^2k+12-14hk-6hk^2$
4) $3a^2b, 5ba^2, 6a^2b^2, 3ab, -9b^2a, 14ba^2, 7ab^2$		

5.3.3 *Adding and Subtracting Multivariate Terms*

Like with single variable terms, we add and subtract multivariate terms using the coefficients of the like terms.

Practice 5.3.3 Adding and Subtracting Multivariate Terms
Simplify.

a) $3x^2y - 6yx - yx^2 + 8xy$

$2x^2y$	$\Rightarrow$	Began with the third-degree terms so my answer will be in standard form. Operated on the coefficients, $3 - 1 = 2$ and wrote the variable factors in alphabetical order using the commutative property.
$2x^2y + 2xy$	$\Rightarrow$	Moved to the second-degree terms, operated on the coefficients, $-6 + 8 = 2$, and wrote the variable factors in alphabetical order.

b) $-2x - 15 + 4xy + 2 - xy + 11x$

$3xy$	$\Rightarrow$	Began with the second-degree terms and operated on the coefficients, $4 - 1 = 3$.
$3xy + 9x$	$\Rightarrow$	Continued with the first-degree terms, $-2 + 11 = 9$.
$3xy + 9x - 13$	$\Rightarrow$	Finished with the constants.

c) $3y - 1 + 4x^2 - 6yx - 9y^2 - 2x + 6xy - 3y + 2x$

$0xy = 0$	$\Rightarrow$	Operated on the second-degree coefficients in *x* and *y*, $-6 + 6 = 0$, there will be no mixed variable term.
$4x^2$	$\Rightarrow$	Moved to the second-degree term in *x*. There was only one.
$4x^2 - 9y^2$	$\Rightarrow$	Moved to the second-degree terms in *y*.
$4x^2 - 9y^2$	$\Rightarrow$	Operated with the first-degree terms in *x*, $-2 + 2 = 0$. There will be no first-degree terms in *x*.
$4x^2 - 9y^2$	$\Rightarrow$	Operated with the first-degree terms in *y*, $3 - 3 = 0$. There will be no first-degree terms in *y*.
$4x^2 - 9y^2 - 1$	$\Rightarrow$	Finished with the constant.

Homework 5.3 Simplify.

5) $3mn - m + 2mn - 2m$ 6) $2 - 2ab - 4ba - 6ba$ 7) $p^2 + 2q^2 - qp^2 - p^2q$

8) $a^2 - b^2 - 4ab + 3b^2 - 7a^2 + 4ab$ 9) $x^2y - 6y^2x^2 + xy^2 + 2x^2y^2$

10) $4y - 11 + 3xy + 14 - y - 7xy - x$ 11) $-8xy^2 + 4y^2x - x^2y + 4y^2x$

12) $-5hk + 7hk^2 + 12hk - 14hk - 6hk^2$ 13) $3a^2b - 5ba^2 + 6a^2b^2 + 3 - 9b^2 + 14b^2a^2 - 5$

5.3.4 Beginning Distribution with Multivariate Polynomials

As you distribute, remember that the product rule only works when bases are the same.

Practice 5.3.4 Beginning Distribution with Multivariate Polynomials
Simplify.

a) $7h(2j^2+3j)$

$7h(2j^2)+7h(3j)$ $\Rightarrow$ Distributed $7h$ to both terms inside the parentheses.

$14hj^2+21hj$ $\Rightarrow$ Simplified each term.

b) $-pq(p^2-4q+q^2)$

$-p^3q+4pq^2-pq^3$ $\Rightarrow$ Multiplied. It's common to leave the product in this order.

c) $xy^2+2y^2-3y(xy+2y)$

$xy^2+2y^2-3xy^2-6y^2$ $\Rightarrow$ Multiplied first.

$-2xy^2-4y^2$ $\Rightarrow$ Collected like terms.

Homework 5.3 Simplify.

14) $2x(x^2+xy+y^2)$
15) $-3ab(-2a-3b)$
16) $-3st^2(3-s^2+2s)$
17) $3a(ab+5b)+2b(a^2-7a)$
18) $w(v^2+4w)-v(vw-4v)$
19) $-2w(-2y-2-wy)-4y(w^2-w)$
20) $12wk(w+3k-4)+6k(w^2-6wk+8w)$
21) $-2x^2y(4-y)-xy(x+2xy)+3x^2(3y-1)$
22) $j(j-2k)-k(k-j)-(j^2-k^2)$

5.3.5 Multiplying Multivariate Binomials

Here's some practice multiplying multivariate binomial factors.

Practice 5.3.5 Multiplying Multivariate Binomials
Simplify.

a) $(x-y)(x-2y)$

$x^2-2xy-xy+2y^2$
$x^2-3xy+2y^2$ $\Rightarrow$ Multiplied and combined like terms.

b) $(2a^2+3b)(2a^2-3b)$

$4a^4-6a^2b+6ba^2-9b^2$ $\Rightarrow$ Multiplied.

$4a^2-9b^2$ $\Rightarrow$ Combined like terms.

c) $(hk-2)(h+hk)$

$h^2k^2+h^2k-2hk-2h$ $\Rightarrow$ Multiplied. None of the terms are like.

Homework 5.3 Simplify.

23) $(v-2y)(3v+4y)$ 24) $(2a-b)(2a+b)$ 25) $(3x-2y)(3x-2y)$ 26) $(4r+r^2)(2s^2+1)$

27) $(ab-2b^2)(b+a)$ 28) $(x^4-y^2)(x^4+y^2)$ 29) $(mn+n)(n+nm)$

5.3.6 *The Difference of Two Squares*

We'll end today's lesson with a common polynomial product known as the **difference of two squares**. The difference results from multiplying the **conjugates** $(m+n)$ and $(m-n)$ in either order. As you can see to the right, the product always simplifies to a difference of two perfect squares.

$$(m+n)(m-n)$$
$$m^2 - \cancel{mn} + \cancel{mn} - n^2$$
$$m^2 - n^2$$

$$(m-n)(m+n)$$
$$m^2 + \cancel{mn} - \cancel{mn} - n^2$$
$$m^2 - n^2$$

Realize that both *m* and *n* stand in place of any expression. So if $(x+5)(x-5)$ are the conjugates then x is the "*m*" and 5 is the "*n*". For the conjugates $(4a-9b)(4a+9b)$ $4a$ is the "*m*" and $9b$ the "*n*".

To quickly multiply $(x+5)(x-5)$ realize that "*m*" is x, "*n*" is 5 and the difference x^2-25, follows. The idea is worked out to the right.

$$(m+n)(m-n) = m^2 - n^2$$
$$(x+5)(x-5) = (x)^2 - (5)^2$$
$$= x^2 - 25$$

To quickly multiply $(4a-9b)(4a+9b)$ realize that $4a$ is the "*m*", $9b$ is the "*n*" and the difference $16a^2-81b^2$ follows. Here's some practice.

$$(m-n)(m+n) = m^2 - n^2$$
$$(4a-9b)(4a+9b) = (4a)^2 - (9b)^2$$
$$= 16a^2 - 81b^2$$

Practice 5.3.6 The Difference of Two Squares

Identify "m" and "n", find the product using the difference of two squares and then multiply the original binomials to verify your work is correct.

a) $(x-6)(x+6)$

$(x-6)(x+6)$	$\Rightarrow$	x is the "*m*" and 6 is the "*n*".
m^2-n^2 $(x)^2-(6)^2$	$\Rightarrow$	Substituted using the difference of two squares.
x^2-36	$\Rightarrow$	Simplified.
$(x-6)(x+6)$ $x^2+6x-6x-36$ x^2-36	$\Rightarrow$	Multiplied the original binomials and simplified to find the result is the same.

b) $(1+5w)(1-5w)$

$(1+5w)(1-5w)$ $\Rightarrow$ 1 is the "m" and $5w$ is the "n".

$m^2 - n^2$

$(1)^2 - (5w)^2$ $\Rightarrow$ Substituted using the difference of two squares and simplified using the power rule for exponents with the second term.

$1 - 25w^2$

$(1+5w)(1-5w)$

$1 - 5w + 5w - 25w^2$ $\Rightarrow$ Multiplied the original binomials and simplified to find the result is the same.

$1 - 25w^2$

c) $(3a^2 + b^2)(3a^2 - b^2)$

$(3a^2 + b^2)(3a^2 - b^2)$ $\Rightarrow$ $3a^2$ is the "m" and b^2 is the "n".

$m^2 - n^2$

$(3a^2)^2 - (b^2)^2$ $\Rightarrow$ Substituted using the difference of two squares and simplified using the power rule for exponents.

$9a^4 - b^4$

$(3a^2 + b^2)(3a^2 - b^2)$

$9a^4 - 3a^2b^2 + 3a^2b^2 - b^4$ $\Rightarrow$ Multiplied the original binomials and simplified to find the result is the same.

$9a^4 - b^4$

Homework 5.3 Identify "m" and "n", find the product using the difference of two squares and then multiply the original binomials to verify your work is correct.

30) $(y-5)(y+5)$ 31) $(2k+1)(2k-1)$ 32) $(7-2p)(7+2p)$ 33) $(3m+n)(3m-n)$

34) $(p^2-1)(p^2+1)$ 35) $(9-6x^2)(9+6x^2)$ 36) $(2k-5h)(2k+5h)$

37) $(x^2+2y^2)(x^2-2y^2)$ 38) $(4a^2-8b^2)(4a^2+8b^2)$ 39) $(x^4+y^4)(x^4-y^4)$

Homework 5.3 Answers

1) $3mn$ and $2nm$ are like. 2) x^2y , $-2yx^2$ and $2x^2y$ are like. 3) –5 and 12 are like.

4) $3a^2b, 5ba^2, 14a^2b$ are like, and $-9b^2a$ and $7ab^2$ are like.

5) $5mn-3m$ 6) $-12ab+2$ 7) $-2p^2q+p^2+2q^2$ 8) $2b^2-6a^2$

9) $-4x^2y^2+x^2y+xy^2$ 10) $-4xy-x+3y+3$ 11) $-x^2y$ 12) hk^2-7hk

13) $20a^2b^2-2a^2b-9b^2-2$ 14) $2x^3+2x^2y+2xy^2$ 15) $6a^2b+9ab^2$

16) $3s^3t^2-6s^2t^2-9st^2$ 17) $5a^2b+ab$ 18) $4v^2+4w^2$

19) $-2w^2y+8wy+4w$ 20) $18kw^2$ 21) $-3x^2$ 22) $-jk$ 23) $3v^2-2vy-8y^2$

24) $4a^2-b^2$ 25) $9x^2-12xy+4y^2$ 26) $2r^2s^2+r^2+8rs^2+4r$

27) $a^2b-ab^2-2b^3$ 28) x^8-y^4 29) $m^2n^2+2mn^2+n^2$

30) y is the "m" and 5 is the "m". y^2-25 31) $2k$ is the "m" and 1 is the "m". $4k^2-1$

32) 7 is the "m" and $2p$ is the "m". $49-4p^2$

33) $3m$ is the "m" and n is the "m". $9m^2-n^2$

34) p^2 is the "m" and 1 is the "m". p^4-1

35) 9 is the "m" and $6x^2$ is the "m". $81-36x^4$

36) $2k$ is the "m" and $5h$ is the "m". $4k^2-25h^2$

37) x^2 is the "m" and $2y^2$ is the "m". x^4-4y^4

38) $4a^2$ is the "m" and $8b^2$ is the "m". $16a^4-64b^4$

39) x^4 is the "m" and y^4 is the "m". x^8-y^8

5.4 Continuing With Polynomial Multiplication

In this section, we'll continue operating on polynomials.

5.4.1 A General Procedure for Multiplying Polynomials

Although FOIL can be helpful, its usefulness is limited to the product of two binomials. As we move beyond two binomial factors it's important to find the product by multiplying each term in the first factor to every term in the second factor. Notice that FOIL is just a special case of this idea. For example, to multiply $(x+4)(x+2)$ using FOIL we'd first multiply x to both x and 2 and then we'd multiply 4 to both x and 2. That is, we'd multiply each term in the first factor to every term in the second factor.

Practice 5.4.1 A General Procedure for Multiplying Polynomials
Simplify.

a) $(x^2-x+1)(x^2+x-1)$

$(\boxed{x^2}-x+1)(x^2+x-1)$ $x^4+x^3-x^2$	$\Rightarrow$	Multiplied x^2 in the first factor to every term in the second factor.
$(x^2\boxed{-x}+1)(x^2+x-1)$ $x^4+x^3-x^2\boxed{-x^3-x^2+x}$	$\Rightarrow$	Multiplied the term $-x$ in the first factor to every term in the second factor. Paid special attention to the signs, especially of the last term $-x(-1)=x$.
$(x^2-x+\boxed{1})(x^2+x-1)$ $x^4+x^3-x^2-x^3-x^2+x\boxed{+x^2+x-1}$	$\Rightarrow$	Multiplied 1 to all three terms in the second factor.
x^4-x^2+2x-1	$\Rightarrow$	Collected like terms in standard form.

b) $2x(3x-1)(x+4)$

$(6x^2-2x)(x+4)$	$\Rightarrow$	Followed the order of operations and began multiplying left to right. Multiplied the factor $2x$ to the factor $(3x-1)$.
$6x^3+24x^2-2x^2-8x$	$\Rightarrow$	Multiplied the resulting binomial to the original third factor.
$6x^3+22x^2-8x$	$\Rightarrow$	Collected like terms.

Homework 5.4 Simplify.

1) $p(p-1)(2p-1)$	2) $(x-1)(x^2+x+1)$	3) $(x^2+2)(x^2-x-2)$
4) $(3x+2)(x^2-x+4)$	5) $(z^2+3z)(2z^2-z)(z^2-3z)$	6) $r(r-1)(r^2-r-1)$
7) $(k-2)(k^2+4)(k+2)$	8) $(a^2-2a+5)(2a^2+2a-3)$	9) $(2p+1)(2p)(2p-p-1)$

5.4.2 *Multiplying Multivariable Polynomials Using the General Procedure*

We can use the same procedure with multivariable polynomials.

Practice 5.4.2 Multiplying Multivariable Polynomials Using the General Procedure
Simplify.

a) $(a+b)(2a+b-1)$

$(a+b)(2a+b-1)$ ⇒ First distributed a to every term in the second factor.

$2a^2+ab-a$

$(a+b)(2a+b-1)$ ⇒ Now distributed b to every term in the second factor.

$2a^2+ab-a+2ab+b^2-b$

$2a^2+3ab+b^2-a-b$ ⇒ Last, collected like terms.

Homework 5.4 Simplify.

10) $a(a-1)(a-b)$ 11) $(w-y)(w^2+y^2)(w+y)$ 12) $(x-y)(x^2+xy+y^2)$
13) $(m-n-1)(m+n+1)$ 14) $(x-2y+1)(x-1)$ 15) $(k+h-4)(k+2)-k(h+k+2)$
16) $a(2a+b)(2a-1)+a(2a+b)$ 17) $(2z^2+3z-y)(2z^2-3z+y)$
18) $g(g-2a)-(g^2+g+2a)$

5.4.3 *The Power of a Polynomial*

Remember, there is <u>no</u> "distribution" of exponents over addition or subtraction. Instead, a natural number exponent tells you how many factors of the base you have.

Practice 5.4.3 The Power of a Polynomial
Simplify.

a) $(2x+3)^2$

$(2x+3)(2x+3)$ ⇒ There are two factors of the base.

$4x^2+6x+6x+9$
$4x^2+12x+9$ ⇒ Distributed and collected like terms.

b) $(x-y)^3$

$(x-y)(x-y)(x-y)$ ⇒ I have three factors of the base.

$(x^2-xy-xy+y^2)(x-y)$
$(x^2-2xy+y^2)(x-y)$ ⇒ Multiplied the first two factors. Kept parentheses since all three terms in the first factor will need to multiply to both terms in the second factor.

$x^3-x^2y-2x^2y+2xy^2+xy^2-y^3$ ⇒ Distributed.

$x^3-3x^2y+3xy^2-y^3$ ⇒ Collected like terms.

Homework 5.4 Simplify.			
19) $(m-5)^2$	20) $(k-1)^3$	21) $(p^2+2)^3$	22) $(y-2)^4$
23) $(2m-3n)^2$	24) $(k^2+h)^3$	25) $(3p+2q)^2-9p^2-4q^2$	26) $(y-x)^2+(y+x)^2$
27) $(2h^2-p)^2+(h^2-p)^2$	28) $\left((y-1)^2-1\right)^2$	29) $(m+n)^2-(m-n)(m+n)$	

5.4.4 Perfect Square Trinomials

It's common in algebra to find the square of a binomial sum or difference. The result is sometimes called a perfect square trinomial.

In general, $(m+n)^2$ leads to the sum $m^2+2mn+n^2$. Memorizing this general form allows you to quickly find the sum of a squared binomial. For instance $(4x+3)^2$ fits the form if you think of 4*x* as your "*m*" and 3 as your "*n*". The sum is to the right.

$$(m+n)^2 = m^2 + 2\,m\,n + n^2$$
$$(4x+3)^2 = (4x)^2 + 2(4x)(3) + (3)^2$$
$$= 16x^2 + 24x + 9$$

The form for squaring a difference, $(m-n)^2 = m^2 - 2mn + n^2$ can be found if you change subtraction to adding an inverse and then use the form for a sum.

Practice 5.4.4 Perfect Square Trinomials
Identify "m" and "n" and use the general form to find the sum or difference.

a) $(2r+6)^2$

$(2r+6)^2$	$\Rightarrow$	2*r* is the "*m*" and 6 is the "*n*".
$(2r)^2+2(2r)(6)+(6)^2$	$\Rightarrow$	Substituted using the general form for a sum.
$4r^2+24r+36$	$\Rightarrow$	Simplified.

b) $(4x-3y)^2$

$(4x-3y)^2$	$\Rightarrow$	$4x$ is the "*m*" and $3y$ is the "*n*".
$m^2-2(m)(n)+n^2$ $(4x)^2-2(4x)(3y)+(3y)^2$	$\Rightarrow$	Substituted for *m* and *n* using the form for the square of a binomial difference.
$16x^2-24xy+9y^2$	$\Rightarrow$	Simplified.

Homework 5.4 Identify "m" and "n" and use the general form to find the sum or difference.

30) $(y+6)^2$	31) $(x-2y)^2$	32) $(3u-v)^2$	33) $(7+5x)^2$	34) $(6a+4b)^2$
35) $(5h-3k)^2$	36) $(-11-9p)^2$	37) $(15k+4)^2$		

Homework 5.4 Answers

1) $2p^3 - 3p^2 + p$ 2) $x^3 - 1$ 3) $x^4 - x^3 - 2x - 4$ 4) $3x^3 - x^2 + 10x + 8$

5) $2z^6 - z^5 - 18z^4 + 9z^3$ 6) $r^4 - 2r^3 + r$ 7) $k^4 - 16$ 8) $2a^4 - 2a^3 + 3a^2 + 16a - 15$

9) $4p^3 - 2p^2 - 2p$ 10) $a^3 - a^2b - a^2 + ab$ 11) $w^4 - y^4$ 12) $x^3 - y^3$

13) $m^2 - n^2 - 2n - 1$ 14) $x^2 - 2xy + 2y - 1$ 15) $2h - 4k - 8$ 16) $4a^3 + 2a^2b$

17) $4z^4 - y^2 + 6yz - 9z^2$ 18) $-2ag - 2a - g$ 19) $m^2 - 10m + 25$ 20) $k^3 - 3k^2 + 3k - 1$

21) $p^6 + 6p^4 + 12p^2 + 8$ 22) $y^4 - 8y^3 + 24y^2 - 32y + 16$ 23) $4m^2 - 12mn + 9n^2$

24) $h^3 + 3h^2k^2 + 3hk^4 + k^6$ 25) $12pq$ 26) $2x^2 + 2y^2$ 27) $5h^4 - 6h^2p + 2p^2$

28) $y^4 - 4y^3 + 4y^2$ 29) $2n^2 + 2mn$ 30) y is the "m" and 6 is the "n", $y^2 + 12y + 36$

31) x is the "m" and $2y$ is the "n", $x^2 - 4xy + 4y^2$

32) $3u$ is the "m" and v is the "n", $9u^2 - 6uv + v^2$

33) 7 is the "m" and $5x$ is the "n", $25x^2 + 70x + 49$

34) $6a$ is the "m" and $4b$ is the "n", $36a^2 + 48ab + 16b^2$

35) $5h$ is the "m" and $3k$ is the "n", $25h^2 - 30hk + 9k^2$

36) -11 is the "m" and $9p$ is the "n", $81p^2 + 198p + 121$

37) $15k$ is the "m" and 4 is the "n", $225k^2 + 120k + 16$

5.5 Factoring Out the GCF

Recall that the distributive property gives us a relationship between sums and products.

Property **– The Distributive Property of Multiplication over Addition**
A product of two factors, one of which is a sum, can be replaced by a sum of two terms. Each term will be the product of the common factor and one of the original terms.
Example: $2(x+3)=2x+2(3)$

In this chapter, up to now, we've usually used the distributive property to rewrite a product as a sum. For example, we rewrote the product $2(x+3)$ as the sum $2x+6$. For the remainder of this chapter we'll take the other point of view and use the distributive property to replace a sum like $2x+6$ with the equivalent product $2(x+3)$.

5.5.1 Rewriting Terms to Factor Out the Numerical GCF

The GCF (Greatest Common Factor) of a polynomial is the largest factor, common to all the terms. To "factor out" the GCF think of each term in the polynomial as a product of two factors, where one of those factors is the GCF. Once all the terms share the GCF you can rewrite the sum as a product using the distributive property.

Practice 5.5.1 Rewriting Terms to Factor Out the Numerical GCF

Rewrite each term as a product with the GCF as the first factor, factor the polynomial and distribute to obtain the original polynomial.

a) $6x+3$

$3(2x)+3(1)$	$\Rightarrow$	The GCF is 3. Rewrote each term as a product where the first factor is the GCF. Paid particular attention to write 3 as $3(1)$.
$3(2x+1)$	$\Rightarrow$	Used the distributive property to factor out the GCF (the 3) which left the terms $2x$ and 1 inside the parentheses.
$6x+3$	$\Rightarrow$	Distributed. The polynomial was the same as the original.

b) $4y+12$

$4(y)+4(3)$	$\Rightarrow$	The GCF is 4. 2 is a common factor but not the greatest common factor. Rewrote each term as a product with the GCF as the first factor.
$4(y+3)$	$\Rightarrow$	Factored out the 4 which left y and 3 inside the parentheses.
$4y+12$	$\Rightarrow$	Multiplied to check if the polynomial is the same as the original.

Homework 5.5 Rewrite each term as a product with the GCF as the first factor, factor the polynomial and distribute to obtain the original polynomial.

1) $3y-9$	2) $6m-6$	3) $10k+50$	4) $12y+8$	5) $60x-30$

5.5.2 Using Prime Factorization to Help Find a Numerical GCF

Hopefully, in the previous problems, the GCF "popped into your head". Prime factorization is a good idea if it's difficult to see the numerical GCF.

Practice 5.5.2 Using Prime Factorization to Help Find a Numerical GCF

Rewrite each term as a product with the GCF as the first factor, factor the polynomial and distribute to obtain the original polynomial.

a) $12y^2+6y+18$

$12=2\times2\times3$ $6=2\times3$ $18=2\times3\times3$	$\Rightarrow$	Prime factored the coefficients and noticed they all share a factor of 2 and a factor of 3 so the GCF is 6. Noticed y is <u>not</u> part of the GCF since only two of the three terms have a factor of y.
$6(2y^2)+6(y)+6(3)$	$\Rightarrow$	Thought of each term as a product where the GCF is one of the factors.
$6(2y^2+y+3)$	$\Rightarrow$	Factored out the GCF.
$12y^2+6y+18$	$\Rightarrow$	Multiplied to verify the polynomial is the same as the original.

b) $63k-42$

$63=3\times3\times7$ $42=2\times3\times7$	$\Rightarrow$	Prime factored the coefficients and noticed both terms share one factor of 3 and one factor of 7. The GCF is 21.
$21(3k)-21(2)$	$\Rightarrow$	Thought of each term as product where the GCF is one of the factors.
$21(3k-2)$	$\Rightarrow$	Factor out the GCF.
$63k-42$	$\Rightarrow$	Multiplied. The polynomials are the same.

Homework 5.5 Rewrite each term as a product with the GCF as the first factor, factor the polynomial and distribute to obtain the original polynomial.

6) $56c^2+14$ 7) $12x^2-32$ 8) $24y^2-16y-36$ 9) $12p^3+8p^2+20$

10) $24a^4-36a^2+60$

5.5.3 Factoring Out a Variable GCF

The product rule for exponents helps us factor out common variable factors. Although mathematically it's not necessary, it's often helpful to start with the polynomial in standard form.

Practice 5.5.3 Factoring Out a Variable GCF

Rewrite each term as a product with the GCF as the first factor, factor the polynomial and distribute to obtain the original polynomial.

a) $-28x+7x^2$

$7x^2-28x$	$\Rightarrow$	Rewrote the polynomial in standard form.
$7x^2-28x$	$\Rightarrow$	Both terms share a factor of 7 and x. The GCF is $7x$.

$7x(x)-7x(4)$	$\Rightarrow$	Thought of each term as the GCF times any factors that are left.
$7x(x-4)$	$\Rightarrow$	Factored out the GCF.
$7x^2-28x$	$\Rightarrow$	Multiplied to make sure the polynomials are the same.

b) $12p^2+4p^4$

$4p^4+12p^2$	$\Rightarrow$	Rewrote the polynomial in standard form.
$4p^2(p^2)+4p^2(3)$	$\Rightarrow$	Both terms share a factor of 4 and p^2 so the GCF is $4p^2$. Thought of each term as the GCF times any factors that are left.
$4p^2(p^2+3)$	$\Rightarrow$	Factored out the GCF.
$4p^4+12p^2$	$\Rightarrow$	Multiplied to check if the sum is the same as the original.

c) $8n^4-4n^3-10n$

$2n(4n^3)-2n(2n^2)-2n(5)$	$\Rightarrow$	All three terms share a factor of 2 and a factor of n so the GCF is $2n$. Thought of each term as the GCF times the remaining factors.
$2n(4n^3-2n^2-5)$	$\Rightarrow$	Factored out the GCF.
$8n^4-4n^3-10n$	$\Rightarrow$	Distributed. The polynomial is the same as the original.

Homework 5.5 Rewrite each term as a product with the GCF as the first factor, factor the polynomial and distribute to obtain the original polynomial.

11) x^2+5x 12) $14p^4+7p^2$ 13) $16x^3-16x^2$ 14) $7m^3-m^2+7m$

15) $2t^3+2t^2-12t$ 16) $24y^2-16y-36$ 17) $12x^7-4x^4+6x^3$

18) $28a^5-12a^4-20a^3$

5.5.4 Factoring a GCF that Contains -1

If the coefficient of the highest-degree term is negative, we usually include a factor of -1 in the GCF. Keep in mind that when you factor out a negative GCF you'll need to rewrite each term inside the parentheses as it's opposite.

Practice 5.5.4 Factoring a GCF that Contains -1

Rewrite each term as a product with the GCF as the first factor (include a factor of -1), factor the polynomial and distribute to obtain the original polynomial.

a) $-x^2+4$

$-1(x^2)+4$	$\Rightarrow$	Thought of $-x^2$ as the product $-1(x^2)$.
$-1(x^2-4)$	$\Rightarrow$	Factored out a GCF of -1. Notice how the "plus" four needed to become a "minus" four.
$-x^2+4$	$\Rightarrow$	Distributed. The sum is the same as the original.

b) $5a^2 - 25a^3 - 15a$

$-25a^3 + 5a^2 - 15a$	$\Rightarrow$	Rewrote in standard form.
$-5a(5a^2 - a + 3)$	$\Rightarrow$	Factored out the GCF of $-5a$. Made each sign inside the parentheses its opposite.
$-25a^3 + 5a^2 - 15a$	$\Rightarrow$	Distributed. The trinomial is the same as the original.

Homework 5.5 Rewrite each term as a product with the GCF as the first factor (include a factor of -1), factor the polynomial and distribute to obtain the original polynomial.

19) $-p + 1$ 20) $-k^2 + 6k - 4$ 21) $-5t^2 - 5t - 15$ 22) $-4y^3 + 2y^2 + 6y$

23) $-8a^3 + 4a^2 - 4a$ 24) $-20t^4 - 18t^2$ 25) $-10a^2 + 15a + 25$

Homework 5.5 Answers

1) The GCF is 3. $3y-3(3)$ factors to $3(y-3)$.

2) The GCF is 6. $6m-6(1)$ factors to $6(m-1)$.

3) The GCF is 10. $10k+10(5)$ factors to $10(k+5)$.

4) The GCF is 4. $4(3y)+4(2)$ factors to $4(3y+2)$.

5) The GCF is 30. $30(2x)-30(1)$ factors to $30(2x-1)$.

6) The GCF is 14. $14(4c^2)+14(1)$ factors to $14(4c^2+1)$.

7) The GCF is 4. $4(3x^2)-4(8)$ factors to $4(3x^2-8)$.

8) The GCF is 4. $4(6y^2)-4(4y)-4(9)$ factors to $4(6y^2-4y-9)$.

9) The GCF is 4. $4(3p^3)+4(2p^2)+4(5)$ factors to $4(3p^3+2p^2+5)$.

10) The GCF is 12. $12(2a^4)-12(3a^2)+12(5)$ factors to $12(2a^4-3a^2+5)$.

11) The GCF is x. $x(x)+5(x)$ factors to $x(x+5)$.

12) The GCF is $7p^2$. $7p^2(2p^2)+7p^2(1)$ factors to $7p^2(2p^2+1)$.

13) The GCF is $16x^2$. $16x^2(x)-16x^2(1)$ factors to $16x^2(x-1)$.

14) The GCF is m. $m(7m^2)-m(m)+m(7)$ factors to $m(7m^2-m+7)$.

15) The GCF is $2t$. $2t(t^2)+2t(t)-2t(6)$ factors to $2t(t^2+t-6)$.

16) The GCF is 4. $4(6y^2)-4(4y)-4(9)$ factors to $4(6y^2-4y-9)$.

17) The GCF is $2x^3$. $2x^3(6x^4)-2x^3(2x)+2x^3(3)$ factors to $2x^3(6x^4-2x+3)$.

18) The GCF is $4a^3$. $4a^3(7a^2)-4a^3(3a)-4a^3(5)$ factors to $4a^3(7a^2-3a-5)$.

19) $-1(p-1)$ 20) $-1(k^2-6k+4)$ 21) $-5(t^2+t+3)$ 22) $-2y(2y^2-y-3)$

23) $-4a(2a^2-a+1)$ 24) $-2t^2(10t^2+9)$ 25) $-5(2a^2-3a-5)$

5.6 Factoring with Guess and Check

Sometimes it's possible to factor a polynomial even if it doesn't have a GCF. One example would be factoring the trinomial $x^2+7x+10$ which doesn't have a GCF and yet it factors to $(x+5)(x+2)$. Multiplying verifies that the original sum and the factored product are the same, $(x+5)(x+2)=x^2+2x+5x+10 = x^2+7x+10$. One way to figure out the factored form of a trinomial is known as guess and check.

5.6.1 A Procedure for Guess and Check

The idea behind guess and check is simple. We assume the trinomial will factor to the product of two binomials and we "guess" what those two binomials might be. To "check" our guess we multiply the two binomials and compare the result to our original polynomial. If the polynomials match, we've found the factored form. If they don't match, we start over with a different guess. With practice, you'll find you can do most of guess and check "in your head" but for now, you might find this procedure helpful.

***Procedure* – Factoring a Trinomial Using Guess and Check**

1. Write the polynomial in standard form and identify the values of *a*, *b* and *c*.
2. Assume the trinomial will factor to two binomials and build a template.
   a) If *c* is positive and *b* is positive use $(\ +\ )(\ +\ )$
   b) If *c* is positive and *b* is negative use $(\ -\ )(\ -\ )$.
   c) If *c* is negative use $(\ +\ )(\ -\ )$ or $(\ -\ )(\ +\ )$.
3. Fill in the "First" terms in both binomials.
4. Substitute factor pairs for the "Last" terms in both binomials and compare the sum or difference of the "Outside" and "Inside" products to the first-degree term. You've found the factored form when these match.
5. After filling in both terms in both binomials, multiply to "check" your work.

Let's begin with step 1 and practice finding the values of *a*, *b* and *c*.

5.6.2 Identifying Trinomial Coefficients and Constants

The first step when factoring a trinomial is to write the polynomial in the standard form ax^2+bx+c and identify the values of *a*, *b* and *c*. The "*a*" is the coefficient of the second-degree term, the "*b*" is the coefficient of the first-degree term and "*c*" is the constant. Notice the form ax^2+bx+c involves addition so it's useful to think of subtraction as adding an opposite. For instance, thinking of $4x^2-x-5$ as $4x^2+-1x+-5$ makes 4 the value of *a*, -1 the value of *b* and -5 the value of *c*. If *b* or *c* is "missing" its value is 0. Let's practice assigning values to trinomial coefficients and constants.

Practice 5.6.2 Identifying Trinomial Coefficients and Constants
Identity a, b and c in the following polynomials.

a) $x^2 - x - 2$

$1x^2 + -1x + -2$	$\Rightarrow$	"Saw" all subtractions as adding an inverse, $a = 1, b = -1, c = -2$.

b) $9y^2 - 4$

$9y^2 + 0y - 4$	$\Rightarrow$	Thought of the coefficient of the first-degree term as 0 so $a = 9, b = 0, c = -4$

c) $-12t + 16 + 2t^2$

$2t^2 - 12t + 16$	$\Rightarrow$	After arranging the polynomial in standard form $a = 2, b = -12, c = 16$

Homework 5.6 Identify a, b and c in the following polynomials.

1) $k^2 - k - 2$ 2) $-4b - 15 + 4b^2$ 3) $-9c^2 + 25$ 4) $-12x + 16x^2 + 2$

5.6.3 Factoring Trinomials Using Guess and Check

The best way to learn guess and check is through some examples.

Practice 5.6.3 Factoring Trinomials Using Guess and Check
Factor using guess and check.

a) $x^2 - 8x + 7$

$a = 1, b = -8, c = 7$	$\Rightarrow$	Began with step 1 of the procedure, noticed the polynomial is in standard form and identified the values of *a*, *b* and *c*.
$(\ -\ )(\ -\ )$	$\Rightarrow$	Moved to step 2 and prepared a template for *c* positive and *b* negative.
$(x - \ )(x - \ )$	$\Rightarrow$	Moved to step 3 and filled in the "first" terms so the product, x^2, would match the first term of the original trinomial.
$(x - 7)(x - 1)$	$\Rightarrow$	Moved to step 4. The only way to get a 7 as the last term, like in the original trinomial, is to replace the "last" terms of the binomials with the factors 1 and 7 (or 7 and 1).
$-x - 7x = -8x$	$\Rightarrow$	The difference of the "outside" and "inside" terms matches the first-degree term in the original trinomial. My guess looks good.
$(x - 7)(x - 1)$ $x^2 - 8x + 7$	$\Rightarrow$	Moved to step 5 and multiplied to make sure my guess gave the original trinomial. (Notice that since multiplication is commutative the product $(x - 1)(x - 7)$ would have been fine also.)

b) $10y + y^2 - 11$		
$y^2 + 10y - 11$ $a = 1, b = 10, c = -11$	$\Rightarrow$	Started with step 1, wrote the trinomial in standard form and identified the values of *a, b* and *c*.
$(\; + \;)(\; - \;)$	$\Rightarrow$	Moved to step 2 and prepared the template for *c* negative. Using $(\; - \;)(\; + \;)$ would have been fine also.
$(y + \;)(y - \;)$	$\Rightarrow$	Moved to step 3 and filled in the first terms in the binomials.
$(y+1)(y-11)$	$\Rightarrow$	Moved to step 4. The last term of the polynomial is -11 so I used the factors 1 and 11 in the last terms of the binomials.
$-11y + y = -10y$	$\Rightarrow$	The difference of the "outside" and "inside" terms <u>isn't</u> $10y$ (the first-degree term in the original trinomial) so I'll need to try a different guess.
$(y+11)(y-1)$	$\Rightarrow$	Rearranged the "last" terms in the binomials.
$-y + 11y = 10y$	$\Rightarrow$	This time the difference of the "outside" and "inside" terms <u>does</u> match the first-degree term in the original trinomial. I've found the factored form is $(y+11)(y-1)$.
$(y+11)(y-1)$ $y^2 + 10y - 11$	$\Rightarrow$	Moved to step 5 and multiplied to make sure the factored form is correct

c) $7k + 6 + k^2$		
$k^2 + 7k + 6$ $a = 1, b = 7, c = 6$	$\Rightarrow$	Started with step 1, wrote the trinomial in standard form and identified the values of *a, b* and *c*.
$(\; + \;)(\; + \;)$	$\Rightarrow$	Moved to step 2 and prepared the template for both *b* and *c* positive.
$(k + \;)(k + \;)$	$\Rightarrow$	Moved to step 3 and filled in the first terms in the binomials.
$(k+2)(k+3)$	$\Rightarrow$	Moved to step 4. The last term of the polynomial is 6 so I guessed 2 and 3 as the last terms of the binomials.
$3k + 2k = 5k$	$\Rightarrow$	The sum of the "outside" and "inside" terms <u>isn't</u> $7k$ so I'll need to try a different guess. Noticed that rearranging the factors to $(k+3)(k+2)$ would again give $5k$ so I'll need to guess different "last" terms.
$(k+1)(k+6)$	$\Rightarrow$	Tried another factorization of 6.
$6k + 1k = 7k$	$\Rightarrow$	This time the sum of the "outside" and "inside" terms <u>is</u> $7k$ so $(k+1)(k+6)$ is a factored form. (Noticed that $(k+6)(k+1)$ would have also been fine.)
$(k+1)(k+6)$ $k^2 + 7k + 6$	$\Rightarrow$	Moved to step 5 and multiplied to make sure the factored form is correct

Homework 5.6 Factor using guess and check.		
5) h^2+4h+4	6) r^2-8r+7	7) $2x+x^2-3$
8) x^2+5x+6	9) $-6t+t^2+8$	10) $y^2-2y-15$
11) k^2-16	12) $y^2-7y+12$	13) $-28-12x+x^2$
14) $p^2-19p+18$	15) $y^2+10y-24$	16) $n^2-3n-40$

5.6.4 Guess and Check When $a \neq 1$

In our past problems the value of a was always 1. Now let's practice with some trinomials where the value of a isn't a 1.

Practice 5.6.4 Guess and Check When $a \neq 1$

Factor using guess and check.

a) $2x^2+9x+10$

$a=2, b=9, c=10$	$\Rightarrow$	Started with step 1 of the procedure.
$(\ +\ )(\ +\ )$	$\Rightarrow$	Moved to step 2 and choose the template for b and c positive.
$(2x+\ )(x+\ )$	$\Rightarrow$	Moved to step 3 and guessed the "first" terms to match the $2x^2$ of the original trinomial. $(x+\ )(2x+\ )$ is also fine.
$(2x+5)(x+2)$	$\Rightarrow$	Moved to step 4 and guessed the "last" terms. Noticed that $4x+5x=9x$ so I did guess the factored form.
$2x^2+4x+5x+10$ $2x^2+9x+10$	$\Rightarrow$	Moved to step 5 and distributed to check my work. (You should start doing this step "in your head".)

b) $3y^2-11y+8$

$a=3, b=-11, c=8$	$\Rightarrow$	Started with step 1 of the procedure.
$(\ -\ )(\ -\ )$	$\Rightarrow$	Moved to step 2 and choose the template for b negative and c positive.
$(3y-\ )(y-\ )$	$\Rightarrow$	Moved to step 3 and guessed the "first" terms to match the $3y^2$ of the original trinomial.
$(3y-2)(y-4)$	$\Rightarrow$	Moved to step 4 and guessed the "last" terms. Noticed that $-12y-2y=-14y$ not $-11y$ so my guess is wrong.
$(3y-4)(y-2)$	$\Rightarrow$	Guessed the "last" terms in a different order. Noticed that $-6y-4y=-10y$ not $-11y$ so my guess is still wrong.
$(3y-8)(y-1)$	$\Rightarrow$	Tried a different factorization for the "last" terms. Noticed that $-3y-8y=-11y$ so my guess is finally correct.
$3y^2-3y-8y+8$ $3y^2-11y+8$	$\Rightarrow$	Moved to step 5 and checked my work.

Homework 5.6 Factor using guess and check.		
17) $2y^2 - 11y + 5$	18) $4n^2 + 5n + 1$	19) $2t^2 + 3t - 14$
20) $3k^2 - 2k - 1$	21) $2m^2 - 7m + 6$	22) $5x^2 - 8x - 4$
23) $3w^2 - 14w + 15$	24) $5x^2 + 19x + 12$	25) $13k^2 + 2k - 11$
26) $2x^2 + 19x + 42$	27) $4k^2 - 9$	28) $5x^2 - 16x + 12$
29) $3y^2 - 22y + 35$	30) $7n^2 + 41n + 30$	

Homework 5.6 Answers

1) $a=1, b=-1, c=-2$ 2) $a=4, b=-4, c=-15$ 3) $a=-9, b=0, c=25$

4) $a=16, b=-12, c=2$ 5) $(h+2)(h+2)$ 6) $(r-1)(r-7)$ 7) $(x+3)(x-1)$

8) $(x+2)(x+3)$ 9) $(t-2)(t-4)$ 10) $(y-5)(y+3)$ 11) $(k-4)(k+4)$

12) $(y-3)(y-4)$ 13) $(x-14)(x+2)$ 14) $(p-1)(p-18)$ 15) $(y+12)(y-2)$

16) $(n+5)(n-8)$ 17) $(2y-1)(y-5)$ 18) $(4n+1)(n+1)$ 19) $(2t+7)(t-2)$

20) $(k-1)(3k+1)$ 21) $(2m-3)(m-2)$ 22) $(5x+2)(x-2)$ 23) $(3w-5)(w-3)$

24) $(5x+4)(x+3)$ 25) $(13k-11)(k+1)$ 26) $(2x+7)(x+6)$ 27) $(2k-3)(2k+3)$

28) $(5x-6)(x-2)$ 29) $(3y-7)(y-5)$ 30) $(7n+6)(n+5)$

5.7 Factoring with the AC Method

In this section, we'll practice factoring trinomials using the AC method. The AC method depends on replacing your original trinomial with a factorable four-term polynomial. So first we'll practice factoring a four-term polynomial using factoring by grouping.

5.7.1 Factoring Out a Binomial GCF

So far, when we've used the distributive property, both our terms had monomial factors. For example, with $3x+3(2)$ both the first and second term shared a common factor of 3, so we were able to replace the original sum with the product $3(x+2)$.

Today our terms will share a binomial factor. For example, with $x(x+1)+2(x+1)$ the first and second term share a factor of $(x+1)$. This allows us to use the distributive property to factor out the common binomial factor. The steps are shown to the right. Here's some practice factoring a binomial GCF.

$$x\mathbf{(x+1)}+2\mathbf{(x+1)}$$
$$\mathbf{(x+1)}(x+2)$$

Practice 5.7.1 Factoring Out a Binomial GCF

Factor out the binomial GCF.

a) $x(x+1)+2(x+1)$

$x\boxed{(x+1)}+2\boxed{(x+1)}$	$\Rightarrow$	Concentrated on the GCF $(x+1)$.
$\boxed{(x+1)}(x+2)$	$\Rightarrow$	After factoring out $(x+1)$, $(x+2)$ is left in parentheses.
$\boxed{(x+1)}x+\boxed{(x+1)}(2)$	$\Rightarrow$	Distributed the binomial $x+1$. Noticed the commutative property allowed me to reorder the factors as $x(x+1)+2(x+1)$ which matches the original sum.

b) $y(y+4)-(y+4)$

$y\boxed{(y+4)}-(1)\boxed{(y+4)}$	$\Rightarrow$	The GCF is $(y+4)$. Realized there is a factor of 1 before the second factor.
$\boxed{(y+4)}(y-1)$	$\Rightarrow$	Factored out the GCF.

c) $7p(5p+4)+(5p+4)$

$7p(5p+4)+1(5p+4)$	$\Rightarrow$	The GCF is $(5p+4)$. Thought of a factor of 1 in the second term.
$(5p+4)(7p+1)$	$\Rightarrow$	Factored out the GCF.

Homework 5.7 Factor out the binomial GCF.

1) $x(2x+3)+4(2x+3)$	2) $n(n-2)+(n-2)$	3) $x(4x-7)-9(4x-7)$
4) $y(4y-1)-(4y-1)$	5) $4r(3r+2)-3(3r+2)$	6) $6x(5x-6)+(5x-6)$

5.7.2 *Factoring by Grouping*

Factoring by grouping factors a four-term polynomial using the idea of a GCF three times. We begin by factoring out the GCF of the first and second terms. Then, we factor out the GCF of the third and fourth terms. Last, we factor out any binomial GCF that's common to both of the remaining factored terms. For these problems only, make sure you don't combine like terms first.

Practice 5.7.2 Factoring by Grouping

Factor by grouping. Make sure not to add or subtract like terms first.

a) $x^2+2x+4x+8$

$\underbrace{x^2+2x}_{x(x+2)}+4x+8$	$\Rightarrow$	The first and second terms have a GCF of x. Factored out the GCF.
$x^2+2x+\underbrace{4x+8}$ $x(x+2)+4(x+2)$	$\Rightarrow$	The third and fourth terms have a GCF of 4, factored out the GCF. Noticed both remaining terms have a common factor of $(x+2)$ which means it's possible to continue with the procedure.
$x\boxed{(x+2)}+4\boxed{(x+2)}$ $\boxed{(x+2)}(x+4)$	$\Rightarrow$	Factored out the GCF of $(x+2)$ which left the terms x and 4 in the second set of parentheses.

b) $m^2-m+m-1$

$\underbrace{m^2-m}_{m(m-1)}+m-1$	$\Rightarrow$	Factoring an m from the first and second terms left $m-1$ inside the parentheses.
$m^2-m+m-1$ $m(m-1)+1(m-1)$	$\Rightarrow$	The only factors common to the third and fourth terms is either 1 or -1. Factoring out a 1 left a common binomial of $(m-1)$.
$(m-1)(m+1)$	$\Rightarrow$	Factored out the GCF of $(m-1)$.

c) $6a^2-8a-9a+12$

$\underbrace{6a^2-8a}_{2a(3a-4)}-9a+12$	$\Rightarrow$	Factored 2a from the first and second terms which left the terms 3a and 4.
$6a^2-8a-\underbrace{9a+12}$ $2a(3a-4)-3(3a-4)$	$\Rightarrow$	Factoring -3 from the third and fourth terms again left the binomial $(3a-4)$. Made sure to change the sign inside the parentheses. Factored out a -3 so the binomials inside the parentheses of both terms were a perfect match.
$(3a-4)(2a-3)$	$\Rightarrow$	Last factored out the GCF of $(3a-4)$.

Homework 5.7 Factor by grouping. Make sure not to add or subtract like terms first.

7) $t^2+t-t-1$	8) $x^2+2x+x+2$	9) $6a^2-3a-8a+4$
10) $w^2-3w-2w+6$	11) $4y^2+24y-3y-18$	12) $12v^2-18v-10v+15$
13) $12x^2+9x+16x+12$	14) $k^2-k-9k+9$	15) $14p^2-49p-4p+14$

5.7.3 Factoring With the AC Method

So far, we've factored trinomials using guess and check. The advantage of guess and check, is that it's easy to remember. You simply guess the binomials and then multiply to check your guess. The disadvantage is that it might take quite a while to make the right guess. To see what I mean, take a couple of minutes and try using guess and check to factor $12x^2 - 31x - 15$.

The AC method is another way to factor trinomials. The advantage is that, with practice, you can use the AC method to quickly factor trinomials like $12x^2 - 31x - 15$. The disadvantage is that you have to remember, and carry out correctly, a series of steps. Here's the procedure.

***Procedure* – The AC Method**
1. Starting with $ax^2 + bx + c$ find the product *ac* and identify the value of *b*. 2. Find a factor pair of *ac* whose sum is *b*. 3. Replace *bx* with the sum of two like terms. Use as coefficients the pair found in step 2. Make sure to include the variable factor. 4. Factor by grouping.

Practice 5.7.3 Factoring With the AC Method

Factor using the AC method.

a) $2x^2 + 9x + 4$

$ac = 2 \times 4 = 8 \quad b = 9$	$\Rightarrow$	Multiplied *a* and *c* to get 8, *b* is 9.
$8 = \boxed{1 \times 8}$ or -1×-8 2×4 or -2×-4	$\Rightarrow$	Found the factor pairs of 8 and then found a pair that adds to 9 (since *b* is 9).
$2x^2 \boxed{+9x} + 4$ $2x^2 + \overbrace{1x + 8x} + 4$	$\Rightarrow$	Rewrote *bx* as the sum of two like terms using as coefficients the pair found in the previous step. Ordering the terms $8x + 1x$ is also fine.
$2x^2 + x + 8x + 4$ $x(2x+1) + 4(2x+1)$ $(2x+1)(x+4)$	$\Rightarrow$	Factored by grouping. If you ordered the first-degree terms $8x + 1x$ you may have gotten $(x+4)(2x+1)$ which, by the commutative property, is the same product.

b) $-13k + k^2 + 12$

$k^2 - 13k + 12$	$\Rightarrow$	Wrote the trinomial in standard form.
$ac = 1 \times 12 = 12, \; b = -13,$	$\Rightarrow$	Multiplied *a* and *c* to get 12, *b* is -13.
$12 = 1 \times 12$ or $\boxed{-1 \times -12}$ 2×6 or -2×-6 3×4 or -3×-4	$\Rightarrow$	Found the factor pairs of 12 and found a pair that adds to *b*. (The first column wasn't necessary since adding two positive numbers gives a positive sum and *b* is negative.)
$k^2 \boxed{-13k} + 12$ $k^2 + \overbrace{-1k - 12k} + 12$	$\Rightarrow$	Rewrote *bk* as the sum of two like terms using as coefficients the pair found in the previous step.

$k^2 - k - 12k + 12$ $k(k-1) - 12(k-1)$ $(k-1)(k-12)$	$\Rightarrow$	Factored by grouping. If you ordered the first-degree terms $-12k - k$ you may have gotten $(k-12)(k-1)$ which is also fine.

c) $6t^2 - t - 2$

$ac = 6 \times -2 = -12 \qquad b = -1$	$\Rightarrow$	Multiplied *a* and *c* to get -12, *b* is -1.
$-12 = -1 \times 12$ or 1×-12 -2×6 or 2×-6 -3×4 or $\boxed{3 \times -4}$	$\Rightarrow$	Since AC is negative, one of the factors must be negative. Found a factor pair of -12 that adds to -1.
$6t^2 \boxed{-t} - 2$ $6t^2 + \overbrace{3t - 4t} - 2$	$\Rightarrow$	Rewrote *bx* as the sum of two like terms using as coefficients the pair found in the previous step. Ordering the terms $-4t + 3t$ is also fine.
$6t^2 + 3t - 4t - 2$ $3t(2t+1) - 2(2t+1)$ $(2t+1)(3t-2)$	$\Rightarrow$	Factored by grouping. If you had ordered the terms $-4t + 3t$ the answer would be $(3t-2)(2t+1)$ which, by the commutative property, is the same.

d) $4r^2 - 25$

$4r^2 + 0r + -25$ $ac = 4 \times -25 = -100, \ b = 0,$	$\Rightarrow$	Wrote the trinomial as a sum in standard form and identified the values of *ac* and *b*.
$-10 + 10$	$\Rightarrow$	"Saw" a factor pair of -100 that adds to 0.
$4r^2 \boxed{+0r} - 25$ $4r^2 + \overbrace{-10r + 10r} - 25$	$\Rightarrow$	Rewrote *bx* as the sum of two like terms using as coefficients the pair found in the previous step.
$4r^2 - 10r + 10r - 25$ $2r(2r-5) + 5(2r-5)$ $(2r-5)(2r+5)$	$\Rightarrow$	Factored by grouping. The answer $(2r+5)(2r-5)$ is also fine.

e) $2x^2 - 19x + 35$

$ac = 2 \times 35 = 70$ $b = -19$	$\Rightarrow$	The product *ac* is 70, and *b* is -19. Since *b* is negative and *ac* is positive the signs of *ac* must both be negative.
$70 = -1 \times -70, \ -2 \times -35,$ $\boxed{-5 \times -14}, \ -7 \times -10$	$\Rightarrow$	Found the factor pairs of 70 using a calculator. Another technique is to prime factor 70 to $2 \times 5 \times 7$ and "see" the factor pairs by choosing different sets of factors. Found the pair that adds to -19
$2x^2 \boxed{-19x} + 35$ $2x^2 \ \overbrace{-5x - 14x} + 35$	$\Rightarrow$	Rewrote *bx* as the sum of two like terms using as coefficients the pair found above. Replacing $-19x$ with $-14x - 5x$ is also fine.
$2x^2 - 5x - 14x + 35$ $x(2x-5) - 7(2x-5)$ $(2x-5)(x-7)$	$\Rightarrow$	Factored by grouping.

Homework 5.7 Factor With the AC method.

16) $h^2+15h+36$	17) $r^2-11r+28$	18) $-40+b^2-3b$
19) $36v^2-1$	20) $2t^2-15t+7$	21) $2w^2+15w+18$
22) $6q^2+7q-5$	23) $7y^2+23y+6$	24) $49p^2-14p+1$
25) $24x+7+9x^2$	26) $18z^2-21z-4$	27) $24x^2-49x+2$
28) $6x^2-7x-3$	29) $2y^2-17y+8$	30) $6t^2-17t+12$
31) $6k^2-23k+15$	32) $28p+15+12p^2$	33) $9h^2-16$
34) $16x^2-24x+9$	35) $15+6y^2-19y$	36) $3y^2+34y-24$
37) $5t^2+34t+45$	38) $9n^2-22n+8$	39) $49k^2-4$
40) $12b^2+59b-5$	41) $12x^2-31x-15$	42) $2k^2-25k+75$
43) $7x^2+32x+16$	44) $36b^2+12b+1$	45) $3x^2+22x-16$
46) $30x^2-19x+2$	47) $36x^2+28x+5$	48) $16m^2+40m+25$

Homework 5.7 Answers

1) $(2x+3)(x+4)$	2) $(n-2)(n+1)$	3) $(4x-7)(x-9)$	4) $(4y-1)(y-1)$
5) $(3r+2)(4r-3)$	6) $(5x-6)(6x+1)$	7) $(t+1)(t-1)$	8) $(x+2)(x+1)$
9) $(2a-1)(3a-4)$	10) $(w-3)(w-2)$	11) $(y+6)(4y-3)$	12) $(2v-3)(6v-5)$
13) $(4x+3)(3x+4)$	14) $(k-9)(k-1)$	15) $(7p-2)(2p-7)$	16) $(h+3)(h+12)$
17) $(r-4)(r-7)$	18) $(b+5)(b-8)$	19) $(6v+1)(6v-1)$	20) $(2t-1)(t-7)$
21) $(2w+3)(w+6)$	22) $(3q+5)(2q-1)$	23) $(y+3)(7y+2)$	24) $(7p-1)^2$
25) $(3x+1)(3x+7)$	26) $(3z-4)(6z+1)$	27) $(24x-1)(x-2)$	28) $(2x-3)(3x+1)$
29) $(2y-1)(y-8)$	30) $(2t-3)(3t-4)$	31) $(6k-5)(k-3)$	32) $(2p+3)(6p+5)$
33) $(3h-4)(3h+4)$	34) $(4x-3)^2$	35) $(2y-3)(3y-5)$	36) $(y+12)(3y-2)$
37) $(5t+9)(t+5)$	38) $(9n-4)(n-2)$	39) $(7k-2)(7k+2)$	40) $(12b-1)(b+5)$
41) $(x-3)(12x+5)$	42) $(2k-15)(k-5)$	43) $(7x+4)(x+4)$	44) $(6b+1)^2$
45) $(x+8)(3x-2)$	46) $(15x-2)(2x-1)$	47) $(18x+5)(2x+1)$	48) $(4m+5)^2$

5.8 Factoring Special Forms

In this section we'll practice factoring some polynomials that show up often. Although we could use the AC method to factor every polynomial in this section, by memorizing these "special" forms you'll be able to factor the polynomial faster.

5.8.1 Factoring the Difference of Two Squares

Earlier, you factored the binomial $p^2 - 9$ by thinking of it as the trinomial $p^2 + 0p - 9$ and using either guess and check or the AC method. It's often faster to factor this special form using the difference of two squares.

As we saw earlier $(m-n)(m+n)$ multiplies to $m^2 - n^2$. This implies that $m^2 - n^2$ factors to $(m-n)(m+n)$. To factor $x^2 - 9$ using the difference of two squares, think of each term as a perfect square $x^2 - 3^2$, and then since $m^2 - n^2 = (m-n)(m+n)$, it must be that $x^2 - 3^2 = (x-3)(x+3)$.

Sometimes it takes a bit of work to rewrite the binomial in a factorable form. For instance $16a^2 - 25$ is the difference of two squares if we use the power of a product rule to think of $16a^2$ as $(4a)^2$ and 25 as 5^2. Now, $4a$ would be our "m", 5 would be our "n" and $(4a)^2 - 5^2 = (4a-5)(4a+5)$. Here's some practice using the difference of two squares.

Practice 5.8.1 Factoring the Difference of Two Squares

Factor using the difference of two squares.

a) $c^2 - 36$		
$c^2 - 6^2$	$\Rightarrow$	Thought of c as the m and 6 as the n.
$(c-6)(c+6)$	$\Rightarrow$	$(c+6)(c-6)$ is also fine since multiplication is commutative.
b) $9w^2 - 1$		
$(3w)^2 - (1)^2$	$\Rightarrow$	Thought of $3w$ as the m and 1 as the n. Noticed $1 = 1^2$
$(3w-1)(3w+1)$	$\Rightarrow$	Followed the pattern for the difference of two squares.
c) $4p^2 - 49$		
$(2p)^2 - (7)^2$	$\Rightarrow$	Visualized both terms as perfect squares.
$(2p+7)(2p-7)$	$\Rightarrow$	Factored using the difference of two squares. Remember, since multiplication is commutative, $(2p-7)(2p+7)$ is also correct.

Homework 5.8 Factor using the difference of two squares.			
1) p^2-4	2) t^2-25	3) y^2-1	4) y^2-100
5) x^2-64	6) $4y^2-1$	7) $9x^2-4$	8) $16a^2-25$
9) $25s^2-36$	10) $16x^2-49$	11) $100x^2-81$	12) $64y^2-49$

5.8.2 Factoring a Perfect Square Trinomial

Earlier you learned that the square of a binomial led to a perfect square trinomial. That is, $(m+n)^2=m^2+2mn+n^2$ and $(m-n)^2=m^2-2mn+n^2$. Now we'll factor a perfect square trinomial by looking at the forms the other way around;

$$m^2+2mn+n^2=(m+n)(m+n)=(m+n)^2$$
$$m^2-2mn+n^2=(m-n)(m-n)=(m-n)^2$$

The key in both cases, is to look for trinomials where the first and last terms are perfect squares and the middle term is twice the product of the bases that are being squared.

Practice 5.8.2 Factoring a Perfect Square Trinomial

Factor using the form for the perfect square trinomial.

a) x^2+6x+9

$x^2+6x+3^2 \Rightarrow$ The first and last terms are perfect squares and the middle term is twice the product of the bases, $2(x)(3)=6x$. The trinomial has the proper form. This step is usually done mentally.

$(x+3)^2 \Rightarrow$ Followed the form with x as the "*m*" and 3 as the "*n*".

b) $4y^2-12y+9$

$(2y)^2-12y+3^2 \Rightarrow$ The first and last terms are perfect squares. Twice the product of the bases is $2(2y)(3)=12y$ which is the middle term. The polynomial has the proper form.

$(2y-3)^2 \Rightarrow$ Followed the form with $2y$ as the *m* and 3 as the *n*.

c) $9x^2+42x+49$

$(3x)^2+42x+7^2 \Rightarrow$ The polynomial has the proper form since the first and last terms are perfect squares and the middle term is twice the product of the bases, $2(3x)(7)=42x$.

$(3x+7)^2 \Rightarrow$ Followed the form with $3x$ as the *m* and 7 as the *n*.

Homework 5.8 Factor using the form for the perfect square trinomial.

13) x^2+2x+1	14) x^2-4x+4	15) $y^2+8y+16$
16) $w^2-16w+64$	17) $s^2+10st+25t^2$	18) $4r^2+4r+1$
19) $4k^2-12k+9$	20) $9a^2-12a+4$	21) $100x^2+20x+1$
22) $4s^2+20s+25$	23) $64h^2-48h+9$	24) $36a^2-84a+49$
25) $49x^2-70x+25$	26) $64k^2+112k+49$	

Homework 5.8 Answers

1) $(p+2)(p-2)$	2) $(t+5)(t-5)$	3) $(y-1)(y+1)$	4) $(y-10)(y+10)$
5) $(x-8)(x+8)$	6) $(2y+1)(2y-1)$	7) $(3x-2)(3x+2)$	8) $(4a+5)(4a-5)$
9) $(5s-6)(5s+6)$	10) $(4x-7)(4x+7)$	11) $(10x-9)(10x+9)$	12) $(8y+7)(8y-7)$
13) $(x+1)^2$	14) $(x-2)^2$	15) $(y+4)^2$	16) $(w-8)^2$
17) $(s+5t)^2$	18) $(2r+1)^2$	19) $(2k-3)^2$	20) $(3a-2)^2$
21) $(10x+1)^2$	22) $(2s+5)^2$	23) $(8h-3)^2$	24) $(6a-7)^2$
25) $(7x-5)^2$	26) $(8k+7)^2$		

5.9 Factoring Out the GCF First

In this section you'll first factor out the GCF and then continue with one of your other factoring techniques.

5.9.1 Factoring Out a Numerical GCF

Factoring out the numerical GCF first makes the remaining polynomial easier to factor.

Practice 5.9.1 Factoring the GCF First

Factor completely. Make sure to factor out the GCF first.

a) $2x^2+8x+6$

$2(x^2+4x+3)$	$\Rightarrow$	Factored out the GCF of 2.
$2((x+1)(x+3))$	$\Rightarrow$	Factored x^2+4x+3 using guess and check.
$2(x+1)(x+3)$	$\Rightarrow$	The outside set of parentheses are not needed.
$(2x+2)(x+3)$ $2x^2+6x+2x+6$ $2x^2+8x+6$	$\Rightarrow$	To check my work, I multiplied left to right. Since multiplication is commutative, I could have also "foiled" $(x+1)(x+3)$ and then multiplied by 2.

b) $12x^2-3$

$3(4x^2-1)$	$\Rightarrow$	Factored out the GCF of 3.
$3(2x-1)(2x+1)$	$\Rightarrow$	Factored the binomial using the difference of two squares.

c) $-t^2+6t-9$

$-1(t^2-6t+9)$	$\Rightarrow$	If *a* is negative it's best to factor out a factor of -1. Made sure the change the signs inside the parentheses correctly.
$-1(t-3)(t-3)$ $-1(t-3)^2$	$\Rightarrow$	Factored the trinomial using the form for a perfect square binomial. Guess and check would have also worked well.

d) $-4k^2+26k-42$

$-2(2k^2-13k+21)$	$\Rightarrow$	Factored out the GCF of -2.
$-2(2k^2-6k-7k+21)$ $-2((k-3)(2k-7))$	$\Rightarrow$	Factored the trinomial using the AC method.
$-2(k-3)(2k-7)$	$\Rightarrow$	The outside set of parentheses wasn't necessary.

Homework 5.9 Factor completely. Make sure to factor out the GCF first.

1) $2y^2-12y+10$	2) $4n^2+8n+4$	3) $-t^2-10t+24$
4) $3k^2-27$	5) $-9p^2-12p-4$	6) $-k^2+100$
7) $-4x^2-38x-70$	8) $4x^2-64$	9) $15y^2-40y+25$
10) $-36y^2+64$	11) $28y^2+84y+63$	12) $-24w^2-58w-20$

5.9.2 Factoring Out a Variable GCF

Sometimes a variable might be included in the GCF.

Practice 5.9.2 Factoring Out a Variable GCF

Factor completely. Make sure to factor out the GCF first.

a) $x^3 - 8x^2 + 7x$

$x\left(x^2 - 8x + 7\right)$	$\Rightarrow$	Factored out the GCF of x.
$x((x-7)(x-1))$ $x(x-7)(x-1)$	$\Rightarrow$	Factored the trinomial using guess and check and removed the nonessential set of parentheses.

b) $6y^3 + 3y^2 - 45y$

$3y\left(2y^2 + y - 15\right)$	$\Rightarrow$	Factored out the GCF of $3y$.
$3y(2y-5)(y+3)$	$\Rightarrow$	Factored the remaining trinomial and removed the nonessential parentheses.

c) $-5t^4 + 20t^2$

$-5t^2\left(t^2 - 4\right)$	$\Rightarrow$	Factored out the GCF of $-5t^2$.
$-5t^2(t-2)(t+2)$	$\Rightarrow$	Factored the binomial using the difference of two squares.

Homework 5.9 Factor completely. Make sure to factor out the GCF first.

13) $h^3 + 4h^2 + 4h$	14) $x^4 + 2x^3 - 3x^2$	15) $2x^3 + 10x^2 + 12x$
16) $3k^3 - 21k^2 - 30k$	17) $k^3 - 9k$	18) $-t^3 + 9t^2 - 20t$
19) $-2n^3 + 12n^2 - 18n$	20) $48y^3 - 3y$	21) $-18h^3 + 50h$
22) $2w^3 + 4w^2 - 70w$	23) $5y^3 - 30y^2 + 25y$	24) $-12w^3 - 68w^2 - 80w$

Homework 5.9 Answers

1) $2(y-1)(y-5)$	2) $4(n+1)^2$	3) $-(t+12)(t-2)$	4) $3(k-3)(k+3)$
5) $-(3p+2)^2$	6) $-(k-10)(k+10)$	7) $-2(2x+5)(x+7)$	8) $4(x-4)(x+4)$
9) $5(y-1)(3y-5)$	10) $-4(3y-4)(3y+4)$	11) $7(2y+3)^2$	12) $-2(12w+5)(w+2)$
13) $h(h+2)^2$	14) $x^2(x+3)(x-1)$	15) $2x(x+3)(x+2)$	16) $3k(k-7)(k-10)$
17) $k(k-3)(k+3)$	18) $-t(t-5)(t-4)$	19) $-2n(n-3)^2$	20) $3y(4y-1)(4y+1)$
21) $-2h(3h-5)(3h+5)$	22) $2w(w+7)(w-5)$	23) $5y(y-5)(y-1)$	
24) $-4w(3w+5)(w+4)$			

5.10 The Zero-Product Method

The zero-product method gives us a quick way to solve a quadratic equation like $x^2 + 5x + 6 = 0$ where the left side factors easily.

5.10.1 The Idea Behind the Zero-Product Method

When my daughter was young she enjoyed asking adults a question like this, "What's seven, times six, times four, times zero, times eleven, times sixteen?" When they said they didn't know, she'd reply laughingly, "It's zero because I multiplied by zero." This is the idea behind the zero-product method.

The Idea Behind the Zero-Product Method
If a product is zero, at least one factor is zero.

Let's use this idea to solve the equation $x^2 + 5x + 6 = 0$.

Notice first that the zero-product method has to do with products. Unfortunately, $x^2 + 5x + 6 = 0$ is a sum of terms. Fortunately, the polynomial on the left side can be replaced with the product $(x+2)(x+3) = 0$. For the product on the left to be 0, either the first factor $(x+2)$ has to be 0, the second factor $(x+3)$ has to be 0 or both factors have to be 0.

For the first factor, $(x+2)$, to be 0, we need to replace x with -2. Now the first factor simplifies to 0, the second factor to 1 and the product, $(0)(1)$ to 0. This means -2 is one solution to the original equation.

$$\begin{gathered}(x+2)(x+3)\\(-2+2)(-2+3)\\(0)(1)\\0\end{gathered}$$

For the second factor, $(x+3)$, to be 0, we need to replace x with -3. Now the first factor simplifies to -1, the second factor to 0 and the product $(-1)(0)$ to 0. This means -3 is a second solution. The solution set for $(x+2)(x+3) = 0$ will be $\{-3, -2\}$ which implies the solution set for $x^2 + 5x + 6 = 0$ must also be $\{-3, -2\}$. Before we practice with the full zero-product method let's make sure you're able to find the solution(s) once the polynomial is factored.

$$\begin{gathered}(x+2)(x+3)\\(-3+2)(-3+3)\\(-1)(0)\\0\end{gathered}$$

Practice 5.10.1 An Introduction to the Zero-Product Method

Solve using the zero-product method.

a) $(y-1)(y+2) = 0$

$y - 1 = 0$ $y = 1$	$\Rightarrow$	If the first factor, $y-1$, becomes 0, the product on the left will be 0. Solved for y to find the first solution is 1.
$y + 2 = 0$ $y = -2$	$\Rightarrow$	If the second factor, $y+2$, becomes 0, the product on the left will again be 0. Solved for y to find another solution is -2.
$\{-2, 1\}$	$\Rightarrow$	The solution set should contain both solutions.

b) $x(x-3)=0$

$x=0$	$\Rightarrow$	If the first factor, *x*, becomes 0 then the left product would be 0. At first, this solution is easy to miss.
$x-3=0$ $x=3$	$\Rightarrow$	When the value of *x* is 3 the second factor becomes 0.
$\{0,3\}$	$\Rightarrow$	There are two solutions.

c) $-2k(k+5)(k+1)=0$

$k=0, k=-5, k=-1$	$\Rightarrow$	Only the factors *k*, $k+5$ and $k+1$ can take on the value 0. -2 can't be a zero, its value is -2.
$\{-5,-1,0\}$	$\Rightarrow$	There are three solutions.

Homework 5.10 Solve using the zero-product method.

1) $(y-3)(y-1)=0$	2) $k(k+4)=0$	3) $9a(a-1)=0$
4) $0=-15(h+9)(h-9)$	5) $0=-k(k-2)(k+7)$	6) $7y(y+11)(y+1)=0$

5.10.2 Working With Complicated Factors

With a little practice, students can often look at $(x+4)(x-2)=0$ and "see" the solutions would be -4 and 2. Things get a little harder if the factors have more than one operation. For instance, it's difficult at first to see what the solutions to $(3x-5)(2x+3)=0$ would be.

To see what value(s) will make a factor equal 0, we can always set the factor equal to 0 and solve. Here are some examples.

Practice 5.10.2 Working With Complicated Factors

Solve using the zero-product method.

a) $(3x-5)(2x+3)=0$

$3x-5=0$ $x=5/3$ or	$2x+3=0$ $x=-3/2$	$\Rightarrow$ Set each factor equal to 0 and solved.
$\{-3/2, 5/3\}$	$\Rightarrow$	The solution set contains both values.

Homework 5.10 Solve using the zero-product method.

7) $(3t+4)(3t-4)=0$	8) $(2f+5)(2f+1)=0$	9) $0=-3w(9-6w)(5w-15)$
10) $(4x-3)(4x+3)(2x-1)=0$	11) $7(3k-75)(-3k+36)=0$	
12) $0=2h(6h-8)(6h+8)$		

5.10.3 Factoring to Use the Zero-Product Method

The zero-product method is helpful as long as the polynomial is a product. Unfortunately, the polynomial often comes to us as a sum. Your job will be to replace the polynomial sum with an equivalent product and then use then solve the equation.

Procedure **– Solving Equations Using the Zero-product Method**
1. Write the equation as a polynomial in standard form set equal to 0. Try to keep the value of the highest degree coefficient positive. 2. Factor the polynomial. 3. Set each variable factor equal to 0 and solve. 4. Check your solutions.

We'll begin with steps 2 through 4.

Practice 5.10.3 Factoring to Use the Zero-Product Method

Solve using the zero-product method.

a) $a^2 - 14a + 48 = 0$

$(a-6)(a-8) = 0$	$\Rightarrow$	Factored the trinomial.
$\{6,8\}$	$\Rightarrow$	Was able to "see" the two solutions were 6 and 8.

b) $0 = k^3 + 6k^2 + 8k$

$0 = k\left(k^2 + 6k + 8\right)$ $0 = k(k+2)(k+4)$	$\Rightarrow$	Factored out the GCF and then factored the trinomial using guess and check.
$\{-4, -2, 0\}$	$\Rightarrow$	Found the three solutions.

c) $-16x^2 + 9 = 0$

$-1\left(16x^2 - 9\right) = 0$ $-1(4x-3)(4x+3) = 0$	$\Rightarrow$	Factored out the GCF of -1 and then used the difference of two squares.
$4x - 3 = 0$, $4x + 3 = 0$ $x = 3/4$, $x = -3/4$	$\Rightarrow$	Set each factor equal to 0 and solved.
$\{-3/4, 3/4\}$	$\Rightarrow$	The solution set.

Homework 5.10 Solve using the zero-product method.

13) $t^2 - 11t + 18 = 0$	14) $x^2 - x = 0$	15) $0 = x^2 - 7x + 12$
16) $0 = 16k^3 - 8k^2 + k$	17) $2x^3 - 18x = 0$	18) $14y^2 + 23y + 3 = 0$
19) $6h^2 - 17h + 12 = 0$	20) $0 = -2w^3 + 15w^2 + 50w$	

5.10.4 The General Procedure for the Zero-Product Method

Now let's practice the entire general procedure.

Practice 5.10.4 The General Procedure for the Zero-Product Method

Solve using the zero-product method.

a) $x^2 = -48 + 16x$

$x^2 - 16x + 48 = 0$	$\Rightarrow$	Started with step 1 and wrote the polynomial in standard form set equal to 0. I kept x^2 on the left so the value of a would be positive.
$(x-4)(a-12) = 0$	$\Rightarrow$	Factored the trinomial.
$\{4,12\}$	$\Rightarrow$	There are two solutions.

b) $4(f+2)(f+1) = 3$

$4f^2 + 12f + 8 = 3$ $4f^2 + 12f + 5 = 0$	$\Rightarrow$	Began with step 1, distributed on the left and then subtracted 3 from both sides so the polynomial was set equal to 0.
$(2f+5)(2f+1) = 0$	$\Rightarrow$	Factored the trinomial.
$f = -5/2$ or $f = -1/2$	$\Rightarrow$	Set each factor equal to 0 and solved. The solution set is $\left\{-5/2, -1/2\right\}$

c) $x^3 = x$

$x^3 - x = 0$	$\Rightarrow$	Wrote the terms in standard form set equal to 0.
$x(x^2 - 1) = 0$ $x(x-1)(x+1) = 0$	$\Rightarrow$	Factored out the GCF and then used the difference of two squares.
$\{-1, 0, 1\}$	$\Rightarrow$	The solution set.

Homework 5.10 Solve using the zero-product method.

21) $x^2 + 10x = 11$ 22) $9y^2 = 16$ 23) $x^2 + 9 = -6x$ 24) $2w^3 = 32w$

25) $4h^2 = -20h - 25$ 26) $18b - 28 = 2b^2$ 27) $t(t+3) = 12 + 4t$

28) $p^3 + 2p = 3p^2$ 29) $14t^2 = -t - 49t^3$ 30) $(2a+3)(a+4) = -3$

31) $(3x-2)^2 = x$ 32) $6x^2 + 4x + 7 = 2x^2 + 6$ 33) $(2y+3)^2 = (3y+2)^2$

34) $y^3 = -3y^2 - 2y$ 35) $16n^3 = 25n$

Homework 5.10 Answers

1) $\{1,3\}$ 2) $\{-4,0\}$ 3) $\{0,1\}$ 4) $\{-9,9\}$ 5) $\{-7,0,2\}$ 6) $\{-11,-1,0\}$

7) $\{-\frac{4}{3}, \frac{4}{3}\}$ 8) $\{-\frac{5}{2}, -\frac{1}{2}\}$ 9) $\{0, \frac{3}{2}, 3\}$ 10) $\{-\frac{3}{4}, \frac{1}{2}, \frac{3}{4}\}$

11) $\{25, 12\}$ 12) $\{-\frac{4}{3}, 0, \frac{4}{3}\}$ 13) $\{2,9\}$ 14) $\{0,1\}$ 15) $\{3,4\}$

16) $\{0, 1/4\}$ 17) $\{-3,0,3\}$ 18) $\{-3/2, -1/7\}$ 19) $\{\frac{4}{3}, \frac{3}{2}\}$ 20) $\{-\frac{5}{2}, 0, 10\}$

21) $\{-11,1\}$ 22) $\{-\frac{4}{3}, \frac{4}{3}\}$ 23) $\{-3\}$ 24) $\{-4,0,4\}$ 25) $\{-\frac{5}{2}\}$

26) $\{2,7\}$ 27) $\{-3,4\}$ 28) $\{0,1,2\}$ 29) $\{-1/7, 0\}$ 30) $\{-3, -\frac{5}{2}\}$

31) $\{\frac{4}{9}, 1\}$ 32) $\{-\frac{1}{2}\}$ 33) $\{-1, 1\}$ 34) $\{-2,-1,0\}$ 35) $\{-5/4, 0, 5/4\}$

5.11 Factoring Multivariate Polynomials

In this section, we'll practice factoring multivariate polynomials.

5.11.1 Factoring Multivariate Polynomials

A multivariate polynomial has more than one kind of letter. For example, $m^2 - n^2$ is a multivariate polynomial since it has the letters *m* and *n*. Even though $m^2 - n^2$ has two letters, it's still the difference of two squares and can be factored to $(m-n)(m+n)$. Here's some practice using our previous techniques to factor multivariate polynomials.

Practice 5.11.1 Factoring Multivariate Polynomials

Factor completely.

a) $ab + a + 2b + 2$

$ab + a + 2b + 2$	$\Rightarrow$	Since there are four terms grouping is the right approach.
$a(b+1) + 2(b+1)$ $(b+1)(a+2)$	$\Rightarrow$	Factored by grouping.

b) $r^2 - 9rs + 14s^2$

$r^2 - 9rs + 14s^2$	$\Rightarrow$	It's a trinomial and $a = 1$. I'll start with guess and check and then switch to AC if necessary.
$(r- \quad)(r- \quad)$	$\Rightarrow$	Filled in the first terms of the binomials.
$(r-7s)(r-2s)$	$\Rightarrow$	Since $-2rs - 7rs = -9rs$ and $(-7s)(-2s) = 14s^2$ made these my "last" terms.
$r^2 - 2rs - 7rs + 14s^2$ $r^2 - 9rs + 14s^2$	$\Rightarrow$	Multiplied to make sure my factored form was the same polynomial that I started with.

c) $10x^2 + 11xy - 6y^2$

$15 \times -4 = -60,\ 15 + -4 = 11$	$\Rightarrow$	Decided to use the AC method and found the factor pair.
$10x^2 + 15xy - 4xy - 6y^2$ $5x(2x+3y) - 2y(2x+3y)$ $(2x+3y)(5x-2y)$	$\Rightarrow$	Factored using the AC method. Made sure my like terms were in the form *xy*.

d) $8p^3 - 50q^2p$

$2p\left(4p^2 - 25q^2\right)$	$\Rightarrow$	Factored out the GCF.
$2p\left((2p)^2 - (7q)^2\right)$	$\Rightarrow$	Noticed the binomial was the difference of two squares.
$2p(2p-7q)(2p+7q)$	$\Rightarrow$	Factored using the difference of two squares.

Homework 5.11 Factor completely.

1) $k^2 - 4kt + 3t^2$ 2) $x^2 - y^2$ 3) $6x^2 + 9xy + 3y^2$ 4) $a^2 + 6ab - 4a - 24b$

5) $16a^2 - b^2$ 6) $4ab - 8a - b + 2$ 7) $81h^2 - 16k^2$ 8) $x^2 - 2xy - 15y^2$

9) $s^2 + 10st + 25t^2$ 10) $-x^2 + xy - 4x + 4y$ 11) $9u^2 - 49v^2$ 12) $4s^2 - 20st + 25t^2$

13) $50w^2 - 2v^2$ 14) $6rt + 18r + 5t + 15$ 15) $12p^2 + 20pq + 7q^2$

16) $36y^3 - 4k^2y$ 17) $36t^3 - 12st^2 + s^2t$ 18) $200x^2 - 40xy + 2y^2$

19) $-32u^2 + 18v^2$ 20) $9n^2 - 43mn + 28m^2$ 21) $-9xy + 6x + 45y^2 - 30y$

22) $10xy^2 - 40xy + 2y^3 - 8y^2$ 23) $-k^2 - 16hk - 64h^2$ 24) $49a^3b - 100ab^3$

Homework 5.11 Answers

1) $(k-t)(k-3t)$ 2) $(x+y)(x-y)$ 3) $3(2x+y)(x+y)$ 4) $(a+6b)(a-4)$

5) $(4a-b)(4a+b)$ 6) $(4a-1)(b-2)$ 7) $(9h-4k)(9h+4k)$ 8) $(x+3y)(x-5y)$

9) $(s+5t)^2$ 10) $-(x-y)(x+4)$ 11) $(3u-7v)(3u+7v)$ 12) $(2s-5t)^2$

13) $2(5w-v)(5w+v)$ 14) $(6r+5)(t+3)$ 15) $(2p+q)(6p+7q)$

16) $4y(3y-k)(3y+k)$ 17) $t(s-6t)^2$ 18) $2(10x-y)^2$ 19) $-2(4u-3v)(4u+3v)$

20) $(7m-9n)(4m-n)$ 21) $(3x-15y)(2-3y)$ 22) $2y(y-4)(5x+y)$

23) $-(8h+k)^2$ 24) $ab(7a-10b)(7a+10b)$

5.12 Polynomial Prime Factorization

In chapter one we discussed the difference between factorization and prime factorization. As you'll recall both 6×2 and $2\times3\times2$ were factorizations of 12 but only $2\times3\times2$ was the prime factorization of 12 since all the factors were prime. Polynomials can also have a prime factorization. For instance, although both $(6x+3)(x+4)$ and $3(2x+1)(x+4)$ are factorizations of $6x^2+27x+12$, only $3(2x+1)(x+4)$ is the prime factorization since the common factor of 3 has been factored out. Here's the procedure we'll use to find polynomial prime factorizations.

Procedure **– Prime Factoring a Polynomial**
1. Factor any G.C.F. 2. Choose the appropriate procedure. a) Binomial – Use the difference of two squares. b) Trinomial – If, after a few tries, the guess and check method hasn't worked, shift to the AC method. c) Four terms – Factor by grouping. 3. Check each polynomial factor and return to step 1 with any factor that isn't prime.

5.12.1 Practicing Step 3 of the Procedure

If someone told you the prime factorization of 12 is 2×6 you'd look at each factor individually and know the answer is wrong because 6 is composite and can be factored further. It's the same for polynomials. Step 3 helps you remember to check each factor of your factored form and decide whether the factor is prime or composite. For example, if you factored $6x^2+27x+12$ to $(6x+3)(x+4)$ you'd need to recognize that $(x+4)$ is prime but $(6x+3)$ is composite since both terms share a common factor of 3. Only after factoring out the 3, $3(2x+1)$ will you have the prime factorization $3(2x+1)(x+4)$. For this next set of problems, we'll assume you've completed steps 1 and 2 and moved on to step 3 to decide if it's possible to factor further.

Practice 5.12.1 Practicing Step 3 of the Procedure

If it's necessary, prime factor the product.

a) $(3x-2)(4x-12)$

$(3x-2)4(x-3)$	$\Rightarrow$	The first factor is prime but the second factor had a GCF of 4 which needs to be factored out. It's true both terms also have a factor of 2 but 4 is the GCF of the two terms.
$4(3x-2)(x-3)$	$\Rightarrow$	Usually a constant factor will be written to the left. All the polynomial factors are now prime. It's not necessary to prime factor the 4

b) $(5a+3)(4a-5)$

$(5a+3)(4a-5) \Rightarrow$ No further factoring is necessary. This is a prime factorization.

c) $(a+8)(9a^2-1)$

$(a+8)(3a-1)(3a+1)$	$\Rightarrow$	The second factor was the difference of two squares and needed to be factored further.

d) $(5x-10)(2x+6)$

$(5x-10)(2x+6)$	$\Rightarrow$	The first factor has a GCF of 5 while the second factor has a GCF of 2.
$5(x-2)2(x+3)$ $5\times 2(x-2)(x+3)$ $10(x-2)(x+3)$	$\Rightarrow$	Factored out and multiplied the GCFs.

Homework 5.12 If it's necessary, prime factor the product.

1) $(9y-18)(y+3)$ 2) $(k-7)(k^2+9)$ 3) $-3(x-1)(6x-3)$ 4) $(2b-6)(5b+5)$

5) $(4m^2-9)(m+4)$ 6) $(8v-12)(4v-6)$ 7) $(14x-7)(36a^2-b^2)$

8) $(9m^2-5)(2m+7)$ 9) $(m^2-16)(n^2-4)$

5.12.2 Prime Factoring a Polynomial

Now let's practice the entire procedure.

Practice 5.12.2 Prime Factoring a Polynomial

Factor completely.

a) x^3+x^2-4x-4

x^3+x^2-4x-4	$\Rightarrow$	Noticed that no factor is common to <u>all</u> the terms.
$x^2(x+1)-4(x+1)$ $(x+1)(x^2-4)$	$\Rightarrow$	Grouping is the right approach for a four-term polynomial. After factoring noticed the second factor was the difference of two squares.
$(x+1)(x-2)(x+2)$	$\Rightarrow$	All factors are now prime so the problem is finished.

b) $7k^2-3k+4$

$7k^2-3k+4$	$\Rightarrow$	There is no GCF and can't see a quick way to factor the trinomial so went to the AC method.
Prime	$\Rightarrow$	No factor pair of 28 added to -3 (-7 and 4 don't multiply to positive 28). The polynomial is already prime.

c) $-8b^2+56b-98$

$-2(4b^2-28b+49)$	$\Rightarrow$	Factored out the GCF. (Included a factor of -1.)
$4b^2-28b+49$ $(2b)^2-2(2b)(7)+7^2$	$\Rightarrow$	Noticed the trinomial had the form for a perfect square trinomial.
$-2(2b-7)^2$	$\Rightarrow$	Factored completely.

d) $9y^3 - z^2y$		
$y(9y^2 - z^2)$	$\Rightarrow$	Factored out the GCF of y. Noticed the remaining binomial is the difference of two squares.
$y(3y - z)(3y + z)$	$\Rightarrow$	Factored completely.

Homework 5.12 Factor completely.

10) $y^3 + 3y^2 - 9y - 27$	11) $6x^2 + 21x + 9$	12) $8p^3 - 16p^2 + 3p - 3$
13) $-8t^2 + 12t + 8$	14) $y^3 + y^2 - 4y - 4$	15) $1 - 64y^2$
16) $-6k^2 + 42k - 72$	17) $4h^2 - 12h + 6$	18) $4k^3 - 16k^2 - 9k + 36$
19) $-3y^3 - 42y^2 - 147y$	20) $9t^3 + 36t$	21) $3a^2 + 16a + 12$
22) $3x^3 - 7x^2 - 27x + 63$	23) $42r^4 + 39r^3 - 36r^2$	24) $4x^4 + 12x^3 + 9x^2$
25) $m^3 + m^2 - m - 1$	26) $18t^2 - 24t + 8$	27) $b^3 - a^2b$
28) $-9w^3 - 54w^2 - 45w$	29) $p^4 - 16$	30) $4x^3 - x^2 - 4xy^2 + y^2$
31) $-6m^3 + 25m^2 - 4m$	32) $t^3 - t^2 - 9t - 9$	33) $-25p^2 - 30p - 9$
34) $a^4 - a^2b^2 - 9a^2 + 9b^2$	35) $8y^4 + 12y^3 - 18y^2 - 27y$	36) $36y^3 + 33y^2 - 45y$

Homework 5.12 Answers

1) $9(y-2)(y+3)$ 2) Already prime factored. There is no sum of two squares form.

3) $-9(x-1)(2x-1)$ 4) $10(b-3)(b+1)$ 5) $(2m-3)(2m+3)(m+4)$ 6) $8(2v-3)^2$

7) $7(2x-1)(6a-b)(6a+b)$ 8) Prime 9) $(m-4)(m+4)(n-2)(n+2)$

10) $(y-3)(y+3)^2$ 11) $3(x+3)(2x+1)$ 12) Prime 13) $-4(2t+1)(t-2)$

14) $(y+1)(y-2)(y+2)$ 15) $-(8y-1)(8y+1)$ 16) $-6(k-3)(k-4)$

17) $2(2h^2-6h+3)$ 18) $(k-4)(2k-3)(2k+3)$ 19) $-3y(y+7)^2$ 20) $9t(t^2+4)$

21) Prime 22) $(3x-7)(x-3)(x+3)$ 23) $3r^2(2r+3)(7r-4)$ 24) $x^2(2x+3)^2$

25) $(m-1)(m+1)^2$ 26) $2(3t-2)^2$ 27) $-b(a-b)(a+b)$ 28) $-9w(w+5)(w+1)$

29) $(p-2)(p+2)(p^2+4)$ 30) $(4x-1)(x-y)(x+y)$ 31) $-m(6m-1)(m-4)$

32) Prime 33) $-(5p+3)^2$ 34) $(a-3)(a+3)(a-b)(a+b)$

35) $y(2y-3)(2y+3)^2$ 36) $3y(4y-3)(3y+5)$

Chapter 6

Roots

6.1 An Introduction to Roots

Earlier, you used $\text{Base}^{\text{exponent}} = \text{Power}$, to simplify a numerical base and exponent to a power. For example, you simplified a base of 3 with an exponent of 2 to the power 9, $3^2 = 9$. In this chapter, we'll look at the relationship $\text{Base}^{\text{exponent}} = \text{Power}$ from a different point of view. You'll be given the exponent and power and you'll be asked to find the base. This is known as finding a root. Before we begin with roots though, it's important to mention irrational and real numbers.

6.1.1 Irrational Numbers and The Real Numbers

Recall that a rational number can be written as the quotient of two integers (as long as the denominator isn't 0). Usually, we've been able to simplify all our values to a rational number. This often won't be the case with roots. Even something as simple as $\sqrt{2}$ can't be written as a fraction. Now you might think a value like $\sqrt{2}$ would only show up under very unusual circumstance but even for something as common as a 1-foot by 1-foot piece of floor tile, if you could exactly measure the distance of the diagonal, that distance would be $\sqrt{2}$ feet.

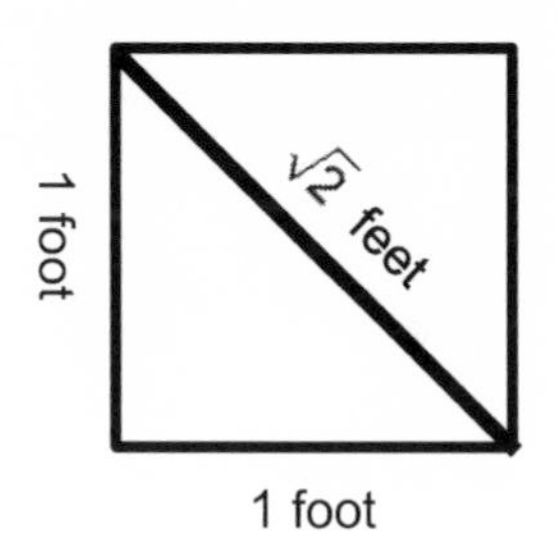

We call numbers like $\sqrt{2}$ (numbers whose value can't be written as the ratio of two integers) "irrational" numbers. Defining the **irrational numbers** is one of those cart and horse situations. In order to define the irrationals, I first have to define the **real numbers**. It's not possible, at this level, to actually define the real numbers. In fact, historically, some of the best mathematicians of their age wrestled with the issue of defining the real numbers for a long time. For now, we'll use this as our definition for the real numbers.

***Definition* – Real Numbers**
The set of numbers which fills in every point on the (real) number line.

Now we can define the irrational numbers.

***Definition* – Irrational Numbers**
The set of real numbers that aren't rational numbers.

Here's a picture that helps show how our numbers sets are related to each other.

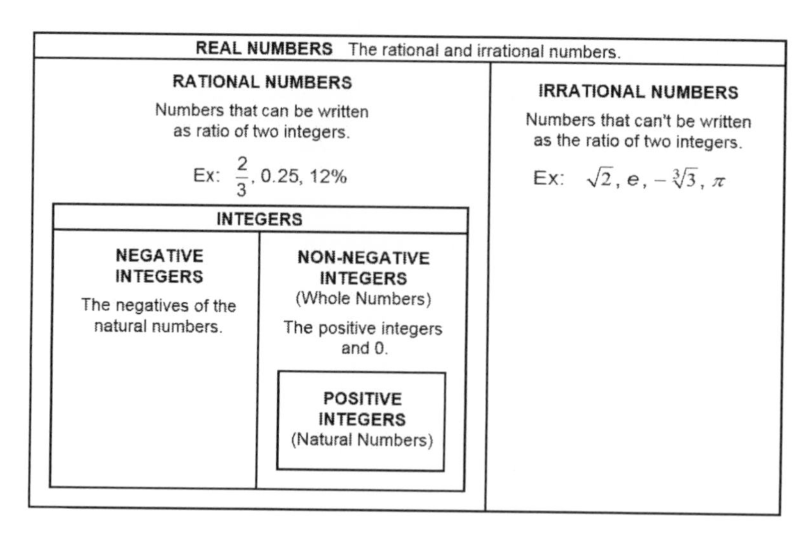

6.1.2 Some Vocabulary for Roots

A radical expression uses a radical symbol, an index and a radicand to give the information needed to find the root. The radical symbol, $\sqrt{\ }$, asks you to find a base. The index, which is written as a superscript on the left side of the radical symbol, $\sqrt[index]{\ }$ gives the original exponent and the radicand, which is the expression under the radical symbol, $\sqrt[index]{radicand}$ gives the original power. When you simplify a radical expression, you are finding a **root**, $\sqrt[index]{radicand} = \text{Root}$. For example, to simplify the radical expression $\sqrt[3]{8}$ you need to find the base, which when cubed, gives eight. Since two cubed is eight $(2^3 = 8)$, $\sqrt[3]{8}$ simplifies to 2.

When the radical symbol doesn't have a visible index $\sqrt{\ }$, it has an index of two, $\sqrt[2]{\ }$, and is usually called a square root. An index of three, $\sqrt[3]{\ }$, is often called a cube root. We don't use special names for a natural number index greater than 3 we just say "fourth" root, "fifth" root etc. Let's practice with some of the vocabulary we'll need for discussing roots.

Practice 6.1.2 Some Vocabulary for Roots

Fill in each blank with term, sum, factor, product, minuend, subtrahend, difference, dividend, divisor, quotient, base, exponent, power, index, radicand or root.

---

a) Before simplifying $\dfrac{4\sqrt{15}}{\sqrt[3]{64}}$,

4 is a(n)_____, $\sqrt{15}$ is a(n)____ and a(n)______, $4\sqrt{15}$ is a(n)____ and the ____, 3 is a(n) ____, 64 is a(n) _____, $\sqrt[3]{64}$ is a(n) ____ and the_____ and $\dfrac{4\sqrt{15}}{\sqrt[3]{64}}$ is a(n)____.

factor, factor, root, product, dividend, index, radicand, root, divisor, quotient

---

Homework 6.1 Fill in each blank with term, sum, factor, product, minuend, subtrahend, difference, dividend, divisor, quotient, base, exponent, power, index, radicand or root.

1) Before simplifying $(\sqrt{9}+\sqrt{36})^2$

9 is a(n) _____, $\sqrt{9}$ is a(n) _____ and a(n)_____, 36 is a(n) _____, $\sqrt{36}$ is a(n) _____ and a(n)_____, $\sqrt{9}+\sqrt{36}$ is a(n) _____, $(\sqrt{9}+\sqrt{36})$ is a(n) _____, 2 is a(n) _____, and $(\sqrt{9}+\sqrt{36})^2$ is a(n) _____.

2) Before simplifying $3+\sqrt{(-3)^2-4(-1)(5)}$,

(-3) is a(n)_____, 2 is a(n)_____, $(-3)^2$ is both a(n)_____ and the_____, 4 is a(n)____, (-1) is a(n)____, (5) is a(n)____, $4(-1)(5)$ is both a(n)_____ and the_____, $(-3)^2-4(-1)(5)$ is both a(n)_____ and the_____, 3 is a(n)_____, $\sqrt{(-3)^2-4(-1)(5)}$ is both a(n)_____ and a(n)_____, and $3+\sqrt{(-3)^2-4(-1)(5)}$ is a(n)_____.

3) Before simplifying $\dfrac{15\sqrt{5}}{\sqrt{25x^2}}$,

15 is a(n) _____, 5 is the _____, $\sqrt{5}$ is a(n) _____ and a(n) _____, $15\sqrt{5}$ is both a(n) _____ and the _____, 25 is a(n)____, x is a(n)____, 2 is a(n)_____, x^2 is both a(n) _____ and a(n) _____, $25x^2$ is both a(n) _____ and the _____, $\sqrt{25x^2}$ is both a(n) _____ and the _____ and $\dfrac{15\sqrt{5}}{\sqrt{25x^2}}$ is a(n) _____.

6.1.3 Simplifying Common Square Roots and Cube Roots

Often roots are irrational numbers. Sometimes though we are able to simplify a root to a rational number. Before we begin simplifying expressions that contain roots, I want to mention a couple points having to do with square roots.

First, consider $\sqrt{4}$. This radical term is asking me to find a base, which when squared, gives the value 4. I hope you can see that $\sqrt{4}=2$ since $2^2=4$. Notice though that -2 also "works", that is $(-2)^2=4$ so why isn't -2 also a square root of 4? Actually, it is. Both 2 and -2 are square roots of 4. When someone asks for "the" square root they're usually asking for the **principal** (positive) square root.

The second point pertains to a radical term like $\sqrt{-4}$ where we need to find a real number, which when squared, gives -4. Since squaring any non-zero real number returns a positive number (for instance both 2 and -2 return 4 when squared) there is no real number, which when squared, gives -4. To simplify $\sqrt{-4}$ we need to include "imaginary" numbers in our number set but doing so would take us too far afield today. So for now, if you're asked to simplify a square root where the radicand is negative, the answer is, "It's not a real number."

To speed your work simplifying radicals, you should memorize the first few whole number square roots;

$$\sqrt{0}=0 \quad \sqrt{1}=1 \quad \sqrt{4}=2 \quad \sqrt{9}=3 \quad \sqrt{16}=4 \quad \sqrt{25}=5$$

$$\sqrt{36}=6 \quad \sqrt{49}=7 \quad \sqrt{64}=8 \quad \sqrt{81}=9 \quad \sqrt{100}=10$$

And the first few integer cube roots.

$$\sqrt[3]{0}=0 \quad \sqrt[3]{1}=1 \quad \sqrt[3]{-1}=-1 \quad \sqrt[3]{8}=2 \quad \sqrt[3]{-8}=-2 \quad \sqrt[3]{27}=3$$

$$\sqrt[3]{-27}=-3 \quad \sqrt[3]{64}=4 \quad \sqrt[3]{-64}=-4 \quad \sqrt[3]{125}=5 \quad \sqrt[3]{-125}=-5$$

Here's some practice simplifying roots.

Practice 6.1.3 Simplifying Common Square Roots and Cube Roots

Describe what the radical term is asking and then simplify. As a check rewrite your answer in exponential form and simplify.

a) $\sqrt{16}$

$\sqrt{16}$	$\Rightarrow$	This is asking, "What positive base when squared gives a result of sixteen?"
$\sqrt{16}=4$	$\Rightarrow$	A base of 4 would work.
$4^2=16$	$\Rightarrow$	Checked the answer by rewriting in exponential form and simplifying.

b) $\sqrt{-25}$

$\sqrt{-25}$	$\Rightarrow$	This is asking, "What base when squared gives a result of negative twenty-five?"
It's not a real number.	$\Rightarrow$	Squaring both 5 and -5 results in a positive 25. There is no real number, which when squared, will result in negative twenty-five.

c) $\sqrt[3]{-27}$

$\sqrt[3]{-27}$	$\Rightarrow$	This is asking, "What base when cubed gives a result of negative twenty-seven?"
$\sqrt[3]{-27}=-3$	$\Rightarrow$	Negative three cubed gives a result of negative twenty-seven.
$(-3)^3=-27$	$\Rightarrow$	Checked the answer by rewriting in exponential form and simplifying.

Homework 6.1 Describe what the radical term is asking and then simplify. As a check rewrite your answer in exponential form and simplify.

4) $\sqrt{25}$	5) $\sqrt[3]{64}$	6) $\sqrt[3]{-1}$	7) $\sqrt{1}$
8) $\sqrt{-81}$	9) $\sqrt{121}$	10) $\sqrt{0}$	11) $\sqrt[3]{-125}$

6.1.4 *Irrational Numbers and the Order of Operations*

Now let's combine roots and other operations. Remember, finding roots is included in line 1 of the order of operations. Also, operations in the radicand are a type of implicit grouping, so operations under the radical symbol need to be completed before finding the root. Also, please keep your answers exact and don't change an irrational root to a rational approximation.

Practice 6.1.4 Irrational Numbers and the Order of Operations

Count the number of operations, name the operations using the correct order and then simplify the expression. Only use a calculator when necessary.

a) $\sqrt{3^2-4(2)(-2)}$

There are five operations. The radicand is an implicit grouping symbol so squaring is first, then multiplication left to right, subtraction is next and taking the square root is last.

$\sqrt{9-4(2)(-2)}$	$\Rightarrow$	Squared first.
$\sqrt{9-(-16)}$	$\Rightarrow$	Multiplied left to right.
$\sqrt{25}$	$\Rightarrow$	Thought of subtracting a negative as addition.
5	$\Rightarrow$	Simplified

b) $\dfrac{6\sqrt{16-9}}{\sqrt{9}}$

There are five operations. The subtraction is first, taking the square root is next, then multiply in the dividend ($6\sqrt{16-9}$ implies 6 is a factor), find the square root in the divisor and find the quotient last.

$\dfrac{6\sqrt{7}}{\sqrt{9}}$	$\Rightarrow$	Subtracted in the radicand. In the dividend, the square root and the multiplication can't be simplified further.
$\dfrac{6\sqrt{7}}{3}$	$\Rightarrow$	Simplified in the divisor.
$\dfrac{\cancel{6}\sqrt{7}}{\cancel{3}}=2\sqrt{7}$	$\Rightarrow$	Reduced the common factor of 3.

c) $\sqrt{(12-10)^2+(9-12)^2}$

There are six operations. Subtracting left to right is first, squaring left to right is next, the addition follows and finally take the square root.

$\sqrt{(2)^2+(-3)^2}$	$\Rightarrow$	Subtracted first left to right.
$\sqrt{4+9}$	$\Rightarrow$	Squared left to right.
$\sqrt{13}$	$\Rightarrow$	Added
$\sqrt{13}$	$\Rightarrow$	This answer is simplified. It's true you can get a decimal approximation using a calculator but this is the exact answer.

Homework 6.1	*Count the number of operations, name the operations using the correct order and then simplify the expression. Only use a calculator when necessary.*

12) $\sqrt{(-2)^2-4(-1)(3)}$ 13) $(\sqrt{4}+\sqrt{9})(\sqrt{4}-\sqrt{9})$ 14) $8\left(\sqrt{\frac{24}{6}}\right)^3$ 15) $\frac{\sqrt[3]{64}}{\sqrt{64}}$

16) $\frac{6+\sqrt{6^2-4(4)(2)}}{2(4)}$ 17) $\sqrt{9-4^2}$ 18) $\frac{14\sqrt{9}+4}{\sqrt{49}}$ 19) $\frac{-9-\sqrt{10^2-4(-3)(-7)}}{2(-3)}$

20) $\sqrt{(3-(-4))^2+(7-11)^2}$ 21) $\left(\sqrt{9}+\sqrt{36}\right)^2$ 22) $\frac{15\sqrt{7}}{\sqrt{25}}$ 23) $\sqrt{\sqrt{16}\sqrt{81}}$

6.1.5 Using the Power of a Product Rule

The power of a product rule for exponents, along with the idea of inverse operations, is often helpful when our base has at least one radical factor.

Square Root and Squaring Are Inverse Operations
For non-negative radicands squaring a square root gives the original radicand. Example: $(\sqrt{3})^2=3$

For instance, if we start with $(2\sqrt{3})^2$ we're able to simplify the power using the power of a product rule and inverse operations, $(2\sqrt{3})^2=2^2(\sqrt{3})^2=4(3)=12$. Here's some practice.

Practice 6.1.5	**Using the Power of a Product Rule** *Simplify using the power of a product rule.*

a) $(2\sqrt{3})^2$

$2^2(\sqrt{3})^2$	$\Rightarrow$	Used the power of a product rule to square each factor of the base.
$4(3)$	$\Rightarrow$	Squared 2 and $(\sqrt{3})^2=3$ by the idea of inverse operations.
12	$\Rightarrow$	Multiplied.

b) $\left(-\sqrt{6}\sqrt{7}\right)^2$

$(-1)^2(\sqrt{6})^2(\sqrt{7})^2$	$\Rightarrow$	Used the power of a product rule to square each factor of the base.
$(6)(7)$	$\Rightarrow$	Simplified.
42	$\Rightarrow$	Multiplied.

Homework 6.1 Simplify using the power of a product rule.

24) $\left(6\sqrt{2}\right)^2$ 25) $\left(-3\sqrt{3}\right)^2$ 26) $\left(\sqrt{8}\sqrt{3}\right)^2$ 27) $-\left(2\sqrt{15}\right)^2$

28) $\left(-2\sqrt{0.5}\right)^2$ 29) $\left(-\sqrt{16}\sqrt{15}\right)^2$

6.1.6 Using the Power of a Quotient Rule

The power of a quotient rule is helpful if we want to work with the power of the dividend and/or the divisor independently.

Practice 6.1.6 Using the Power of a Quotient Rule

Simplify using the power of a quotient rule.

a) $\left(\frac{\sqrt{3}}{2}\right)^2$

$\frac{(\sqrt{3})^2}{2^2}$	$\Rightarrow$	Used the power of a quotient rule.
$\frac{3}{4}$	$\Rightarrow$	Used the idea of inverse operators in the dividend and squared the divisor.

b) $\left(-\frac{\sqrt{2}}{\sqrt{3}}\right)^2$

$\left(-1\times\frac{\sqrt{2}}{\sqrt{3}}\right)^2$	$\Rightarrow$	Decided to process the negative as a factor of negative one.
$(-1)^2\times\frac{(\sqrt{2})^2}{(\sqrt{3})^2}$	$\Rightarrow$	Used the power of a product rule followed by the power of a quotient rule for the second factor.
$\frac{2}{3}$	$\Rightarrow$	Squaring negative one gives positive one and used the inverse idea in both the numerator and denominator.

Homework 6.1 Simplify using the power of a quotient rule.

30) $\left(\frac{\sqrt{2}}{2}\right)^2$ 31) $\left(-\frac{\sqrt{3}}{2}\right)^2$ 32) $\left(\frac{5}{\sqrt{5}}\right)^2$ 33) $-\left(\frac{\sqrt{6}}{3}\right)^2$ 34) $\left(\frac{\sqrt{6}}{\sqrt{3}}\right)^2$ 35) $\left(\frac{-\sqrt{20}}{\sqrt{5}}\right)^2$

Homework 6.1 Answers

1) radicand, root, term, radicand, root, term, sum, base, exponent, power

2) base, exponent, power, minuend, factor, factor, factor, product, subtrahend, difference, radicand, term, root, term, sum

3) factor, radicand, root, factor, product, dividend, factor, base, exponent, factor, power, product, radicand, root, divisor, quotient

4) What base when squared results in 25?, 5

5) What base when cubed results in 64?, 4

6) What base when cubed results in -1?, -1

7) What base when squared results in 1?, 1

8) What base when squared results in -81?, Not a real number.

9) What squared base results in 121?, 11 10) What squared base results in 0?, 0

11) What base when cubed results in -125?, -5

12) There are five operations. In the radicand squaring is first followed by multiplication left to right, then the subtraction and finally the square root. The answer is 4.

13) There are seven operations. Inside the parentheses each square root is simplified and the addition and subtraction are finished. Last the product is found. The answer is -5.

14) There are four operations. The division is first, the square root is next, cubing next and multiplication by 8 last. The answer is 64.

15) There are three operations. The roots are first, the division last. The answer is 1/2.

16) There are eight operations. Starting in the radicand the squaring is first, followed by the two multiplications and then the subtraction. Still in the numerator find the square root and then add the six. In the denominator multiply and then finally divide. The answer is 1.

17) There are three operations. In the radicand the squaring is first followed by the subtraction and then finding the square root. Because the radicand becomes -7 this is not a real number.

18) There are five operations. In the dividend the sum is first, the square root is next and the product last. In the divisor simplify the square root and the quotient is last. The final answer is $2\sqrt{13}$.

19) There are eight operations. In the radicand the squaring is first, followed by the two multiplications and then the subtraction. Still in the dividend subtract the square root from negative nine. In the divisor, multiply and then divide. The answer is $^{13}/_{6}$.

20) There are six operations. In the radicand the subtractions inside the parentheses are first. The results are squared and added. The square root is last. The answer is $\sqrt{65}$.

21) There are four operations. Inside the parentheses simplify the square roots left to right, add next and finally square the sum. The answer is 81.

22) There are four operations. In the dividend the square root can't be simplified so the numerator is already simplified. In the divisor the square root of 25 is 5. After reducing the common factor of 5 in the dividend and divisor the final answer is $3\sqrt{7}$.

23) There are four operations. The two square roots in the radicand are first followed by the multiplication. Taking the larger square root is last. The answer is 6.

24) 72 25) 27 26) 24 27) -60 28) 2 29) 240 30) 1/2 31) 3/4

32) 5 33) - 2/3 34) 2 35) 4

6.2 Applying Square Roots

In this section, we'll practice applying formulas and functions that involve square roots.

6.2.1 The Pythagorean Theorem

Two lines are perpendicular if they meet in a ninety-degree angle. A right triangle has two perpendicular sides known as the legs and a third longest side known as the hypotenuse. Notice, to the right, the little box in the lower left corner of the triangle. This tells us that the lines meet in a 90° angle.

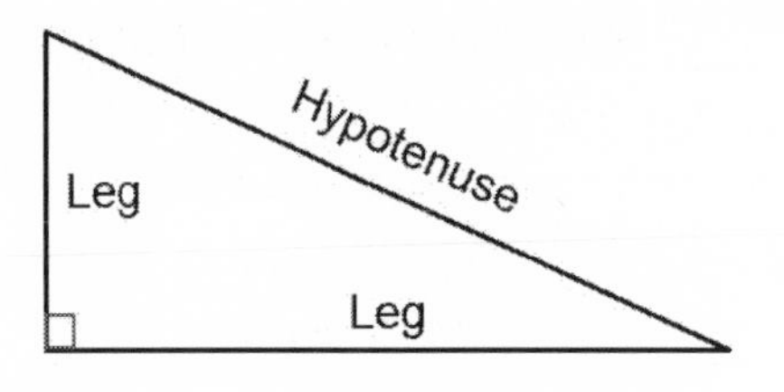

The Pythagorean Theorem describes a relationship between the lengths of the sides of a right triangle. When describing the Pythagorean Theorem, the tradition is to use the letter *a* to stand for the length of one leg, *b* to stand for the length of the other leg and *c* to stand for the length of the hypotenuse. It doesn't matter which leg length you decide to make *a* and which you decide to make *b*. Our use of the letters *a*, *b* and *c* for the Pythagorean Theorem has nothing to do with the way we used *a*, *b* and *c* for the values of the coefficients and constant when we factored polynomials. Here's the Pythagorean Theorem.

The Pythagorean Theorem
Given a right triangle, the sum of the squares of the lengths of the legs equals the square of the length of the hypotenuse. Using algebra, if a and b are the lengths of the legs of a right triangle and c is the length of the hypotenuse, then $a^2+b^2=c^2$.

The formula $a=\sqrt{c^2-b^2}$ helps us find an unknown side length if we know the length of the other side and the hypotenuse. The formula $c=\sqrt{a^2+b^2}$ helps us find an unknown hypotenuse length if we know the lengths of both legs. Let's practice answering some questions using the Pythagorean Theorem.

Practice 6.2.1 The Pythagorean Theorem

Find the length of the unknown side.

a)

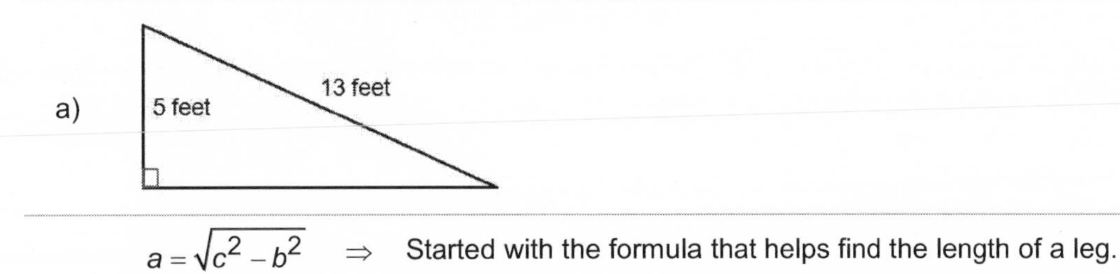

$a=\sqrt{c^2-b^2}$ $\Rightarrow$ Started with the formula that helps find the length of a leg.

$a=\sqrt{13^2-5^2}$ $\Rightarrow$ Substituted the known values for *c* and *b*.

$a = \sqrt{144}$
$a = 12$ $\Rightarrow$ Simplified.

The unknown leg length is 12 feet. $\Rightarrow$ Answered the question.

b) The two legs of a right triangle have lengths 2 miles and 4 miles. Find the length of the hypotenuse. Find both the exact and an estimated answer.

$c = \sqrt{a^2 + b^2}$ $\Rightarrow$ Started with the formula that helps us find the length of the hypotenuse.

$c = \sqrt{2^2 + 4^2}$ $\Rightarrow$ Substituted values for *a* and *b*.

$c = \sqrt{20} = 2\sqrt{5}$ miles
$c \approx 4.5$ miles $\Rightarrow$ The exact length, $2\sqrt{5}$ miles, is approximately 4.5 miles.

c) A 20-foot ladder is set against the wall of a house with the base of the ladder 6 feet away from the wall. If the top of the ladder just reaches the bottom of the second story window, how far above the ground is the bottom of the window?

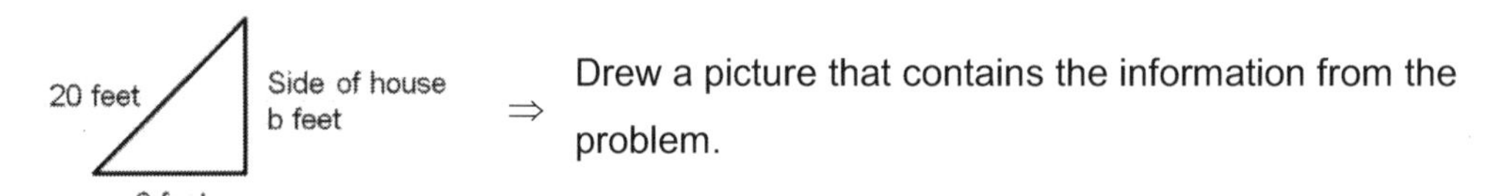

$\Rightarrow$ Drew a picture that contains the information from the problem.

$a = \sqrt{c^2 - b^2}$
$a = \sqrt{20^2 - 6^2}$ $\Rightarrow$ Chose the formula for an unknown leg length and substituted the given values.

$a = \sqrt{364} \approx 19$ feet $\Rightarrow$ Simplified and found the estimated answer.

Homework 6.2: Find the length of the unknown side. For problems 1-6 include both an exact and an estimated answer if appropriate.

1) 2)

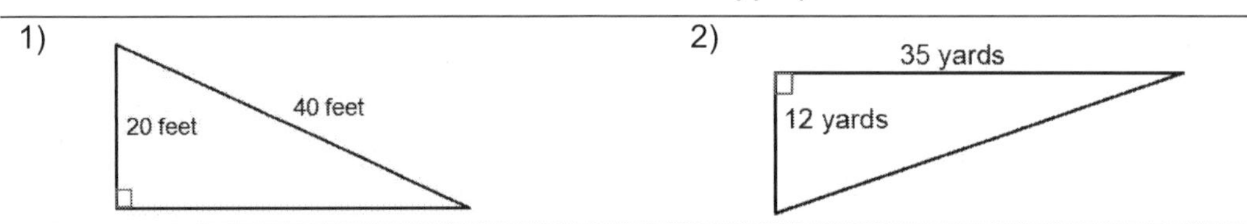

3) One leg is 20 miles the other leg is 21 miles.

4) The hypotenuse is 14 meters and one leg is 4 meters.

5) One leg is 2 cm shorter than the 19 cm hypotenuse.

6) One leg is 0.75 feet, the other leg is 1.2 feet.

7) The Pythagorean Theorem can help us decide if something we build is "square" (the corners are 90 degrees). You've built a deck that's 12 feet by 16 feet. If the deck is square what should the diagonal distance from corner to corner through the center of the deck be?

8) A square cabinet has a side length of 32 inches. If the cabinet is truly "square" what should the diagonal distance be?

9) A daughter is moving out of her parent's home. She wants to take her queen-sized box spring which is 60 inches wide. The dimensions of the back of her parent's SUV is 42 inches wide by 35 inches high. Will the box spring fit in the SUV if it's turned? (A box spring is stiff, not flexible.)

10) A student wants to estimate the height of a tree to the nearest foot. The tree is supported by a rope that's staked into the ground and tied about a quarter of the way up the trunk. She paces off the distance from the base of the tree to the stake and finds it's about 12 feet. She assumes the rope from the tree to the stake is about 15 feet long. What is the estimated height of the tree?

11) A right triangular window is above a staircase which makes measuring the hypotenuse unsafe. It's easy to measure the bottom of the window, it's 12 feet. By extending the measuring tape straight up it's also possible to find that the length of the other leg is 5 feet. What is the length of the hypotenuse?

6.2.2 The Distance Formula

Recall that we use an ordered pair to describe a location on a two-dimensional graph. The pair is "ordered" because the first number (the "x" value) describes how far we should move left or right from the origin and the second number (the "y" value) describes how far we should move up or down from the origin. The origin itself is considered the point (0,0).

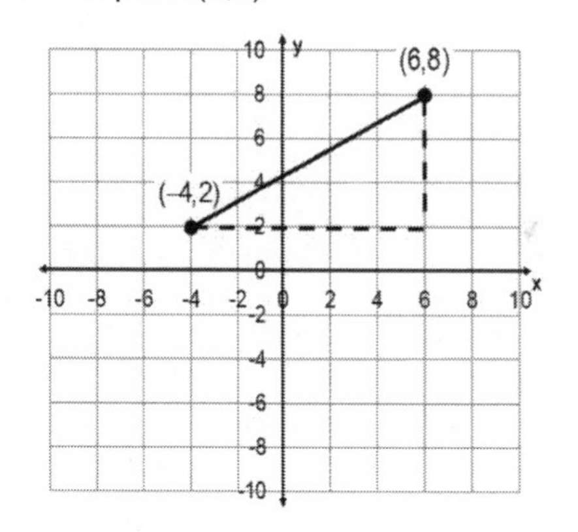

The distance formula finds the "straight line" distance from one point to another in two-dimensional space. On the graph to the right the distance from (–4,2) to (6,8) would be the solid line. To find this distance we imagine the line is the hypotenuse of a right triangle and the two dotted lines are the legs of the triangle. Since we're looking for the length of the hypotenuse we use the formula $c=\sqrt{a^2+b^2}$ where a is the difference in our x values and b is the difference in our y values. If x_1 and x_2 are the two x values and y_1 and y_2 are the two y values our formula becomes $d=\sqrt{(x_2-x_1)^2+(y_2-y_1)^2}$ where d stands for the straight line distance. Here's some practice with the distance formula.

Practice 6.2.2 The Distance Formula

Find the exact distances.

a) *From the point* $(2,-3)$ *to* $(-1,-4)$.

(x_1,y_1) (x_2,y_2) $(2,-3)$ $(-1,-4)$	$\Rightarrow$	Decided which point will be (x_1,y_1) and which (x_2,y_2). Switching which pair is which will not affect the outcome.

$d=\sqrt{(x_2-x_1)^2+(y_2-y_1)^2}$ $d=\sqrt{(-1-2)^2+(-4-(-3))^2}$	$\Rightarrow$ Substituted the values for the appropriate variables.
$d=\sqrt{9+1}=\sqrt{10}$	$\Rightarrow$ Simplified. The distance is $\sqrt{10}$ units.

Homework 6.2 Find the exact distances.

12) From $(8,0)$ to $(0,6)$.

13) From $(-10,6)$ to $(-4,2)$.

14) From $(8,-8)$ to $(8,0)$.

15) From $(-8,-4)$ to $(-2,-6)$.

16) From $(-10,6)$ to $(0,6)$.

17) From $(-2,-6)$ to $(-10,6)$.

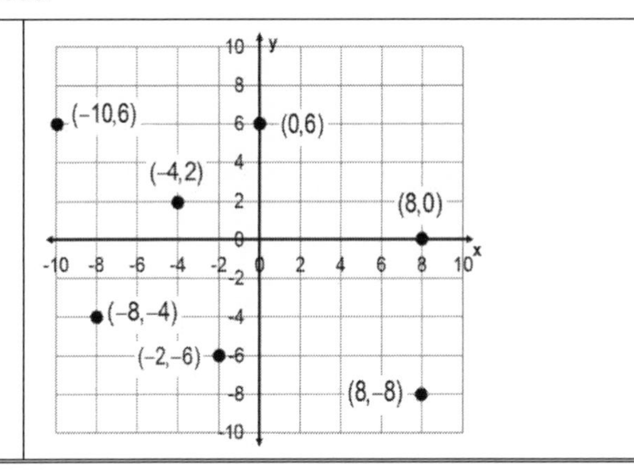

6.2.3 Assorted Square Root Functions

Let's finish this section by applying an assortment of square root functions.

18) The function $y=2\sqrt{x}+13$ predicts the percent of one-person households in the United States since 1960. x is the number of years since 1960.
   a) Approximately what percent of household were one-person households in 1960?
   b) Approximately what percent of household were one-person households in 2000?
   c) Predict the percent of one-person households in 2020.
   d) Use your calculator to try to find the percent of households in 1959 (year -1). Why was there a problem?

19) A person who makes "vintage" bicycles finds the function $y=\sqrt{156{,}025-5x}+605$ can predict the price they should charge, y, to make a profit of x dollars as long as the profit is less than \$31,205.
   a) The maximum profit they can make is \$31,205. What price should they charge to make this happen?
   b) If the person raises their price, they will probably reduce demand and be able to make fewer bicycles. If the person is satisfied with a profit of \$25,000 what should their price be?
   c) Solve $156{,}025-5x<0$ and discuss what the value tells you?

20) The function $y = \sqrt{0.2x - 108.2} + 4.2$ can predict the number of years after 1970 that someone attended a four-year college if you supply x, the cost of tuition and fees that year (in dollars).

a) If someone spent $2,000 what year were they attending college?

b) If someone spent $6,000 what year were they attending college?

c) Predict the first year someone will have to spend $10,000 for one year of college?

d) Can you figure out the cost below which the model no longer works?

21) The function $y = \sqrt{864.6 - 27x} + 13.3$ will estimate how many years after 1980 it was (y) if you supply (x), the percent of 12th grade boys who've smoked in the last 30 days.

a) In the last 50 years the highest percent was 32%, what year was this?

b) Use the function to approximate the year the percentage will first drop to 20%?

Homework 6.2 Answers

1) $20\sqrt{3}$ ft. $\approx$ 34.6ft. 2) 37 yds. 3) 29 miles 4) $6\sqrt{5}$ m $\approx$ 13.4 m

5) $6\sqrt{2}$ cm. $\approx$ 8.5cm. 6) The length of the hypotenuse is 1.4 ft

7) 20 ft. 8) A little more than 45 inches. 9) No. The opening is about 54 inches.

10) About 36 feet 11) 13 feet 12) 10 13) $2\sqrt{13}$ 14) 8 15) $2\sqrt{10}$

16) 10 17) $4\sqrt{13}$

18) a) About 13% b) Around 26.5% c) About 28.5% d) $\sqrt{-1}$ is not a real number.

19) a) $605 b) $781

c) The answer tells me when the radicand will become negative and not be a real number. This will happen when $x > 31{,}205$. This is why it originally says the profit has to be less than $31,205.

20) a) Around 1991 b) Around 2007 c) About 2018

d) This will be the same issue as the previous problem so I solved $0.2x - 108.2 < 0$ and found the model gives a negative radicand for costs below $541

21) a) Around 1994 b) In 2011

6.3 A Product Rule for Square Roots

For the next few sections, we'll practice some of the skills necessary for simplifying and operating on square roots.

6.3.1 The Product Rule for Square Roots

A square root isn't considered simplified if there's a perfect square factor in the radicand. For instance, $\sqrt{8}$ is not considered simplified because we can think of $\sqrt{8}$ as $\sqrt{4\times 2}$ and 4 is a perfect square. We'll simplify square roots like $\sqrt{8}$ using the product rule for square roots.

The Product Rule for Square Roots
If all factors of a radicand are non-negative then the square root of the product is the product of the square roots of the factors. Example: $\sqrt{4\times 2}=\sqrt{4}\times\sqrt{2}$

Here's some practice simplifying square roots using the product rule.

Practice 6.3.1 The Product Rule for Square Roots
Simplify.

a) $\sqrt{12}$

$\sqrt{4\times 3}$	$\Rightarrow$	Factored the radicand so one factor is the largest perfect square.
$\sqrt{4}\sqrt{3}$	$\Rightarrow$	Used the product rule for square roots.
$2\sqrt{3}$	$\Rightarrow$	Simplified $\sqrt{4}$ to 2. Recalled that $2\sqrt{3}$ is a product built from the factors 2 and $\sqrt{3}$.

b) $5\sqrt{18}$

$5\sqrt{9\times 2}$	$\Rightarrow$	Factored the radicand so one factor is the largest perfect square.
$5\sqrt{9}\sqrt{2}$	$\Rightarrow$	Used the product rule.
$5(3)\sqrt{2}$	$\Rightarrow$	The square root of 9 is 3.
$15\sqrt{2}$	$\Rightarrow$	Multiplied.

Homework 6.3 Simplify.

1) $\sqrt{50}$ 2) $3\sqrt{20}$ 3) $-2\sqrt{18}$ 4) $-\sqrt{45}$

6.3.2 Reviewing Prime Factorization

Hopefully, as you worked with the previous radicands, a perfect square factor, "popped into your head". What if nothing pops? Then prime factoring will often show the way. Let's spend a little time reviewing prime factoring a natural number.

Homework 6.3 Prime factor the following numbers.

5) 42 6) 90 7) 126 8) 252 9) 405

6.3.3 *Simplifying Using Prime Factorization*

Now that you've reviewed prime factorization, let's use the idea to simplify the radical term $\sqrt{126}$. When I look at $\sqrt{126}$ I don't see a largest perfect square factor. If I prime factor the radicand though $\sqrt{2\times3\times3\times7}$ I notice two factors of 3 which is 9, this gives me the idea of using the commutative property and the associative property to reorder and regroup the factors $\sqrt{(3\times3)\times(2\times7)}$ so I have a perfect square factor in the radicand, $\sqrt{9\times14}$. Using the product rule, we can simplify the root, $\sqrt{9\times14}=\sqrt{9}\sqrt{14}=3\sqrt{14}$.

Careful!	**Use the largest perfect square**
	Sometimes you may not notice the largest perfect square. For instance, you may see $\sqrt{32}$ as $\sqrt{4\times8}$ which simplifies to $2\sqrt{8}$. Notice though that $\sqrt{8}$ can itself be simplified so you'll need to continue the process. To avoid the extra work always look for the largest perfect square factor. With $\sqrt{32}$ the best factorization would be $\sqrt{16\times2}=\sqrt{16}\sqrt{2}=4\sqrt{2}$. Seeing the factor of 16 is easier when you're working from the prime factored form for $\sqrt{32}$ which is $\sqrt{2\times2\times2\times2\times2}$.

Practice 6.3.3 Simplifying Using Prime Factorization

Simplify using prime factorization.

a) $\sqrt{108}$

$\sqrt{2\times2\times3\times3\times3}$	$\Rightarrow$	Prime factored the radicand.
$\sqrt{4\times9\times3}$	$\Rightarrow$	Noticed a factor of 4 and 9 in the radicand. $\sqrt{36\times3}$ also works well.
$\sqrt{4}\sqrt{9}\sqrt{3}$	$\Rightarrow$	Used the product rule.
$2(3)\sqrt{3}$	$\Rightarrow$	Simplified
$6\sqrt{3}$	$\Rightarrow$	Multiplied

b) $-4\sqrt{135}$

$-4\sqrt{3\times3\times3\times5}$	$\Rightarrow$	Noticed a factor of 9.
$-4\sqrt{9\times15}$	$\Rightarrow$	Combined factors to build a perfect square.
$-4\sqrt{9}\sqrt{15}$	$\Rightarrow$	Used the product rule.
$-4(3)\sqrt{15}=-12\sqrt{15}$	$\Rightarrow$	Simplified.

Homework 6.3 Simplify using prime factorization.

10) $\sqrt{84}$	11) $-3\sqrt{126}$	12) $\sqrt{252}$	13) $4\sqrt{147}$
14) $\sqrt{297}$	15) $-5\sqrt{196}$	16) $\sqrt{294}$	17) $4\sqrt{280}$

6.3.4 Simplifying and Reducing

Sometimes, after simplifying a square root, you're able to reduce common factors.

Practice 6.3.4 Simplifying and Reducing

Simplify.

a) $\frac{\sqrt{8}}{2}$

$\frac{\sqrt{4\times 2}}{2}=\frac{2\sqrt{2}}{2}$ ⇒ Simplified the numerator.

$\frac{\cancel{2}\sqrt{2}}{\cancel{2}}=\sqrt{2}$ ⇒ Reduced the common factor of 2. The final answer is $\sqrt{2}$.

b) $\frac{\sqrt{20}}{\sqrt{5}}$

$\frac{\sqrt{4\times 5}}{\sqrt{5}}=\frac{2\sqrt{5}}{\sqrt{5}}$ ⇒ Simplified the numerator.

$\frac{2\cancel{\sqrt{5}}}{\cancel{\sqrt{5}}}=2$ ⇒ Reduced the common factor of $\sqrt{5}$.

Homework 6.3 Simplify.

18) $\frac{3}{\sqrt{36}}$ 19) $\frac{\sqrt{18}}{3}$ 20) $\frac{\sqrt{12}}{\sqrt{3}}$ 21) $\frac{3\sqrt{8}}{2}$ 22) $\frac{\sqrt{28}}{\sqrt{7}}$ 23) $\frac{3\sqrt{25}}{\sqrt{9}}$

24) $\frac{\sqrt{45}}{6\sqrt{5}}$ 25) $\frac{-4\sqrt{27}}{\sqrt{16}}$ 26) $\frac{2\sqrt{63}}{3\sqrt{28}}$

6.3.5 Multiplying Radical Factors

Remember that with a rule we are given two points of view. With the product rule, we've only taken the point of view that we're going to rewrite a product as its factors. By taking the other point of view we can use the product rule to multiply factors to build a product.

Practice 6.3.5 Multiplying Radical Factors

Simplify.

a) $\sqrt{6}\sqrt{6}$

$\sqrt{6\times 6}=\sqrt{36}$ ⇒ Used the product rule for square roots.

6 ⇒ Simplified the square root of 36.

b) $\sqrt{6}\sqrt{2}$

$\sqrt{6\times 2}=\sqrt{12}$ ⇒ Used the product rule for square roots.

$\sqrt{4}\sqrt{3}$ ⇒ Thought of $\sqrt{12}$ as $\sqrt{4\times 3}$ and used the product rule.

$2\sqrt{3}$ ⇒ Simplified.

c) $\sqrt{18}\sqrt{6}$		
$\sqrt{2\times3\times3}\ \sqrt{2\times3}$ $\sqrt{2\times3\times3\times2\times3}$	$\Rightarrow$	Instead of multiplying 18 and 6 together, and then factoring, I decided to factor first and then multiply.
$\sqrt{4\times9\times3}$	$\Rightarrow$	Combined factors to make the largest perfect squares. $\sqrt{36\times3}$ also works well.
$\sqrt{4}\sqrt{9}\sqrt{3}$	$\Rightarrow$	Used the product rule.
$2(3)\sqrt{3}$	$\Rightarrow$	Simplified. $\sqrt{36\times3}=\sqrt{36}\sqrt{3}=6\sqrt{3}$ gets us there a little faster.
$6\sqrt{3}$	$\Rightarrow$	Multiplied.

d) $(-4\sqrt{35})(2\sqrt{20})$		
$-4(2)(\sqrt{35}\sqrt{20})$	$\Rightarrow$	Used the commutative and associative properties to combine non-radical and radical factors.
$-8\sqrt{5\times7\times2\times2\times5}$	$\Rightarrow$	Multiplied the non-radical factors and multiplied the radical factors in prime factored form.
$-8\sqrt{4\times25\times7}$	$\Rightarrow$	Found the perfect square factors.
$-8\sqrt{4}\sqrt{25}\sqrt{7}$	$\Rightarrow$	Used the product rule.
$-8(2)(5)\sqrt{7}$	$\Rightarrow$	Simplified.
$-80\sqrt{7}$	$\Rightarrow$	Multiplied.

Homework 6.3 Simplify.

27) $\sqrt{2}\sqrt{10}$ 28) $\sqrt{6}\sqrt{12}$ 29) $\sqrt{12}\sqrt{15}$ 30) $-3\sqrt{14}\sqrt{21}$

31) $7\sqrt{8}\sqrt{12}$ 32) $\sqrt{10}\sqrt{35}$ 33) $(2\sqrt{30})(-3\sqrt{10})$ 34) $\sqrt{6}\sqrt{5}\sqrt{15}$

35) $(-2\sqrt{15})(-\sqrt{35})$ 36) $(2\sqrt{8})(3\sqrt{10})$ 37) $(-6\sqrt{42})(5\sqrt{14})$

Homework 6.3 Answers

1) $5\sqrt{2}$ 2) $6\sqrt{5}$ 3) $-6\sqrt{2}$ 4) $-3\sqrt{5}$ 5) $2\times3\times7$ 6) $2\times3\times3\times5$

7) $2\times3\times3\times7$ 8) $2\times2\times3\times3\times7$ 9) $3\times3\times3\times3\times5$ 10) $2\sqrt{21}$ 11) $-9\sqrt{14}$

12) $6\sqrt{7}$ 13) $28\sqrt{3}$ 14) $3\sqrt{33}$ 15) 70 16) $7\sqrt{6}$ 17) $8\sqrt{70}$ 18) $1/2$

19) $\sqrt{2}$ 20) 2 21) $3\sqrt{2}$ 22) 2 23) 5 24) $1/2$ 25) $-3\sqrt{3}$ 26) 1

27) $2\sqrt{5}$ 28) $6\sqrt{2}$ 29) $6\sqrt{5}$ 30) $-21\sqrt{6}$ 31) $28\sqrt{6}$ 32) $5\sqrt{14}$

33) $-60\sqrt{3}$ 34) $15\sqrt{2}$ 35) $10\sqrt{21}$ 36) $24\sqrt{5}$ 37) $-420\sqrt{3}$

6.4 The Product Rule with Variable Roots

In this section, we'll practice with radicands that have variables. We'll assume that values chosen for the variables will always keep the radicand positive or zero.

6.4.1 Simplifying Variable Radicands

As with the square roots of constants, squaring and square root are inverse operations.

Squaring and Square Roots are Inverse Operations
For nonnegative radicands squaring and taking a square root are inverse operations.   Algebra: $\sqrt{x^2} = x \quad x \ge 0$

We can combine this rule with the product rule for square roots to help simplify radical expressions.

The Product Rule for Square Roots
If all factors of a radicand are non-negative then the square root of the product is the product of the square roots of the factors.   Algebra: For real numbers x and y $\sqrt{xy} = \sqrt{x}\sqrt{y}$, $x \ge 0, y \ge 0$

Let's use these properties to simplify $\sqrt{x^5}$.

First, I'll use the product rule for exponents, $x^5 = x^{2+2+1} = x^2\,x^2\,x$, to rewrite the radicand as factors that are, and factors that are not, a perfect square.	$\sqrt{x^2\,x^2\,x}$
Next, I'll use the product rule for square roots to separate the factors that are perfect squares from the factors that aren't perfect squares.	$\sqrt{x^2}\,\sqrt{x^2}\,\sqrt{x}$
Now, I'll simplify the factors where the inverse rule is applicable.	$x\,x\sqrt{x}$
Last, I'll use the product rule for exponents to simplify the factors of x outside the radical symbol. Here's some practice simplifying radicands with variable factors.	$x^2\sqrt{x}$

Practice 6.4.1 Simplifying Variable Radicands
Simplify.

a) $2a\sqrt{a^3}$

$2a\sqrt{a^2\,a}$	$\Rightarrow$	Used the product rule to think of the a^3 in the radicand as $a^2 \times a$.
$2a\sqrt{a^2}\,\sqrt{a}$	$\Rightarrow$	Used the product rule for radicals.
$2a(a)\sqrt{a}$ $2a^2\sqrt{a}$	$\Rightarrow$	Simplified using inverse operations and the product rule.

b) $\sqrt{x^3y^4}$

$\sqrt{x^2\,x\,y^2\,y^2}$	$\Rightarrow$	The base *x* had one perfect square factor while the base *y* had two factors that were perfect squares since y^4 is $y^2\,y^2$.
$\sqrt{x^2}\sqrt{y^2}\sqrt{y^2}\sqrt{x}$	$\Rightarrow$	Rewrote using the product rule and reordered the factors using the commutative property.
$x\,y\,y\sqrt{x}$	$\Rightarrow$	Used inverse operations with the first three factors.
$xy^2\sqrt{x}$	$\Rightarrow$	Simplified using the product rule for exponents.

Homework 6.4 Simplify.

1) $\sqrt{a^5b^3}$ 2) $3m\sqrt{m^7}$ 3) $\sqrt{m^2n^4s^3}$ 4) $pq^2\sqrt{p^6q^3}$

6.4.2 Using the Index and Exponents to Simplify Square Roots

Here's a different way to simplify $\sqrt{x^5}$. To use this approach, you need to keep in mind that with square roots, the index of 2 is usually implicit. That is, $\sqrt{x^5}$ is really $\sqrt[2]{x^5}$.

Since every factor of x^2 in the radicand is one factor of *x* "outside" the radical symbol we can use division to count the number of perfect square factors that will "get out" of the radical symbol. For instance, to simplify $\sqrt[2]{x^5}$ I could divide the index into the exponent $2\overline{)5}$ (quotient 2). This gives a quotient of 2 (the exponent of *x* outside the radical symbol) with remainder 1 (the exponent of *x* in the radicand) so $\sqrt{x^5}$ would be $x^2\sqrt{x^1}$ or just $x^2\sqrt{x}$ since we usually don't make exponents of 1 explicit. Here's the procedure.

Procedure – Simplifying Square Roots Using the Index and Exponents
1. Combine all similar prime variable bases in the radicand. 2. For each prime base divide the index into the exponent. 3. Use the quotient as the exponent of the base "outside" the radical symbol. 4. Use the remainder as the exponent of the base "inside" the radical symbol.

Let's practice using this idea to simplify radical expressions.

Practice 6.4.2 Using the Index and Exponents to Simplify Square Roots
Simplify.

a) $\sqrt{m^9}$

$m^4\sqrt{m}$	$\Rightarrow$	Divided the index of 2 into the exponent 9. The quotient 4 is the exponent of the base outside the radical symbol and the remainder 1 is the exponent of the base in the radicand.

b) $\sqrt{x^3y^4}$

$x\sqrt{xy^4}$	$\Rightarrow$	There are two prime bases x and y. For the base of x divided 2 into 3. The quotient was 1 with a remainder of 1 so 1 is the exponent on x both outside and in the radicand.
$xy^2\sqrt{xy^0}$ $xy^2\sqrt{x(1)}$ $xy^2\sqrt{x}$	$\Rightarrow$	For the base y divided 2 into 4. The quotient is 2 with remainder 0 so the exponent of y outside is 2 and the exponent of y in the radicand is 0. Remembered that by the zero-exponent rule y^0 is a factor of 1.

Think like an expert

You now have two ways of simplifying radicals. Remember don't try to settle on "one" way or the "best" way. Being comfortable with multiple points of view is one key to mathematical power.

Homework 6.4 Simplify.

5) $\sqrt{w^3}$	6) $\sqrt{a^5b^3}$	7) $3m\sqrt{m^7}$	8) $-\sqrt{x^4y^{10}}$
9) $xy\sqrt{x^3y^6}$	10) $-8\sqrt{m^{12}n^6p^{18}}$	11) $pq^2\sqrt{p^7q^7}$	12) $t^4\sqrt{w^6t^4}$

6.4.3 Simplifying When Radicands Contain Both Constants and Variables

Now let's simplify radicands that contain both numbers and variables.

Practice 6.4.3 Simplifying When Radicands Contain Both Constants and Variables

Simplify.

a) $\sqrt{63x^6}$

$\sqrt{9\times 7x^6}$ $3\sqrt{7x^6}$	$\Rightarrow$	Prime factored 63 to $3\times3\times7$ and recognized the perfect square 9. Simplified the numerical part of the radicand.
$3x^3\sqrt{7x^0}$ $3x^3\sqrt{7}$	$\Rightarrow$	Divided 2 into 6 to find the exponent on x outside the radical symbol is 3, (the quotient) and the exponent on x in the radicand is 0, (the remainder). By the zero-exponent rule $x^0=1$ (assuming x itself isn't 0).

b) $2k\sqrt{18k^3t^3}$

$2k(3)\sqrt{2k^3t^3}$ $6k\sqrt{2k^3t^3}$	$\Rightarrow$	Saw $\sqrt{18k^3t^3}$ as $\sqrt{9\times 2k^3t^3}$ and then as $3\sqrt{2k^3t^3}$. Multiplied the factors 2 and 3.
$6k(k)(t)\sqrt{2kt}$ $6k^2t\sqrt{2kt}$	$\Rightarrow$	For the bases of k and t divided 2 into 3 to find the exponent outside is 1 and the exponent inside is also 1. Then wrote the two factors of k as k^2.

Homework 6.4 Simplify

13) $\sqrt{4k^7}$	14) $5h\sqrt{8h}$	15) $\sqrt{48x^9}$	16) $3ab\sqrt{9b^4}$
17) $-2x\sqrt{12x^4}$	18) $2y\sqrt{50y^6}$	19) $y\sqrt{49x^2y^2}$	20) $9x\sqrt{45x^4}$
21) $\sqrt{24y^{14}z^4}$	22) $-3\sqrt{81a^8}$	23) $-6\sqrt{25a^2b^5}$	24) $3k\sqrt{225k^4h^6}$

6.4.4 Multiplying Square Roots That Have Variable Factors

Even though these radicands have variables we can still use the product rule to combined two or more square root factors into a single square root product.

Practice 6.4.4 Multiplying Square Roots That Have Variable Factors

Simplify.

a) $\sqrt{15y^2}\sqrt{6y}$

$\sqrt{3\times5\times2\times3\times y^2\times y}$ $\sqrt{9\times10\times y^3}$	$\Rightarrow$	Factored 15 and 6 to find the perfect square factor of 9. Used the product rule for exponents and added the exponents on y.
$3y\sqrt{10y}$	$\Rightarrow$	Simplified using the earlier ideas.

b) $\sqrt{14a^3b}\sqrt{7a^5b}$

$\sqrt{2\times7\times7a^8b^2}$ $\sqrt{49\times2\times a^8b^2}$	$\Rightarrow$	Used factoring to find the perfect square factor of 49. Used the product rule with the bases of a and b.
$7a^4b\sqrt{2}$	$\Rightarrow$	Simplified $\sqrt{49}$ to 7. Divided 2 into 8 to find the exponents for base a, and divided 2 into 2 to find the exponents for base b.

c) $(3\sqrt{2y})(-2\sqrt{6y^3})$

$-6\sqrt{2\times2\times3\times y^4}$ $-6\sqrt{4\times3\times y^4}$	$\Rightarrow$	Multiplied "outside" factors. Notice there is no distribution since there is no addition or subtraction. Combined radicand factors using the product rule.
$-6(2)y^2\sqrt{3}$	$\Rightarrow$	Simplified $\sqrt{4}$ to 2 and the factor of y^4 inside the radical symbol to y^2 outside the radical symbol.
$-12y^2\sqrt{3}$	$\Rightarrow$	Multiplied -6×2.

Homework 6.4 Simplify.		
25) $\sqrt{4x^3}\sqrt{x^5}$	26) $\sqrt{x^5 y}\sqrt{x^2 y^3}$	27) $3\sqrt{5a^3}\sqrt{15a^2}$
28) $\sqrt{3k^2}\sqrt{6k}\sqrt{10k^3}$	29) $\sqrt{21yz^2}\sqrt{3yz^2}$	30) $\sqrt{8ab^2}\sqrt{14a^5}\sqrt{3b}$
31) $(3\sqrt{2pq})(2\sqrt{8pq^3})$	32) $(-6\sqrt{3xy^2})(2\sqrt{6xy})$	33) $\sqrt{7x^3y}\sqrt{xy^8}\sqrt{28x}$

Homework 6.4 Answers

1) $a^2b\sqrt{ab}$	2) $3m^4\sqrt{m}$	3) $mn^2s\sqrt{s}$	4) $p^4q^3\sqrt{q}$	5) $w\sqrt{w}$	6) $a^2b\sqrt{ab}$
7) $3m^4\sqrt{m}$	8) $-x^2y^5$	9) $x^2y^4\sqrt{x}$	10) $-8m^6n^3p^9$	11) $p^4q^5\sqrt{pq}$	
12) t^6w^3	13) $2k^3\sqrt{k}$	14) $10h\sqrt{2h}$	15) $4x^4\sqrt{3x}$	16) $9ab^3$	17) $-4x^3\sqrt{3}$
18) $10y^4\sqrt{2}$	19) $7xy^2$	20) $27x^3\sqrt{5}$	21) $2y^7z^2\sqrt{6}$	22) $-27a^4$	23) $-30ab^2\sqrt{b}$
24) $45k^3h^3$	25) $2x^4$	26) $x^3y^2\sqrt{x}$	27) $15a^2\sqrt{3a}$	28) $6k^3\sqrt{5}$	29) $3yz^2\sqrt{7}$
30) $4a^3b\sqrt{21b}$	31) $24pq^2$	32) $-36xy\sqrt{2y}$	33) $14x^2y^4\sqrt{xy}$		

6.5 Adding and Subtracting Square Roots

In this lesson, we'll continue operating on square roots.

6.5.1 Identifying Like Radical Terms

As we've often discussed, using the distributive property to add and subtract requires we have like terms. Radical terms are like if they have the same index and the same radicand. For instance, $2\sqrt{3}$ and $5\sqrt{3}$ are like since they both have an index of 2 and a radicand of 3. Here's some practice identifying like radical terms.

Practice 6.5.1 Identifying Like Radical Terms
Identity which terms are like.

a) $\sqrt{2}\,,\,2\sqrt{2}\,,\,2\sqrt{3}\,,\,-\sqrt{2}$ $\Rightarrow$ $\sqrt{2}$, $2\sqrt{2}$ and $-\sqrt{2}$ are like terms.

b) $5\sqrt{7}\,,\,3\sqrt{5}\,,\,7\sqrt{5}\,,\,3\sqrt{7}$ $\Rightarrow$ $5\sqrt{7}$ and $3\sqrt{7}$ are alike as are $3\sqrt{5}$ and $7\sqrt{5}$.

Homework 6.5 Identify which terms are like.

1) $-7\sqrt{3}\,,\,7\sqrt{5}\,,\,12\sqrt{5},\,3\sqrt{7}$

2) $-6\sqrt{2}\,,\,-2\sqrt{6}\,,\,-\sqrt{6}\,,\,2\sqrt{6}$

3) $6\sqrt{15}\,,\,12\sqrt{14}\,,\,4\sqrt{14}\,,\,14\sqrt{12},\,14\sqrt{15}$

4) $-4\sqrt{10}\,,\,-2\sqrt{10}\,,10\sqrt{10}\,,\,\sqrt{10}$

6.5.2 Adding and Subtracting Square Roots

Like with polynomials, the goal with radicals is to recognize the like terms and then simplify the coefficients. There isn't a "standard form" if the radicands are constants, but I tend to write the larger radicands to the left.

Practice 6.5.2 Adding and Subtracting Square Roots Automatically
Simplify.

a) $3\sqrt{2}-12\sqrt{2}+4\sqrt{2}$

$-9\sqrt{2}+4\sqrt{2}$ $\Rightarrow$ Subtract the first two like terms.

$-5\sqrt{2}$ $\Rightarrow$ Added the remaining like terms.

b) $8\sqrt{7}+4\sqrt{5}-3\sqrt{5}+\sqrt{7}$

$8\sqrt{7}+4\sqrt{5}-3\sqrt{5}+\sqrt{7}$ $\Rightarrow$ There are two types of like terms $\sqrt{5}$ and $\sqrt{7}$.

$9\sqrt{7}$ $\Rightarrow$ Combined the terms containing $\sqrt{7}$ first.

$9\sqrt{7}+\sqrt{5}$ $\Rightarrow$ Then combined the terms containing $\sqrt{5}$.

Homework 6.5 Simplify.

5) $8\sqrt{7}-5\sqrt{7}$ 6) $2\sqrt{2}-5\sqrt{2}-7\sqrt{2}$ 7) $-6\sqrt{10}-6\sqrt{10}+12\sqrt{10}$ 8) $7\sqrt{5}-9\sqrt{7}+\sqrt{7}$

9) $-7\sqrt{15}+17\sqrt{15}-11\sqrt{15}$ 10) $4\sqrt{5}-9\sqrt{2}+3\sqrt{2}-6\sqrt{5}$ 11) $\sqrt{3}-2\sqrt{3}+3\sqrt{2}+\sqrt{3}$

6.5.3 Simplifying Before Adding and Subtracting Square Roots

At first, it would seem we can't add $\sqrt{12}+\sqrt{75}$ since the radicands are different. But $\sqrt{12}$ can be simplified to $2\sqrt{3}$ while $\sqrt{75}$ can be simplified to $5\sqrt{3}$. By simplifying first, we see they are like terms and can be added, $\sqrt{12}+\sqrt{75}=2\sqrt{3}+5\sqrt{3}=7\sqrt{3}$.

Simplifying radical terms first is covered in line 1 of the order of operations.

Practice 6.5.3 Simplifying Before Adding and Subtracting Square Roots
Simplify.

a) $2\sqrt{45}-\sqrt{20}$

$2\sqrt{9\times5}-\sqrt{4\times5}$	$\Rightarrow$	Found perfect square factors in the radicand.
$2(3)\sqrt{5}-2\sqrt{5}$	$\Rightarrow$	"Brought out" the factor of 3 and the factor of 2.
$6\sqrt{5}-2\sqrt{5}$ $4\sqrt{5}$	$\Rightarrow$	Multiplied the factors in the first term and subtracted.

b) $\sqrt{28}+\sqrt{7}-2\sqrt{63}$

$\sqrt{4\times7}+\sqrt{7}-2\sqrt{9\times7}$	$\Rightarrow$	Recognized the perfect square factors in the radicand.
$2\sqrt{7}+\sqrt{7}-6\sqrt{7}$ $-3\sqrt{7}$	$\Rightarrow$	Simplified all three terms and simplified.

Homework 6.5 Simplify.

12) $6\sqrt{5}+\sqrt{20}$	13) $\sqrt{50}-\sqrt{8}$	14) $\sqrt{27}+3\sqrt{12}$
15) $-2\sqrt{6}-3\sqrt{24}$	16) $\sqrt{60}-\sqrt{15}-\sqrt{135}$	17) $\sqrt{2}\sqrt{5}+\sqrt{8}\sqrt{5}$
18) $\sqrt{80}-\sqrt{125}-\sqrt{108}+\sqrt{75}$	19) $\dfrac{\sqrt{7}+\sqrt{28}}{3}$	20) $\dfrac{\sqrt{5}+\sqrt{45}}{\sqrt{5}-\sqrt{45}}$

6.5.4 Identifying Like Square Roots

Terms with variable radicands are like if they have the same index and the same radicand. For instance, $2\sqrt{y}+5\sqrt{y}$ are alike since both have the index 2 and radicand y. Here's some practice identifying like radical terms.

Practice 6.5.4 Identifying Like Square Roots
Identity which terms are like.

a) $\sqrt{a}$, $\sqrt{3a}$, $3\sqrt{a}$, $-3\sqrt{a}$	$\Rightarrow$	The like terms are $\sqrt{a}$, $3\sqrt{a}$ and $-3\sqrt{a}$.
b) $2\sqrt{y}$, $3\sqrt{xy}$, $-2\sqrt{yx}$, $2\sqrt{x}$	$\Rightarrow$	The terms $3\sqrt{xy}$ and $-2\sqrt{yx}$ are like since, by the commutative property, $yx=xy$.
c) $-3\sqrt{p}$, $5\sqrt{2p}$, $-\sqrt{3p}$	$\Rightarrow$	None of the terms are like.

Homework 6.5 Identify which terms are like.	
21) $3\sqrt{7y}$, $7\sqrt{y}$, $12\sqrt{7y}$	22) $-2\sqrt{a}$, $2\sqrt{2}$, $-\sqrt{2a}$, $2\sqrt{2a}$
23) $6\sqrt{t}$, $-2\sqrt{2t}$, $2\sqrt{t}$, $-2\sqrt{t}$	24) $4\sqrt{3m}$, $-\sqrt{2m}$, $7\sqrt{mn}$, $3\sqrt{nm}$

6.5.5 Adding and Subtracting Variable Square Roots Automatically

As usual, using the distributive property with roots depends on having like terms. For instance, $2\sqrt{y}+3\sqrt{y}$ can be simplified to $5\sqrt{y}$ since there's a common factor of $\sqrt{y}$. In practice, we usually identify like terms and then add and subtract coefficients.

Practice 6.5.5 Adding and Subtracting Variable Square Roots Automatically
Simplify.

a) $7\sqrt{d}+\sqrt{d}+2\sqrt{d}$

$8\sqrt{d}+2\sqrt{d}$	$\Rightarrow$	Added like terms left to right. Remembered $\sqrt{d}$ has coefficient 1.
$10\sqrt{d}$	$\Rightarrow$	Added like terms.

b) $5\sqrt{2x}-9\sqrt{2x}+4\sqrt{2x}$

$-4\sqrt{2x}+4\sqrt{2x}$	$\Rightarrow$	Subtracted like terms left to right.
0	$\Rightarrow$	The remaining terms were opposites and added to 0. There is no longer a need for the square root symbol.

c) $3\sqrt{xy}+12\sqrt{yx}-16\sqrt{xy}$

$-\sqrt{xy}$	$\Rightarrow$	Simplified the coefficients for terms in $\sqrt{xy}$ $(3+12-16=-1)$. Recalled that by the commutative property $\sqrt{xy}=\sqrt{yx}$.

Homework 6.5 Simplify.

25) $3\sqrt{7y}+8\sqrt{7y}-12\sqrt{7y}$ 26) $-\sqrt{ab}+5\sqrt{ba}-\sqrt{ba}$ 27) $12\sqrt{xy}+10\sqrt{yx}-15\sqrt{yx}$

28) $-3\sqrt{2p}+6\sqrt{2p}+2\sqrt{2p}-3\sqrt{2p}$ 29) $18\sqrt{mn}+12\sqrt{nm}-40\sqrt{nm}+6\sqrt{mn}$

30) $\frac{2}{3}\sqrt{14x}-\frac{1}{2}\sqrt{14x}$

6.5.6 Simplifying Variable Roots Before Adding and Subtracting

Terms that have both variable and radical factors are like if both the variable part and the radical part are the same. For example, $2y\sqrt{y}$ and $-y\sqrt{y}$ are like since both have the factor y and the radical factor $\sqrt{y}$. On the other hand, $3x\sqrt{x}$ and $3\sqrt{x}$ aren't like since only the first expression has a factor of x outside the radical symbol. Here's some practice with simplifying both variable and radical factors before adding or subtracting.

Practice 6.5.6 Simplifying Variable Roots Before Adding and Subtracting

Simplify.

a) $5\sqrt{8n}-9\sqrt{2n}$

$5\sqrt{4\times 2n}-9\sqrt{2n}$

$5(2)\sqrt{2n}-9\sqrt{2n}$ $\Rightarrow$ Simplified the first term

$10\sqrt{2n}-9\sqrt{2n}$

$\sqrt{2n}$ $\Rightarrow$ The like terms can be subtracted.

b) $\sqrt{12b}+\sqrt{27b}$

$\sqrt{4(3b)}+\sqrt{9(3b)}$

$2\sqrt{3b}+3\sqrt{3b}$ $\Rightarrow$ This time both terms can be simplified.

$5\sqrt{3b}$ $\Rightarrow$ Added like terms.

c) $-2\sqrt{x^3}+7x\sqrt{x}$

$-2x\sqrt{x}+7x\sqrt{x}$ $\Rightarrow$ Simplified the first term. I now have like terms of $x\sqrt{x}$.

$5x\sqrt{x}$ $\Rightarrow$ Simplified the like terms.

d) $w^2\sqrt{4w}-\sqrt{w^5}$

$2w^2\sqrt{w}-\sqrt{w^5}$ $\Rightarrow$ Simplified the first term.

$2w^2\sqrt{w}-w^2\sqrt{w}$ $\Rightarrow$ Simplified the second term. I now have like terms of $w^2\sqrt{w}$.

$w^2\sqrt{w}$ $\Rightarrow$ Subtracted like terms.

Homework 6.5 Simplify.

31) $2\sqrt{32p}+4\sqrt{18p}$

32) $2\sqrt{45xy}-\sqrt{80yx}$

33) $-6\sqrt{5z}+8\sqrt{5z}+\sqrt{20z}$

34) $-2\sqrt{12mn}+3\sqrt{3mn}+\sqrt{27mn}$

35) $\sqrt{32a}-\sqrt{8a}-\sqrt{50a}+\sqrt{18a}$

36) $2\sqrt{2c^3}+3c\sqrt{2c}$

37) $2\sqrt{x^4y}+5x^2\sqrt{y}$

38) $6\sqrt{x^3}-3x\sqrt{9x}$

39) $t\sqrt{5t^4}-\sqrt{5t^6}$

40) $3y\sqrt{24y}+\sqrt{6y^3}+2y\sqrt{6y}$

41) $4y\sqrt{3y}-\sqrt{12y^3}-\sqrt{27y^3}$

Homework 6.5 Answers

1) $7\sqrt{5}$ and $12\sqrt{5}$ are like.

2) $-2\sqrt{6}$ and $-\sqrt{6}$ and $2\sqrt{6}$ are like.

3) $6\sqrt{15}$ and $14\sqrt{15}$ are like as are $12\sqrt{14}$ and $4\sqrt{14}$

4) All the terms are like.

5) $3\sqrt{7}$ 6) $-10\sqrt{2}$ 7) 0 8) $7\sqrt{5}-8\sqrt{7}$ 9) $-\sqrt{15}$ 10) $-6\sqrt{2}-2\sqrt{5}$

11) $3\sqrt{2}$ 12) $8\sqrt{5}$ 13) $3\sqrt{2}$ 14) $9\sqrt{3}$ 15) $-8\sqrt{6}$ 16) $-2\sqrt{15}$

17) $3\sqrt{10}$ 18) $-\sqrt{3}-\sqrt{5}$ 19) $\sqrt{7}$ 20) -2 21) $3\sqrt{7y}$, $12\sqrt{7y}$ are like.

22) $-\sqrt{2a}$, $2\sqrt{2a}$ are like. 23) $6\sqrt{t}$, $2\sqrt{t}$ and $-2\sqrt{t}$ are like 24) $7\sqrt{mn}$, $3\sqrt{nm}$ are like.

25) $-\sqrt{7y}$ 26) $3\sqrt{ab}$ 27) $7\sqrt{xy}$ 28) $2\sqrt{2p}$ 29) $-4\sqrt{mn}$

30) $\frac{1}{6}\sqrt{14x}$ 31) $20\sqrt{2p}$ 32) $2\sqrt{5xy}$ 33) $4\sqrt{5z}$ 34) $2\sqrt{3mn}$

35) 0 36) $5c\sqrt{2c}$ 37) $7x^2\sqrt{y}$ 38) $-3x\sqrt{x}$ 39) 0

40) $9y\sqrt{6y}$ 41) $-y\sqrt{3y}$

6.6 Distribution with Square Roots

In this lesson, we'll practice distributing radical factors.

6.6.1 Distributing a Single Radical Factor

Even though the unlike terms inside the parentheses of $\sqrt{2}(1+\sqrt{6})$ can't be added, we're still able to multiply using the distributive property.

Practice 6.6.1 Distributing a Single Radical Factor
Simplify.

a) $\sqrt{5}(2+\sqrt{5})$

$2\sqrt{5}+\sqrt{5}\sqrt{5}$	$\Rightarrow$	Distributed the factor of $\sqrt{5}$. Although, $\sqrt{5}(2)$ is fine mathematically, we usually use the commutative property to write an integer factor to the left of a radical factor.
$2\sqrt{5}+\sqrt{25}$ $2\sqrt{5}+5$	$\Rightarrow$	Used the product rule to combine the two factors of $\sqrt{5}$ in the second term and then simplified.

b) $\sqrt{x}(1+\sqrt{x})$

$\sqrt{x}+\sqrt{x^2}$	$\Rightarrow$	Distributed the factor of $\sqrt{x}$ to both terms inside the parentheses. Remember that by the product rule $\sqrt{x}\sqrt{x}=\sqrt{x^2}$.
$\sqrt{x}+x$ $x+\sqrt{x}$	$\Rightarrow$	Simplified $\sqrt{x^2}$. It's common to write a radical term second.

c) $-\sqrt{6}(\sqrt{2}-\sqrt{3})$

$-\sqrt{12}+\sqrt{18}$	$\Rightarrow$	Distributed the factor of $-\sqrt{6}$ to both terms $-\sqrt{6}\sqrt{2}+\sqrt{6}\sqrt{3}$ then used the product rule to make each radicand a single product.
$-2\sqrt{3}+3\sqrt{2}$	$\Rightarrow$	Simplified each term.

d) $\sqrt{2y}(\sqrt{y}-\sqrt{2})$

$\sqrt{2y^2}-\sqrt{4y}$	$\Rightarrow$	Distributed the factor of $\sqrt{2y}$ to the unlike terms inside the parentheses.
$y\sqrt{2}-2\sqrt{y}$	$\Rightarrow$	Simplified both terms. The terms are unlike so they can't be subtracted.

Homework 6.6 Simplify.

1) $\sqrt{2}(5+\sqrt{6})$ 2) $\sqrt{3}(\sqrt{3}-\sqrt{12})$ 3) $\sqrt{x}(\sqrt{x}-2)$ 4) $-\sqrt{y}(\sqrt{y}-4)$

5) $-\sqrt{2}(\sqrt{8}-\sqrt{2})$ 6) $\sqrt{15}(\sqrt{3}-\sqrt{5})$ 7) $\sqrt{6h}(\sqrt{6h}+\sqrt{3h})$ 8) $\sqrt{3y}(\sqrt{3y}+\sqrt{3})$

9) $\sqrt{6}(\sqrt{2}+\sqrt{3}+\sqrt{6})$ 10) $\sqrt{2x}(\sqrt{2}-\sqrt{2x}+\sqrt{x})$ 11) $\sqrt{10w}(\sqrt{5w}+\sqrt{2w})$

6.6.2 Continuing with Distribution

In this topic we'll multiply two factors each of which has two unlike terms. Like before, make sure you multiply each term in the first factor to every term in the second factor.

Practice 6.6.2 Continuing with Distribution

Simplify.

a) $(\sqrt{2}+\sqrt{3})(\sqrt{2}-\sqrt{3})$

$\sqrt{2}(\sqrt{2})-\sqrt{2}(\sqrt{3})+\sqrt{3}(\sqrt{2})-\sqrt{3}(\sqrt{3})$	$\Rightarrow$	Multiplied. (If you had trouble with signs, try writing subtractions as adding an inverse.)
$\sqrt{4}-\sqrt{6}+\sqrt{6}-\sqrt{9}$	$\Rightarrow$	Found the products.
$2-3=-1$	$\Rightarrow$	Simplified. Noticed this is the difference of two squares.

b) $(1-\sqrt{6})(\sqrt{3}-\sqrt{2})$

$1(\sqrt{3})-1(\sqrt{2})-\sqrt{6}(\sqrt{3})+\sqrt{6}(\sqrt{2})$	$\Rightarrow$	Multiplied. Paid attention to the sign of the last term.
$\sqrt{3}-\sqrt{2}-\sqrt{18}+\sqrt{12}$	$\Rightarrow$	Multiplied.
$\sqrt{3}-\sqrt{2}-3\sqrt{2}+2\sqrt{3}$	$\Rightarrow$	Simplified each term.
$3\sqrt{3}-4\sqrt{2}$	$\Rightarrow$	Combined like terms.

c) $(x+\sqrt{2})(x+1)$

$x^2+x+\sqrt{2}x+\sqrt{2}$	$\Rightarrow$	Multiplied each term in the first factor to every term in the second factor. None of the terms are like.

d) $(2\sqrt{x}+\sqrt{3})(\sqrt{x}-\sqrt{3})$

$2\sqrt{x^2}-2\sqrt{3x}+\sqrt{3x}-\sqrt{9}$	$\Rightarrow$	Distributed each term in the first factor.
$2x-\sqrt{3x}-3$	$\Rightarrow$	Simplified and collected like terms.

Homework 6.6 Simplify.

12) $(1-\sqrt{2})(2+\sqrt{2})$	13) $(4+\sqrt{6})(4+\sqrt{6})$	14) $(\sqrt{5}+x)(\sqrt{5}-x)$
15) $(\sqrt{3m}-\sqrt{6})(\sqrt{m}+\sqrt{2})$	16) $(\sqrt{8}+\sqrt{3})(\sqrt{2}+\sqrt{3})$	17) $(\sqrt{7}-\sqrt{3})(\sqrt{7}-\sqrt{3})$
18) $(\sqrt{x}-2\sqrt{x})(\sqrt{x}-\sqrt{2x})$	19) $(\sqrt{a}+\sqrt{b})(\sqrt{a}-\sqrt{b})$	20) $(3\sqrt{5}-\sqrt{3})(\sqrt{3}-\sqrt{5})$
21) $(k\sqrt{h}-\sqrt{h})(\sqrt{h}-k\sqrt{h})$	22) $(2y\sqrt{y}+\sqrt{y})(\sqrt{y}+y\sqrt{y})$	23) $(\sqrt{3}+k\sqrt{2k})(k\sqrt{2k}-\sqrt{3})$

6.6.3 *The Power of a Two-Term Radical Expression*

When first trying to simplify an expression like $(\sqrt{3}-\sqrt{2})^2$, it's common for students to <u>incorrectly</u> "distribute" the exponent and get $(\sqrt{3})^2-(\sqrt{2})^2=3-2=1$. To get the right answer, remember that a natural number exponent is telling us how many times to use the base as a factor. To simplify $(\sqrt{3}-\sqrt{2})^2$, you'll need to find the product $(\sqrt{3}-\sqrt{2})(\sqrt{3}-\sqrt{2})$. Here's some practice.

Practice 6.6.3 The Power of a Two-Term Radical Expression
Simplify.

a) $(1+\sqrt{3})^3$

$(1+\sqrt{3})(1+\sqrt{3})$

$1+2\sqrt{3}+3$ ⇒ Began with two of the three factors and simplified.

$4+2\sqrt{3}$

$(4+2\sqrt{3})(1+\sqrt{3})$

$4+6\sqrt{3}+2(3)$ ⇒ Multiplied the product from the first two factors to the third factor and simplified.

$10+6\sqrt{3}$

b) $(\sqrt{3}-\sqrt{2})^2$

$(\sqrt{3}-\sqrt{2})(\sqrt{3}-\sqrt{2})$

$\sqrt{9}-\sqrt{6}-\sqrt{6}+\sqrt{4}$ ⇒ Multiplied two factors of the base.

$3-\sqrt{6}-\sqrt{6}+2$ ⇒ Simplified radical terms.

$5-2\sqrt{6}$ ⇒ Collected like terms.

c) $(\sqrt{y}+\sqrt{2})^2$

$(\sqrt{y}+\sqrt{2})(\sqrt{y}+\sqrt{2})$

$\sqrt{y^2}+\sqrt{2y}+\sqrt{2y}+\sqrt{2^2}$ ⇒ Multiplied two factors of the base and simplified.

$y+2+2\sqrt{2y}$

Homework 6.6 Simplify.

24) $(2+\sqrt{2})^2$	25) $(x+\sqrt{x})^2$	26) $(\sqrt{5}-\sqrt{15})^2$
27) $(1+\sqrt{2})^3$	28) $(x\sqrt{3}-\sqrt{3})^2$	29) $(\sqrt{x}-\sqrt{y})^2$
30) $(\sqrt{3}-\sqrt{6})^3$	31) $(x+\sqrt{2})^3$	32) $(2-\sqrt{x})^3$

Homework 6.6 Answers

1) $5\sqrt{2}+2\sqrt{3}$ 2) -3 3) $x-2\sqrt{x}$ 4) $-y+4\sqrt{y}$ 5) -2

6) $3\sqrt{5}-5\sqrt{3}$ 7) $6h+3h\sqrt{2}$ 8) $3y+3\sqrt{y}$ 9) $2\sqrt{3}+3\sqrt{2}+6$

10) $x\sqrt{2}+2\sqrt{x}-2x$ 11) $2w\sqrt{5}+5w\sqrt{2}$ 12) $-\sqrt{2}$ 13) $8\sqrt{6}+22$

14) $5-x^2$ 15) $m\sqrt{3}-2\sqrt{3}$ 16) $7+3\sqrt{6}$ 17) $10-2\sqrt{21}$

18) $x\sqrt{2}-x$ 19) $a-b$ 20) $4\sqrt{15}-18$ 21) $-hk^2+2hk-h$

22) $2y^3+3y^2+y$ 23) $2k^3-3$ 24) $6+4\sqrt{2}$ 25) $x^2+2x\sqrt{x}+x$

26) $20-10\sqrt{3}$ 27) $7+5\sqrt{2}$ 28) $3x^2-6x+3$ 29) $x+y-2\sqrt{xy}$

30) $21\sqrt{3}-15\sqrt{6}$ 31) $x^3+3\sqrt{2}x^2+6x+2\sqrt{2}$ 32) $-x\sqrt{x}+6x-12\sqrt{x}+8$

6.7 The Quotient Rule for Square Roots

In this section, we'll simplify quotients using the quotient rule for square roots.

The Quotient Rule for Square Roots
If all radicands are non-negative, and the denominator is not 0, then the square root of a quotient is the quotient of the square roots of the radicand.   Example: For non-negative real numbers *x* and positive real numbers *y*, $\sqrt{\frac{x}{y}}=\frac{\sqrt{x}}{\sqrt{y}}$

Notice we can take two points of view with the quotient rule. Let's begin by replacing the expression on the left with the one on the right.

6.7.1 The Quotient Rule for Square Roots – Part 1

The quotient rule for square roots sometimes helps us rewrite the quotient so we can use some of our previous ideas for simplifying square roots.

Practice 6.7.1 The Quotient Rule for Square Roots – Part 1

Simplify using the quotient rule for square roots.

a) $\sqrt{\frac{3}{4}}$

$\frac{\sqrt{3}}{\sqrt{4}}$ ⇒ Used the quotient rule for square roots.

$\frac{\sqrt{3}}{2}$ ⇒ Simplified the denominator. The numerator can't be simplified further.

b) $3\sqrt{\frac{8}{9}}$

$\frac{3}{1}\left(\frac{\sqrt{8}}{\sqrt{9}}\right)$ ⇒ Used the quotient rule for square roots and thought of 3 as a fraction.

$\frac{\cancel{3}}{1}\left(\frac{\sqrt{8}}{\cancel{3}}\right)=\sqrt{8}$ ⇒ Simplified the denominator and reduced.

$2\sqrt{2}$ ⇒ Simplified $\sqrt{8}$.

c) $\sqrt{\frac{4x^2}{y^4}}$

$\frac{\sqrt{4x^2}}{\sqrt{y^4}}$ ⇒ Used the quotient rule for square roots.

$\frac{2x}{y^2}$ ⇒ Simplified.

d) $\frac{k}{3}\sqrt{\frac{18}{k^2}}$

$\frac{k}{3}\frac{\sqrt{18}}{\sqrt{k^2}}$ ⇒ Used the quotient rule with the second factor.

$\frac{k}{3}\frac{3\sqrt{2}}{k}$ ⇒ Simplified the square roots.

$\frac{\cancel{k}}{\cancel{3}}\frac{\cancel{3}\sqrt{2}}{\cancel{k}}=\sqrt{2}$ ⇒ Reduced common factors of 1. The answer is $\sqrt{2}$.

Homework 6.7 Simplify using the quotient rule for square roots.

1) $\sqrt{\frac{1}{16}}$ 2) $\sqrt{\frac{5}{9}}$ 3) $\sqrt{\frac{7}{b^4}}$ 4) $\sqrt{\frac{27}{4}}$ 5) $\sqrt{\frac{a^3}{36}}$

6) $2\sqrt{\frac{11}{36}}$ 7) $r\sqrt{\frac{49}{r^2}}$ 8) $\sqrt{\frac{25}{49c^4}}$ 9) $-\frac{1}{3}\sqrt{\frac{9}{m^{10}}}$ 10) $-5\sqrt{\frac{12}{25}}$

11) $-k\sqrt{\frac{40}{k^4}}$ 12) $\frac{-4}{5}\sqrt{\frac{25}{16}}$ 13) $\frac{y}{6}\sqrt{\frac{9}{y^2}}$ 14) $7\sqrt{\frac{45}{49}}$ 15) $\sqrt{\frac{3a^5}{b^4}}$

16) $\sqrt{\frac{12y^3}{x^2}}$ 17) $-15\sqrt{\frac{18}{25}}$ 18) $-\frac{y}{4}\sqrt{\frac{16x^2}{y^2}}$ 19) $\frac{q}{2p}\sqrt{\frac{8p^2}{q^2}}$

6.7.2 The Quotient Rule for Square Roots – Part 2

Recall the quotient rule tells us that under certain conditions $\sqrt{\frac{x}{y}}=\frac{\sqrt{x}}{\sqrt{y}}$. So far, we've taken the point of view that we'll replace the expression on the left with the one on the right. Now let's practice with some problems where the first step will be to replace the expression on the right with the one on the left. Notice I specifically said, "the first step". That's because, after the first step, you may want to switch to the other point of view within the same problem.

Practice 6.7.2 The Quotient Rule for Square Roots – Part 2
Simplify using the quotient rule for square roots.

a) $\frac{\sqrt{40}}{\sqrt{2}}$

$\sqrt{\frac{40}{2}}$ ⇒ Used the quotient rule for square roots.

$\sqrt{20}$ ⇒ Reduced in the radicand.

$2\sqrt{5}$ ⇒ Simplified $\sqrt{4\times 5}$.

b) $\dfrac{\sqrt{k^8}}{\sqrt{k^5 t^2}}$

$\sqrt{\dfrac{k^8}{k^5 t^2}}$ $\Rightarrow$ Used the quotient rule for square roots.

$\sqrt{\dfrac{k^3}{t^2}}$ $\Rightarrow$ Used the quotient rule for exponents $\left(k^{8-5} = k^3\right)$.

$\dfrac{\sqrt{k^3}}{\sqrt{t^2}} = \dfrac{k\sqrt{k}}{t}$ $\Rightarrow$ Used the quotient rule for square roots again and then simplified.

c) $\dfrac{2\sqrt{6k^2}}{\sqrt{24}}$

$\dfrac{2}{1}\sqrt{\dfrac{6k^2}{24}} = \dfrac{2}{1}\sqrt{\dfrac{k^2}{4}}$ $\Rightarrow$ Used the quotient rule for square roots and reduced the common factor of 6.

$\dfrac{2}{1}\dfrac{\sqrt{k^2}}{\sqrt{4}} = \dfrac{\not{2}}{1}\left(\dfrac{k}{\not{2}}\right) = k$ $\Rightarrow$ Used the quotient rule from the other point of view, simplified the square roots and reduced the common factor of 2.

d) $\dfrac{3\sqrt{2x^6}}{\sqrt{18x}}$

$\dfrac{3}{1}\sqrt{\dfrac{2x^6}{18x}}$ $\Rightarrow$ Used the quotient rule.

$\dfrac{3}{1}\sqrt{\dfrac{x^5}{9}}$ $\Rightarrow$ Divided.

$\dfrac{3}{1}\dfrac{\sqrt{x^5}}{\sqrt{9}}$ $\Rightarrow$ Used the quotient rule again.

$\dfrac{\not{3}x^2\sqrt{x}}{\not{3}} = x^2\sqrt{x}$ $\Rightarrow$ Simplified and reduced the common factor of 3.

Homework 6.7 Simplify using the quotient rule for square roots.

20) $\dfrac{\sqrt{15}}{\sqrt{3}}$ 21) $\dfrac{-\sqrt{14}}{\sqrt{7}}$ 22) $\dfrac{\sqrt{n^5}}{\sqrt{n^3}}$ 23) $\dfrac{\sqrt{12x^3}}{\sqrt{3x}}$ 24) $\dfrac{\sqrt{150}}{\sqrt{3}}$ 25) $\dfrac{2\sqrt{x}}{\sqrt{x^7}}$

26) $\dfrac{\sqrt{5}}{\sqrt{45}}$ 27) $\dfrac{-3\sqrt{2y}}{\sqrt{18}}$ 28) $\dfrac{\sqrt{54a}}{\sqrt{2a^5}}$ 29) $\dfrac{12\sqrt{6t^5}}{\sqrt{24t^5}}$ 30) $\dfrac{3\sqrt{5b}}{\sqrt{45b^3}}$

Homework 6.7 Answers

1) $\frac{1}{4}$	2) $\frac{\sqrt{5}}{3}$	3) $\frac{\sqrt{7}}{b^2}$	4) $\frac{3\sqrt{3}}{2}$	5) $\frac{a\sqrt{a}}{6}$	6) $\frac{\sqrt{11}}{3}$
7) 7	8) $\frac{5}{7c^2}$	9) $\frac{-1}{m^5}$	10) $-2\sqrt{3}$	11) $\frac{-2\sqrt{10}}{k}$	12) -1
13) $\frac{1}{2}$	14) $3\sqrt{5}$	15) $\frac{a^2\sqrt{3a}}{b^2}$	16) $\frac{2y\sqrt{3y}}{x}$	17) $-9\sqrt{2}$	18) $-x$
19) $\sqrt{2}$	20) $\sqrt{5}$	21) $-\sqrt{2}$	22) n	23) $2x$	24) $5\sqrt{2}$
25) $\frac{2}{x^3}$	26) $\frac{1}{3}$	27) $-\sqrt{y}$	28) $\frac{3\sqrt{3}}{a^2}$	29) 6	30) $\frac{1}{b}$

6.8 Rationalizing the Denominator

It's common, when first working with square roots, for a student to assume they've done something wrong if their answer is $\frac{3}{\sqrt{6}}$ and the textbook answer is $\frac{\sqrt{6}}{2}$. Changing $\frac{3}{\sqrt{6}}$ to the equivalent expression $\frac{\sqrt{6}}{2}$ uses a technique known as rationalizing the denominator. In this case the original irrational denominator $\sqrt{6}$ has been replaced by the rational denominator 2.

6.8.1 Rationalizing a Constant Denominator

To rationalize the denominator, without changing the value of the fraction, we'll multiply by 1 (the multiplicative identity). For example, multiplying $\frac{1}{\sqrt{3}}$ by 1 in the form $\frac{\sqrt{3}}{\sqrt{3}}$ leads to a rational denominator, $\frac{1}{\sqrt{3}} = \frac{1}{\sqrt{3}} \times 1 = \frac{1}{\sqrt{3}} \times \frac{\sqrt{3}}{\sqrt{3}} = \frac{1 \times \sqrt{3}}{\sqrt{3}\sqrt{3}} = \frac{\sqrt{3}}{\sqrt{9}} = \frac{\sqrt{3}}{3}$.

From now on we'll assume simplifying a root implies a rationalized denominator.

Practice 6.8.1 Rationalizing a Constant Denominator

Rationalize the denominator and simplify if possible.

a) $\frac{2}{\sqrt{2}}$

$\frac{2}{\sqrt{2}} \times \frac{\sqrt{2}}{\sqrt{2}} = \frac{2\sqrt{2}}{\sqrt{4}}$	$\Rightarrow$	Multiplied the numerator and denominator by $\sqrt{2}$. Noticed the denominator became $\sqrt{4}$ by the product rule.
$\frac{2\sqrt{2}}{2}$	$\Rightarrow$	Simplified the denominator.
$\frac{\cancel{2}\sqrt{2}}{\cancel{2}}$ $\sqrt{2}$	$\Rightarrow$	Reduced the common factor of 2.

b) $\frac{6\sqrt{2}}{\sqrt{3}}$

$\frac{6\sqrt{2}}{\sqrt{3}} \times \frac{\sqrt{3}}{\sqrt{3}} = \frac{6\sqrt{6}}{\sqrt{9}}$	$\Rightarrow$	Multiplied by 1 in the form $\frac{\sqrt{3}}{\sqrt{3}}$.
$\frac{6\sqrt{6}}{3} = 2\sqrt{6}$	$\Rightarrow$	Simplified the denominator and reduced the common factor of 3.

c) $\dfrac{\sqrt{2}}{2\sqrt{5}}$

$\dfrac{\sqrt{2}}{2\sqrt{5}} \times \dfrac{\sqrt{5}}{\sqrt{5}} = \dfrac{\sqrt{10}}{2\sqrt{25}}$ ⇒ Multiplied by 1. Noticed the factor of 2 in the denominator is already rational and doesn't have to be accounted for in the factor of 1.

$\dfrac{\sqrt{10}}{2(5)} = \dfrac{\sqrt{10}}{10}$ ⇒ Simplified the denominator.

Homework 6.8 Rationalize the denominator and simplify if possible.

1) $\dfrac{5}{\sqrt{10}}$ 2) $\dfrac{2}{\sqrt{6}}$ 3) $\dfrac{12\sqrt{2}}{\sqrt{3}}$ 4) $\dfrac{-\sqrt{3}}{3\sqrt{5}}$ 5) $\dfrac{3\sqrt{2}}{\sqrt{15}}$ 6) $\dfrac{\sqrt{21}}{6\sqrt{6}}$

7) $-\dfrac{2\sqrt{2}}{\sqrt{10}}$ 8) $\dfrac{-7\sqrt{2}}{4\sqrt{14}}$

6.8.2 Rationalizing a Variable Denominator

We can use the same process to rationalize a variable denominator.

Practice 6.8.2 Rationalizing a Variable Denominator
Rationalize the denominator and simplify.

a) $\dfrac{2}{\sqrt{x}}$

$\dfrac{2}{\sqrt{x}} \times \dfrac{\sqrt{x}}{\sqrt{x}} = \dfrac{2\sqrt{x}}{\sqrt{x^2}}$ ⇒ Multiplied by 1 in the form $\dfrac{\sqrt{x}}{\sqrt{x}}$.

$\dfrac{2\sqrt{x}}{x}$ ⇒ Simplified the denominator.

b) $\dfrac{25}{\sqrt{5x}}$

$\dfrac{25}{\sqrt{5x}} \times \dfrac{\sqrt{5x}}{\sqrt{5x}} = \dfrac{25\sqrt{5x}}{\sqrt{25x^2}}$ ⇒ Multiplied the original fraction by 1.

$\dfrac{25\sqrt{5x}}{5x} = \dfrac{5\sqrt{5x}}{x}$ ⇒ Simplified the denominator and reduced the common factor of 5.

Homework 6.8 Rationalize the denominator and simplify.

9) $\dfrac{\sqrt{8}}{\sqrt{k}}$ 10) $\dfrac{x}{\sqrt{x}}$ 11) $\dfrac{h}{\sqrt{2h}}$ 12) $\dfrac{6m}{\sqrt{6m}}$ 13) $\dfrac{b\sqrt{15}}{\sqrt{5b}}$ 14) $-\dfrac{h\sqrt{2}}{2\sqrt{h}}$

15) $\dfrac{k\sqrt{6}}{2\sqrt{2k}}$ 16) $\dfrac{-3\sqrt{7y}}{y\sqrt{21}}$

6.8.3 Rationalizing After the Quotient Rule

Let's put the quotient rule and rationalizing the denominator together.

Practice 6.8.3 Rationalizing After the Quotient Rule

Simplify using the quotient rule as the first step.

a) $\sqrt{\frac{1}{3}}$

$\frac{\sqrt{1}}{\sqrt{3}} = \frac{1}{\sqrt{3}}$	$\Rightarrow$	Began with the quotient rule.
$\frac{1}{\sqrt{3}} \times \frac{\sqrt{3}}{\sqrt{3}} = \frac{\sqrt{3}}{\sqrt{9}}$	$\Rightarrow$	Continued with rationalizing the denominator.
$\frac{\sqrt{3}}{3}$	$\Rightarrow$	The final answer.

b) $2\sqrt{\frac{3}{8}}$

$2\frac{\sqrt{3}}{\sqrt{8}} = \cancel{2}\frac{\sqrt{3}}{\cancel{2}\sqrt{2}} = \frac{\sqrt{3}}{\sqrt{2}}$	$\Rightarrow$	Began with the quotient rule, simplified $\sqrt{8} = 2\sqrt{2}$ and reduced the common factor of 2.
$\frac{\sqrt{3}}{\sqrt{2}} \times \frac{\sqrt{2}}{\sqrt{2}} = \frac{\sqrt{6}}{\sqrt{4}}$	$\Rightarrow$	Rationalized the denominator.
$\frac{\sqrt{6}}{2}$	$\Rightarrow$	Simplified.

---

c) $\frac{\sqrt{7}}{\sqrt{14x}}$

$\sqrt{\frac{7}{14x}} = \sqrt{\frac{1}{2x}}$	$\Rightarrow$	Began with the quotient rule and reduced the common factor of 7.
$\frac{\sqrt{1}}{\sqrt{2x}} \times \frac{\sqrt{2x}}{\sqrt{2x}} = \frac{\sqrt{2x}}{\sqrt{4x^2}}$	$\Rightarrow$	Used the quotient rule again and then multiplied to rationalize the denominator.
$\frac{\sqrt{2x}}{2x}$	$\Rightarrow$	Simplified the denominator.

Homework 6.8 Simplify using the quotient rule as the first step.

17) $\sqrt{\frac{2}{7}}$ 18) $\sqrt{\frac{49}{x}}$ 19) $2\sqrt{\frac{3}{10}}$ 20) $-4k\sqrt{\frac{3}{2k}}$ 21) $\frac{-5\sqrt{2}}{\sqrt{10}}$ 22) $\frac{-7\sqrt{2}}{4\sqrt{14}}$

23) $\frac{\sqrt{9y}}{\sqrt{18y}}$ 24) $\frac{b\sqrt{15}}{\sqrt{5b}}$ 25) $-\frac{\sqrt{7}}{y\sqrt{14}}$ 26) $\frac{k\sqrt{4k}}{\sqrt{12k}}$ 27) $-\frac{3\sqrt{7y}}{y\sqrt{21}}$ 28) $\frac{x\sqrt{10}}{\sqrt{5x}}$

Homework 6.8 Answers

1) $\frac{\sqrt{10}}{2}$	2) $\frac{\sqrt{6}}{3}$	3) $4\sqrt{6}$	4) $-\frac{\sqrt{15}}{15}$	5) $\frac{\sqrt{30}}{5}$	6) $\frac{\sqrt{14}}{12}$	7) $\frac{-2\sqrt{5}}{5}$
8) $-\frac{\sqrt{7}}{4}$	9) $\frac{2\sqrt{2k}}{k}$	10) $\sqrt{x}$	11) $\frac{\sqrt{2h}}{h}$	12) $\sqrt{6m}$	13) $\sqrt{3b}$	14) $-\frac{\sqrt{2h}}{2}$
15) $\frac{\sqrt{3k}}{2}$	16) $\frac{-\sqrt{3y}}{y}$	17) $\frac{\sqrt{14}}{7}$	18) $\frac{7\sqrt{x}}{x}$	19) $\frac{\sqrt{30}}{5}$	20) $-2\sqrt{6k}$	21) $-\sqrt{5}$
22) $-\frac{\sqrt{7}}{4}$	23) $\frac{\sqrt{2}}{2}$	24) $\sqrt{3b}$	25) $-\frac{\sqrt{2}}{2y}$	26) $\frac{k\sqrt{3}}{3}$	27) $-\frac{\sqrt{3y}}{y}$	28) $\sqrt{2x}$

6.9 Preparing for the Quadratic Formula

In the next chapter we'll be using the quadratic formula to quickly solve quadratic equations.

Definition – The Quadratic Formula
The solutions to the quadratic equation in standard form $ax^2+bx+c=0$ are given by the quadratic formula $x=\dfrac{-b\pm\sqrt{b^2-4ac}}{2a}$.

Although the formula quickly solves a quadratic equation, simplifying the answer(s) can take a bit of work. In this lesson, we'll practice some of the skills you'll need when you simplify quadratic answers.

6.9.1 Rewriting Quotients

As you can see, the expression $\dfrac{-b\pm\sqrt{b^2-4ac}}{2a}$ is a quotient, so our last operation will be a division. Before dividing we'll have to simplify the dividend which includes a new symbol the "plus-minus" symbol, $\pm$. The idea is that there will be two answers, one where we add and one where we subtract. Here's some practice using the $\pm$ symbol with quotients.

Practice 6.9.1 Rewriting Quotients

Simplify both answers.

a) $x=\dfrac{2\pm 6}{2}$

$x=\dfrac{2+6}{2}$ or $x=\dfrac{2-6}{2}$

$x=\dfrac{8}{2}$ $\quad$ $x=\dfrac{-4}{2}$ $\Rightarrow$ One dividend begins as a sum and the other as a difference. Simplified following the order of operations.

$x=4$ $\quad$ $x=-2$

$\{-2,4\}$ $\Rightarrow$ The solution set.

b) $x=\dfrac{-1\pm 4}{-15}$

$x=\dfrac{-1+4}{-15}$ or $x=\dfrac{-1-4}{-15}$

$x=\dfrac{3}{-15}$ $\quad$ $x=\dfrac{-5}{-15}$ $\Rightarrow$ Wrote both answers and simplified.

$x=-\dfrac{1}{5}$ $\quad$ $x=\dfrac{1}{3}$

$\left\{-\frac{1}{5},\frac{1}{3}\right\}$ $\Rightarrow$ The solution set.

Homework 6.9 Simplify both answers.

1) $x = \dfrac{3 \pm 5}{4}$ 2) $x = \dfrac{-4 \pm 4}{-2}$ 3) $x = \dfrac{-36 \pm 9}{9}$ 4) $x = \dfrac{-7 \pm 12}{-18}$

6.9.2 Rewriting Quotients Involving Radicals

In practice, you'll often have to simplify a square root before you simplify the quotient.

Practice 6.9.2 Rewriting Quotients Involving Radicals

Simplify both answers.

a) $x = \dfrac{-15 \pm \sqrt{25}}{10}$

$x = \dfrac{-15+5}{10}$ or $x = \dfrac{-15-5}{10}$

$x = \dfrac{-10}{10} = 1$ $x = \dfrac{-20}{10} = -2$ $\Rightarrow$ Both answers have three operations. Square root is first, then adding or subtracting and division is last.

$\{-2, 1\}$ $\Rightarrow$ The solution set.

b) $x = \dfrac{2 \pm \sqrt{8}}{4}$

$x = \dfrac{2 \pm 2\sqrt{2}}{4}$ $\Rightarrow$ Simplified the square root of 8. Will wait to reduce common factors until the dividend isn't a sum or difference.

$x = \dfrac{2}{4} \pm \dfrac{2\sqrt{2}}{4}$ $\Rightarrow$ Rewrote the quotient as two terms. Made sure to keep the common denominator.

$x = \dfrac{\not{2}}{\not{4}} \pm \dfrac{\not{2}\sqrt{2}}{\not{4}}$

$x = \dfrac{1}{2} \pm \dfrac{\sqrt{2}}{2}$ $\Rightarrow$ Reduced the common factors of 1.

$\left\{\dfrac{1}{2} - \dfrac{\sqrt{2}}{2}, \dfrac{1}{2} + \dfrac{\sqrt{2}}{2}\right\}$ $\Rightarrow$ The solution set.

$\left\{\dfrac{1 \pm \sqrt{2}}{2}\right\}$ $\Rightarrow$ Sometimes people prefer writing the solution set as a single quotient.

Homework 6.9 Simplify both answers.

5) $x = \dfrac{7 \pm \sqrt{16}}{6}$ 6) $x = \dfrac{-9 \pm \sqrt{49}}{-2}$ 7) $x = \dfrac{-14 \pm \sqrt{0}}{7}$ 8) $x = \dfrac{6 \pm \sqrt{18}}{3}$

9) $x = \dfrac{-5 \pm \sqrt{75}}{10}$ 10) $x = \dfrac{-4 \pm \sqrt{18}}{12}$ 11) $x = \dfrac{9 \pm \sqrt{12}}{-12}$ 12) $x = \dfrac{10 \pm \sqrt{20}}{8}$

6.9.3 Simplifying Quadratic Answers

Earlier you practiced substituting values for *a*, *b* and *c* and evaluating the discriminant b^2-4ac. Let's combine this idea with your current work to simplify some quadratic answers.

Practice 6.9.3 Simplifying Quadratic Answers

Simplify the answers.

a) $x=\dfrac{-(-1)\pm\sqrt{(-1)^2-4(1)(-20)}}{2(1)}$

$x=\dfrac{1\pm\sqrt{1-(-80)}}{2}$ ⇒ Began simplifying the radicand.

$x=\dfrac{1\pm\sqrt{81}}{2}$ ⇒ Continued simplifying the radicand.

$x=\dfrac{1\pm 9}{2}$ ⇒ Simplified $\sqrt{81}$.

$x=5$ or $x=-4$

$\{-4,5\}$ ⇒ Simplified and wrote the solution set.

b) $x=\dfrac{-(-6)\pm\sqrt{(-6)^2-4(1)(7)}}{2(1)}$

$x=\dfrac{6\pm\sqrt{36-(28)}}{2}$ ⇒ Began simplifying.

$x=\dfrac{6\pm\sqrt{8}}{2}=\dfrac{6\pm 2\sqrt{2}}{2}$ ⇒ Continued simplifying the radicand and simplified $\sqrt{8}$.

$x=\dfrac{6}{2}\pm\dfrac{2\sqrt{2}}{2}$ ⇒ Wrote as two fractions and reduced.

$x=3\pm\sqrt{2}$

$\{3-\sqrt{2},3+\sqrt{2}\}$ ⇒ The solution set.

Homework 6.9 Simplify the answers.

13) $x=\dfrac{-(-11)\pm\sqrt{(-11)^2-4(2)(5)}}{2(2)}$

14) $x=\dfrac{-(-2)\pm\sqrt{(-2)^2-4(1)(-1)}}{2(1)}$

15) $a=\dfrac{-(2)\pm\sqrt{(2)^2-4(1)(-2)}}{2(1)}$

16) $y=\dfrac{-(12)\pm\sqrt{(12)^2-4(-12)(1)}}{2(-12)}$

17) $t=\dfrac{-(4)\pm\sqrt{(4)^2-4(1)(-1)}}{2(1)}$

18) $x=\dfrac{-(12)\pm\sqrt{(12)^2-4(4)(9)}}{2(4)}$

19) $x=\dfrac{-(-6)\pm\sqrt{(-6)^2-4(9)(-5)}}{2(9)}$

20) $x=\dfrac{-(-14)\pm\sqrt{(-14)^2-4(7)(4)}}{2(7)}$

6.9.4 *Checking Answers for a Quadratic Equation*

Now let's practice checking some answers.

Practice 6.9.4 Checking Answers for a Quadratic Equation

Check the supplied answer.

---

a) Show that $\sqrt{5}$ is a solution for $3t^2 - 4 = 6 + t^2$.

$3(\sqrt{5})^2 - 4$	$6 + (\sqrt{5})^2$	$\Rightarrow$	Substituted the answer for every occurrence of the variable.
$3(5) - 4$ $15 - 4$ 11	$6 + 5$ 11	$\Rightarrow$	Simplified and found both expressions simplify to the same value as expected.

---

b) Show that $-2\sqrt{7}$ is a solution for $x^2 = 2x^2 - 28$.

$(-2\sqrt{7})^2$	$2(-2\sqrt{7})^2 - 28$	$\Rightarrow$	Substituted the answer for every occurrence of the variable.
$(-2)^2(\sqrt{7})^2$ 4×7 28	$2\left[(-2)^2(\sqrt{7})^2\right] - 28$ $2[4 \times 7] - 28$ $2[28] - 28$ 28	$\Rightarrow$	Simplified using the power of a product rule for exponents and found both expressions simplify to the same value as expected.

---

c) Show that $3 - \sqrt{2}$ is a solution for $x^2 - 6x + 7 = 0$.

$(3 - \sqrt{2})^2 - 6(3 - \sqrt{2}) + 7$	0	$\Rightarrow$	Substituted the answer for every occurrence of the variable.
$(3 - \sqrt{2})(3 - \sqrt{2}) - 6(3 - \sqrt{2}) + 7$ $3^2 - 2(3)(\sqrt{2}) + (\sqrt{2})^2 - 18 + 6\sqrt{2} + 7$ $9 - 6\sqrt{2} + 2 - 18 + 6\sqrt{2} + 7$ 0	0	$\Rightarrow$	Made sure not to "distribute" the exponent in the first term. Simplified using the form for the square of a binomial difference.

---

d) Show that $\frac{\sqrt{3}}{2}$ is a solution for $3y^2 = y^2 + \frac{3}{2}$.

$3\left(\frac{\sqrt{3}}{2}\right)^2$	$\left(\frac{\sqrt{3}}{2}\right)^2 + \frac{3}{2}$	$\Rightarrow$	Substituted the answer for every occurrence of the variable.
$3\frac{(\sqrt{3})^2}{2^2}$	$\frac{(\sqrt{3})^2}{2^2} + \frac{3}{2}$	$\Rightarrow$	Used the power of a quotient rule for exponents.
$3 \times \frac{3}{4}$ $\frac{9}{4}$	$\frac{3}{4} + \frac{3}{2}$ $\frac{3}{4} + \frac{6}{4}$ $\frac{9}{4}$	$\Rightarrow$	Both expressions simplify to the same value.

Homework 6.9 Check the supplied answer.

21) Show that $\sqrt{2}$ is a solution for $-n^2 - 2 = n^2 - 6$.

22) Show that $-3\sqrt{2}$ is a solution for $2t^2 - 18 = t^2$.

23) Show that $1+\sqrt{5}$ is a solution for $x^2 - 2x - 4 = 0$.

24) Show that $1-\sqrt{2}$ is a solution for $a^2 - 2a - 1 = 0$.

25) Show that $1/2+\sqrt{3}$ is a solution for $4x^2 - 4x = 11$.

26) Show that $-\sqrt{6}/3$ is a solution for $-y^2 + 2 = 2y^2$.

27) Show that $\sqrt{21}/3$ is a solution for $0 = 2(3 - x^2) - (x^2 - 1)$.

28) Show that $1+4\sqrt{3}$ is a solution for $y^2 - 2y - 47 = 0$.

29) Show that $-1-\sqrt{3}$ is a solution for $k^2 = 2 - 2k$.

30) Show that $2\sqrt{2}$ is a solution for $-6(m^2 + 2) + 4(m^2 - 5) = -48$.

31) Show that $2-3\sqrt{5}$ is a solution for $41 + 4x = x^2$.

32) Show that $\frac{3}{2}+\frac{\sqrt{5}}{2}$ is a solution for $k^2 + 1 = 3k$.

33) Show that $\frac{-\sqrt{10}}{5}$ is a solution for $\frac{1}{2} = -2(5k^2 - 4) - \frac{7}{2}$.

34) Show that $\frac{2\sqrt{5}}{3}$ is a solution for $9t^2 = 3t^2 + \frac{40}{3}$.

35) Show that $1+3\sqrt{3}$ is a solution for $2x + 26 + 2x^2 = 3x^2$.

Homework 6.9 Answers

1) $\left\{-\frac{1}{2}, 2\right\}$ 2) $\{0,4\}$ 3) $\{-5,-3\}$ 4) $\left\{-\frac{5}{18}, \frac{19}{18}\right\}$ 5) $\left\{\frac{1}{2}, \frac{11}{6}\right\}$

6) $\{1,8\}$ 7) $\{-2\}$ 8) $\{2-\sqrt{2}, 2+\sqrt{2}\}$ 9) $\left\{-\frac{1}{2}-\frac{\sqrt{3}}{2}, -\frac{1}{2}+\frac{\sqrt{3}}{2}\right\}$

10) $\left\{-\frac{1}{3} \pm \frac{\sqrt{2}}{4}\right\}$ 11) $\left\{-\frac{3}{4} \pm \frac{\sqrt{3}}{6}\right\}$ 12) $\left\{\frac{5}{4} \pm \frac{\sqrt{5}}{4}\right\}$ 13) $\left\{\frac{1}{2}, 5\right\}$ 14) $\{1 \pm \sqrt{2}\}$

15) $\{-1 \pm \sqrt{3}\}$ 16) $\left\{\frac{1}{2} \pm \frac{\sqrt{3}}{3}\right\}$ 17) $\{-2 \pm \sqrt{5}\}$ 18) $\left\{-\frac{3}{2}\right\}$ 19) $\left\{\frac{1}{3} \pm \frac{\sqrt{6}}{3}\right\}$

20) $\left\{1 \pm \frac{\sqrt{21}}{7}\right\}$ 21) $-4, -4$ 22) $18, 18$ 23) $0, 0$ 24) $0, 0$ 25) $11, 11$ 26) $\frac{4}{3}, \frac{4}{3}$

27) $0, 0$ 28) $0, 0$ 29) $4+2\sqrt{3}, 4+2\sqrt{3}$ 30) $-48, -48$ 31) $49-12\sqrt{5} = 49-12\sqrt{5}$

32) $\frac{9}{2}+\frac{3\sqrt{5}}{2}, \frac{9}{2}+\frac{3\sqrt{5}}{2}$ 33) $\frac{1}{2}, \frac{1}{2}$ 34) $20, 20$ 35) $84+18\sqrt{3}, 84+18\sqrt{3}$

Chapter 7

Quadratic Functions

7.1 The Square Root Method

As you've seen, after using the quadratic formula, simplifying the solution can be a bit of work. In certain situations, the square root method allows you to find quadratic solutions faster.

7.1.1 The Square Root Method

If, after writing your quadratic equation in standard form, it happens that the value of b is 0, then the square root method is a good choice for quickly solving the equation.

***Procedure* – The Square Root Method**
If $x^2 = k$ then $x = \sqrt{k}$ or $x = -\sqrt{k}$. We assume $k \geq 0$. Comments: The idea that x equals $\sqrt{k}$ or $-\sqrt{k}$ is often written $x = \pm\sqrt{k}$.

Let's practice solving some quadratic equations using the square root method.

Practice 7.1.1 The Square Root Method
Solve using the square root method.

a) $x^2 = 12$

$x = \sqrt{12}$ or $x = -\sqrt{12}$	$\Rightarrow$	Solved for x using the square root method.
$x = 2\sqrt{3}$ or $x = -2\sqrt{3}$	$\Rightarrow$	Simplified the answers.
$\{-2\sqrt{3}, 2\sqrt{3}\}$	$\Rightarrow$	The solution set.

Homework 7.1 Solve using the square root method.

1) $a^2 = 36$	2) $x^2 = 3$	3) $y^2 = 32$	4) $k^2 = 45$

7.1.2 Using Division to Isolate the Variable

If our quadratic equation has the form $ax^2 = b$ we'll need to divide before using the square root method. For example, to solve $2x^2 = 1$ my first step should be to divide both expressions by 2 and isolate x^2. When it's necessary, remember to rationalize the denominator.

Practice 7.1.2 Using Division to Isolate the Variable
Solve using the square root method.

a) $-3x^2 = -24$

$x^2 = 8$	$\Rightarrow$	Divided both sides by -3. Made sure the value on the right became positive.
$x = \sqrt{8}$ or $x = -\sqrt{8}$ $x = 2\sqrt{2}$ or $x = -2\sqrt{2}$	$\Rightarrow$	Used the square root method and simplified the answers.

b) $3x^2 = 2$	
$x^2 = \frac{2}{3}$	$\Rightarrow$ Divided both sides by 3.
$x = \pm\sqrt{\frac{2}{3}} = \pm\frac{\sqrt{2}}{\sqrt{3}}$	$\Rightarrow$ Used the square root method and then the quotient rule to write the solution as the quotient of square roots.
$x = \pm\frac{\sqrt{2}}{\sqrt{3}} \times \frac{\sqrt{3}}{\sqrt{3}}$	$\Rightarrow$ Multiplied by 1 to rationalize the denominator.
$x = \pm\frac{\sqrt{6}}{\sqrt{9}} = \pm\frac{\sqrt{6}}{3}$	$\Rightarrow$ Simplified the answers.

Homework 7.1 Solve using the square root method.

5) $2n^2 = 72$ 6) $-a^2 = -18$ 7) $-12y^2 = -8$ 8) $4 = 5x^2$ 9) $18k^2 = 6$

7.1.3 Writing Quadratic Equations in the Form $x^2 = k$

We'll often need to use our techniques for solving linear equations to rewrite an equation into the form $x^2 = k$.

Practice 7.1.3 Writing Quadratic Equations in the Form $x^2 = k$

Solve using the square root method.

a) $2y^2 + 2 = 14 - y^2$	
$2y^2 + 2 + y^2 = 14 - y^2 + y^2$ $3y^2 + 2 = 14$	$\Rightarrow$ Isolated the variable term on the side with the larger coefficient.
$3y^2 + 2 - 2 = 14 - 2$ $3y^2 = 12$	$\Rightarrow$ Isolated the constant term on the other side.
$\frac{3y^2}{3} = \frac{12}{3}$ $y^2 = 4$	$\Rightarrow$ Divided and simplified.
$y = \pm\sqrt{4}$	$\Rightarrow$ Solved using the square root method.
$y = 2$ or $y = -2$	$\Rightarrow$ Simplified the radical term.
$2(-2)^2 + 2$ $\quad$ $14 - (-2)^2$ $8 + 2$ $\quad$ $14 - 4$ 10 $\quad$ 10	$\Rightarrow$ Checked one of the answers.

b) $(2x-3)^2 = 3x^2 - 12x + 27$		
$4x^2 - 12x + 9 = 3x^2 - 12x + 27$	$\Rightarrow$	Distributed to simplify the left side.
$x^2 + 9 = 27$ $x^2 = 18$	$\Rightarrow$	Isolated the variable term on the left since the coefficient is greater. After adding $12x$ to both expressions the first-degree term goes to 0.
$x = \pm\sqrt{18}$ $x = \pm 3\sqrt{2}$	$\Rightarrow$	Solved using the square root method.
$x = -3\sqrt{2}$ or $x = 3\sqrt{2}$	$\Rightarrow$	There are two answers.

Homework 7.1 Solve using the square root method.

10) $4a^2 + 12 = 36$ 11) $-x^2 + 12 = x^2 - 4$ 12) $8 + 3t^2 + 4t = 2(2t+5)$

13) $5r^2 - 41 = 13 + 2r^2$ 14) $-2(2y^2 - y - 3) = 5 - 2y + y^2$ 15) $-3(p+2)^2 + 12p = -13$

16) $6(x-1)^2 + 6(x^2 - 1) = (2x-2)^2$

7.1.4 Solving Equations of the Form $(x-a)^2 = k$

It's common for students to solve an equation like $(x-1)^2 + 3 = 5$ by first "FOILing" $(x-1)^2$. Although this method will work, there's a faster approach if you realize the equation can be written in the form $(x-a)^2 = k$ and then solve for x using the square root method.

$(x-1)^2 + 3 = 5$	$\Rightarrow$	The original equation
$(x-1)^2 = 2$	$\Rightarrow$	Subtracted 3 from both expressions
$x - 1 = \pm\sqrt{2}$	$\Rightarrow$	Used the square root method with $x-1$ as the "x".
$x - 1 + 1 = \pm\sqrt{2} + 1$ $x = 1 \pm \sqrt{2}$	$\Rightarrow$	Added 1 to both sides to isolate the variable. The terms are not like so you can't add or subtract. It's common to put the constant term to the left of the radical term.
$\{1-\sqrt{2}, 1+\sqrt{2}\}$	$\Rightarrow$	The solution set

Here's some practice with the idea.

Practice 7.1.4 Solving Equations of the Form $(x-a)^2 = k$

Solve using the square root method.

a) $6(y-4)^2 = 6$		
$\frac{6(y-4)^2}{6} = \frac{6}{6}$ $(y-4)^2 = 1$	$\Rightarrow$	Divided both expressions using the property of equality and simplified.
$y - 4 = \pm\sqrt{1}$	$\Rightarrow$	Used the square root method.
$y - 4 = \pm 1$	$\Rightarrow$	The square root of 1 is 1.

$$y-4+4=4\pm 1$$

$y=4\pm 1$ ⇒ Solved for *y*.

$$y=5 \text{ or } y=3$$

b) $2(3b+2)^2-11=25$

$2(3b+2)^2=36$	⇒	Added 11 to both expressions and simplified.
$(3b+2)^2=18$	⇒	Divided both expressions by 2 to isolate the variable factor.
$3b+2=\pm\sqrt{18}$ $3b+2=\pm 3\sqrt{2}$	⇒	Used the square root method and simplified the radical.
$3b=-2\pm 3\sqrt{2}$	⇒	Subtracted 2 from both expressions and simplified.
$b=\dfrac{-2\pm 3\sqrt{2}}{3}$ $b=\dfrac{-2}{3}\pm\sqrt{2}$	⇒	Divided both expressions by 3, replaced the expression on the right with two terms with a common denominator and reduced the common factor of 3 in the second term.

Homework 7.1 Solve using the square root method.

17) $(x+3)^2=4$ 18) $(q-1)^2=2$ 19) $3(m+4)^2=48$

20) $-(2x-1)^2=-8$ 21) $(4x+3)^2-1=4$ 22) $8(r-4)^2+6=8$

23) $15(3m-5)^2-5=-2$ 24) $-7+6(2w+3)^2=3$ 25) $\left(c-\frac{5}{2}\right)^2=\frac{1}{4}$

7.1.5 Realizing a Quadratic Solution is Not a Real Number

Sometimes solutions for a quadratic equation aren't real numbers. For example, $x^2=-4$ has no solution in the real numbers since the left expression is a perfect square, which means it will be positive or zero, while the right expression is constantly negative.

As you solve these equations keep your eye on the "big picture". If, during the process you realize there's no solution in the real numbers, state the fact. Otherwise, find any solutions.

Practice 7.1.5 Realizing a Quadratic Solution is Not a Real Number

Solve using the square root method.

a) $4x^2-5=-13$

$4x^2=-8$ $x^2=-2$	⇒	Added 5 to both sides and divided by 4. Noticed the left side is positive or 0 while the right side is negative.
No solution in the real numbers.	⇒	Using the square root method would have given $x=\pm\sqrt{-2}$ which is not a real number.

b) $3 - y^2 = 1$		
$-y^2 = -2$	$\Rightarrow$	Subtracted 3 from both sides.
$y^2 = 2$	$\Rightarrow$	Divided by -1. This time the equation can be solved.
$y = \pm\sqrt{2}$	$\Rightarrow$	Used the square root method to find my answers.

Homework 7.1 Solve using the square root method.

26) $4k^2 + 20 = 6k^2 + 23$	27) $4x^2 - 2(x^2 + 1) = 14$	28) $-5(C - 4)^2 - 7 = 4$
29) $6 + 3(2h - 5)^2 = 27$	30) $1 = 2 - (N^2 - 3)$	31) $-4(y - 3)^2 + 6 = 10$
32) $4(2 - h^2) + 7 = 19$	33) $2 - 9(2 - B)^2 = -16$	34) $-3 = 15 - 2(3 - 5y)^2$

Homework 7.1 Answers

1) $\{-6,6\}$
2) $\{-\sqrt{3},\sqrt{3}\}$
3) $\{-4\sqrt{2},4\sqrt{2}\}$
4) $\{-3\sqrt{5},3\sqrt{5}\}$
5) $\{-6,6\}$
6) $\{-3\sqrt{2},3\sqrt{2}\}$
7) $\left\{-\frac{\sqrt{6}}{3},\frac{\sqrt{6}}{3}\right\}$
8) $\left\{-\frac{2\sqrt{5}}{5},\frac{2\sqrt{5}}{5}\right\}$
9) $\left\{-\frac{\sqrt{3}}{3},\frac{\sqrt{3}}{3}\right\}$
10) $\{-\sqrt{6},\sqrt{6}\}$
11) $\{-2\sqrt{2},2\sqrt{2}\}$
12) $\left\{-\frac{\sqrt{6}}{3},\frac{\sqrt{6}}{3}\right\}$
13) $\{-3\sqrt{2},3\sqrt{2}\}$
14) $\{-\frac{1}{5},1\}$
15) $\left\{-\frac{\sqrt{3}}{3},\frac{\sqrt{3}}{3}\right\}$
16) $\{-\frac{1}{2},1\}$
17) $\{-5,-1\}$
18) $\{1-\sqrt{2},1+\sqrt{2}\}$
19) $\{-8,0\}$
20) $\left\{\frac{1}{2}-\sqrt{2},\frac{1}{2}+\sqrt{2}\right\}$
21) $\left\{\frac{-3-\sqrt{5}}{4},\frac{-3+\sqrt{5}}{4}\right\}$
22) $\left\{\frac{7}{2},\frac{9}{2}\right\}$
23) $\left\{\frac{5}{3}-\frac{\sqrt{5}}{15},\frac{5}{3}+\frac{\sqrt{5}}{15}\right\}$
24) $\left\{-\frac{3}{2}-\frac{\sqrt{15}}{6},-\frac{3}{2}+\frac{\sqrt{15}}{6}\right\}$
25) $\{2,3\}$
26) No solution in the real numbers.
27) $\{-2\sqrt{2},2\sqrt{2}\}$
28) No solution in the real numbers.
29) $\left\{\frac{5-\sqrt{7}}{2},\frac{5+\sqrt{7}}{2}\right\}$
30) $\{-2,2\}$
31) No solution in the real numbers.
32) No solution in the real numbers.
33) $\{2-\sqrt{2},2+\sqrt{2}\}$
34) $\{0,\frac{6}{5}\}$

7.2 The Quadratic Formula

It was easy to solve a linear equation like $x-1=0$ because, after adding 1 to both expressions and simplifying, $x=1$, we could "isolate" the variable term and see the solution is 1. If we add a second-degree term in *x* to the left side, $x^2+x-1=0$, it's surprising how difficult solving this new quadratic equation becomes. Unlike the linear equation, where there was a single variable to isolate, we now have two unlike terms in *x* which can't be combined. Because of the unlike terms, it's no longer possible to isolate a variable term using addition or subtraction. Today you'll practice solving an equation like $x^2+x-1=0$ using the quadratic formula.

7.2.1 Identifying a, b and c in a Quadratic Equation

To use the quadratic formula, we must always write the quadratic equation in the form $ax^2+bx+c=0$ and identify the coefficients and the constant. Your work with the AC method pays off here since the values for *a*, *b* and *c* are the same.

Definition – The Quadratic Formula
The solutions to the quadratic equation in standard form $ax^2+bx+c=0$ are given by the quadratic formula $x=\dfrac{-b\pm\sqrt{b^2-4ac}}{2a}$.

Let's practice writing a quadratic equation in the form $ax^2+bx+c=0$ and identifying *a*, *b* and *c*.

Practice 7.2.1 Identifying *a*, *b* and *c* in a Quadratic Equation

Write the polynomial in the form $ax^2+bx+c=0$ *and identify a, b, and c.*

a) $x^2-2=x$

$x^2-2-x=x-x$ $x^2-x-2=0$	$\Rightarrow$	Used the property of equality so all variables are on the left then wrote the expression in standard form.
$a=1,\ b=-1,\ c=-2$	$\Rightarrow$	Identified *a*, *b* and *c*. Remembered to think of the equation as $x^2+-1x+-2=0$.

b) $-3x^2=2x+26+2x^2$

$-3x^2+3x^2=2x+26+2x^2+3x^2$ $0=2x+26+5x^2$ $0=5x^2+2x+26$	$\Rightarrow$	Used the property of equality to move all variable terms to the right side then wrote the right side in standard form.
$a=5,\ b=2,\ c=26$	$\Rightarrow$	Identified *a*, *b* and *c*.
$-5x^2-2x-26=0$	$\Rightarrow$	Some students prefer moving all terms to the left side. In that case $a=-5,\ b=-2,\ c=-26$.

Homework 7.2 Write the polynomial in the form $ax^2+bx+c=0$ and identify a, b, and c.			
1) $5-11x=-2x^2$	2) $6y^2+4y+7=2y^2+6$	3) $x(x+1)=4(x^2-1)$	4) $4m^2=9$

7.2.2 Beginning to Use the Quadratic Formula

Now let's solve some quadratic equations using the quadratic formula.

Procedure – Solving Quadratic Equations Using the Quadratic Formula
1. Write the equation in the form $ax^2+bx+c=0$ and identify *a, b* and *c*.
2. Substitute the values of *a, b* and *c* into the quadratic formula $x=\frac{-b\pm\sqrt{b^2-4ac}}{2a}$.
3. Simplify your answer(s).
4. Check your answer(s).

Here's some practice.

Practice 7.2.2 Beginning to Use the Quadratic Formula

Solve using the quadratic formula.

a) $x^2-20=8x$

$x^2-8x-20=0$ ⇒ Subtracted $8x$ from both expressions and wrote the polynomial on the left in standard form.

$a=1,\ b=-8, c=-20$ ⇒ Identified *a, b* and *c*. Remembered to think of the polynomial as a sum which makes both *b* and *c* negative.

$$x=\frac{-(-8)\pm\sqrt{(-8)^2-4(1)(-20)}}{2(1)}$$ ⇒ Substituted for *a, b* and *c* in the quadratic formula. It's best to substitute into parentheses. I now have both answers to the original equation.

$$x=\frac{8\pm\sqrt{64-(-80)}}{2}$$

$$x=\frac{8\pm\sqrt{144}}{2}$$

$$x=\frac{8\pm12}{2}$$

$$x=\frac{8+12}{2} \text{ or } x=\frac{8-12}{2}$$

$$x=\frac{20}{2} \text{ or } x=\frac{-4}{2}$$

$$x=10 \text{ or } x=-2$$

⇒ Getting answers for the original equation was easy, simplifying those answers will take a bit of work. My simplified answers are 10 and -2.

$(10)^2-20$	$8(10)$	$(-2)^2-20$	$8(-2)$
$100-20$	80	$4-20$	-16
80		-16	

⇒ Checked both answers using the original equation. The solution set is $\{-2,10\}$.

b) $2w^2 = 2w + 1$

$2w^2 - 2w - 1 = 0$	$\Rightarrow$ Wrote the equation in the form $ax^2 + bx + c = 0$.
$a = 2,\ b = -2,\ c = -1$	$\Rightarrow$ Identified *a, b* and *c*.
$x = \dfrac{-(-2) \pm \sqrt{(-2)^2 - 4(2)(-1)}}{2(2)}$	$\Rightarrow$ Substituted for *a, b* and *c* in the quadratic formula.
$x = \dfrac{2 \pm \sqrt{4 - (-8)}}{4}$ $x = \dfrac{2 \pm \sqrt{12}}{4}$	$\Rightarrow$ Simplified.
$x = \dfrac{2 \pm 2\sqrt{3}}{4}$	$\Rightarrow$ Simplified $\sqrt{12}$.
$x = \dfrac{2}{4} + \dfrac{2\sqrt{3}}{4}$ or $x = \dfrac{2}{4} - \dfrac{2\sqrt{3}}{4}$	$\Rightarrow$ The original quotient can be rewritten as two fractions each with the common denominator.
$x = \dfrac{1}{2} + \dfrac{\sqrt{3}}{2}$ or $x = \dfrac{1}{2} - \dfrac{\sqrt{3}}{2}$	$\Rightarrow$ Reduced common factors.
$x = \dfrac{1+\sqrt{3}}{2}$ or $x = \dfrac{1-\sqrt{3}}{2}$	$\Rightarrow$ Sometimes the answers will be written as a single fraction.
$2\left(\dfrac{1+\sqrt{3}}{2}\right)^2 \quad 2\left(\dfrac{1+\sqrt{3}}{2}\right)+1$ $2\left(\dfrac{1+2\sqrt{3}+3}{4}\right) \quad 1+\sqrt{3}+1$ $\dfrac{4+2\sqrt{3}}{2} \quad 2+\sqrt{3}$ $2+\sqrt{3}$	$\Rightarrow$ The check for one of the answers. Remembered to not "distribute" the exponent with the expression on the left, but instead, to first use the quotient rule for exponents and then multiply two factors of the base, $2\left(\dfrac{(1+\sqrt{3})^2}{2^2}\right) = 2\left(\dfrac{(1+\sqrt{3})(1+\sqrt{3})}{4}\right)$.

Homework 7.2 Solve using the quadratic formula.

5) $m^2 - 2m - 24 = 0$ 6) $k^2 - 6k = -1$ 7) $x^2 - 2x - 1 = 0$ 8) $6x^2 - 6 = 5x$

9) $2x^2 - 11x + 5 = 0$ 10) $4x^2 + 12x + 9 = 0$ 11) $2p^2 = -2p + 7$

12) $a^2 + 2a - 2 = 0$ 13) $12x^2 = 12x + 1$ 14) $y^2 + 4y = 1$

15) $6x^2 - 6x + 5 = -3x^2 + 10$ 16) $7y^2 = 5$ 17) $2x + 26 + 2x^2 = 3x^2$

7.2.3 Uses of the Discriminant

You can use the discriminant, $b^2 - 4ac$, to predict the number and kind of solutions. If the discriminant is positive, you will have two real solutions. If the discriminant is 0, you will have one real solution. And if the discriminant is negative, you will have no real solutions since a square root with a negative radicand is not a real number.

Uses of the Discriminant
The discriminant $b^2 - 4ac$ can be used to predict the number and kind of solutions for a quadratic equation. 1. If $b^2 - 4ac > 0$ you will have two real solutions. 2. If $b^2 - 4ac = 0$ you will have one real solution. 3. If $b^2 - 4ac < 0$ you will have no real solutions.

You might think that finding the discriminant as the first step in the solution process will increase your work. Actually, it's just the opposite. If the discriminant is negative you can, for now, stop the process and answer, "No solution in the real numbers." If the discriminant is positive or 0, simply replace the discriminant with the value and continue simplifying the answers. Let's practice using the discriminant as the first step in the solution process.

Practice 7.2.3 Uses of the Discriminant

Find the discriminant as the first step in the solution process and discuss the implications of your value.

a) $h - 1 = 2h^2$

$-2h^2 + h - 1 = 0$	$\Rightarrow$	Wrote the equation in the form $ax^2 + bx + c = 0$.
$a = -2,\ b = 1, c = -1$	$\Rightarrow$	Identified *a, b* and *c*.
$(1)^2 - 4(-2)(-1)$	$\Rightarrow$	Substituted for *a, b* and *c* in the discriminant.
-7	$\Rightarrow$	The discriminant is negative there are no solutions in the real numbers. For now, there is no reason to continue the process.
No solution in the real numbers.	$\Rightarrow$	Included an answer.

b) $2x^2 - 6x + 1 = 0$

$a = 2,\ b = -6, c = 1$	$\Rightarrow$	Identified *a, b* and *c*.
$(-6)^2 - 4(2)(1) = 28$	$\Rightarrow$	Substituted for *a, b* and *c* in the discriminant. The discriminant is positive. There will be two real solutions.
$x = \dfrac{-(-6) \pm \sqrt{28}}{2(2)}$	$\Rightarrow$	Continued with the quadratic formula using the discriminant found previously.
$x = \dfrac{6 \pm 2\sqrt{7}}{4}$	$\Rightarrow$	Simplified $\sqrt{28}$.
$x = \dfrac{3}{2} + \dfrac{\sqrt{7}}{2}$ or $x = \dfrac{3}{2} - \dfrac{\sqrt{7}}{2}$ $x = \dfrac{3 + \sqrt{7}}{2}$ or $x = \dfrac{3 - \sqrt{7}}{2}$	$\Rightarrow$	Simplified the answers. Wrote the answers in both forms to stay in practice. As expected, there are two answers in the real numbers.

Homework 7.2 Find the discriminant as the first step in the solution process and discuss the implications of your value.

18) $-3k^2-3k+2=0$	19) $n-4n^2=9$	20) $(x+1)=(3x+2)^2$
21) $\frac{1}{2}x^2+x-2=0$	22) $3y^2-7y=-15$	23) $z^2-z+\frac{1}{4}=0$
24) $3(m^2+m)-4m=m+1$	25) $-2x^2-7=-14x$	26) $8p^2+\frac{1}{7}p+1=0$
27) $2x-3(2-x^2)=-4-(x+x^2)$	28) $1-6x=-9x^2$	29) $-80p^2+23{,}000=0$
30) $0=-16t^2+88t$	31) $-1374=6.7x^2-158x$	32) $1.7t^2-42t=1.16t^2+10t+218$
33) $0=-0.2p^2+242p-42{,}000$	34) $408=-25.5d^2$	35) $0.6(t-26.5)^2=2965$

Homework 7.2 Answers

1) $2x^2 - 11x + 5 = 0;\ a = 2\ \ b = -11\ \ c = 5$ 2) $4y^2 + 4y + 1 = 0;\ a = 4\ \ b = 4\ \ c = 1$

3) $-3x^2 + x + 4 = 0;\ a = -3\ \ b = 1\ \ c = 4$ 4) $4m^2 + 0m - 9 = 0;\ a = 4\ \ b = 0\ \ c = -9$

5) $\{-4, 6\}$ 6) $\{3 - 2\sqrt{2}, 3 + 2\sqrt{2}\}$ 7) $\{1 - \sqrt{2}, 1 + \sqrt{2}\}$ 8) $\{-\tfrac{2}{3}, \tfrac{3}{2}\}$

9) $\{\tfrac{1}{2}, 5\}$ 10) $\{-\tfrac{3}{2}\}$ 11) $\left\{\frac{-1 \pm \sqrt{15}}{2}\right\}$ 12) $\{-1 \pm \sqrt{3}\}$ 13) $\left\{\frac{1}{2} - \frac{\sqrt{3}}{3}, \frac{1}{2} + \frac{\sqrt{3}}{3}\right\}$

14) $\{-2 - \sqrt{5}, -2 + \sqrt{5}\}$ 15) $\left\{\frac{1}{3} \pm \frac{\sqrt{6}}{3}\right\}$ 16) $\left\{-\frac{\sqrt{35}}{7}, \frac{\sqrt{35}}{7}\right\}$ 17) $\{1 - 3\sqrt{3}, 1 + 3\sqrt{3}\}$

18) The discriminant is 33, there are two real solutions. $\left\{-\frac{1}{2} - \frac{\sqrt{33}}{6}, -\frac{1}{2} + \frac{\sqrt{33}}{6}\right\}$.

19) The discriminant is -143, there are no solutions in the real numbers.

20) The discriminant is 13, there are two real solutions. $\left\{-\frac{11}{18} - \frac{\sqrt{13}}{18}, -\frac{11}{18} + \frac{\sqrt{13}}{18}\right\}$.

21) The discriminant is 5, there are two real solutions. $\{-1 - \sqrt{5}, -1 + \sqrt{5}\}$.

22) The discriminant is -131, there are no solutions in the real numbers.

23) The discriminant is 0, there is one real solution. $\{\tfrac{1}{2}\}$.

24) The discriminant is 16, the two real solutions are $\{-\tfrac{1}{3}, 1\}$.

25) The discriminant is 140, the two real solutions are $\left\{\frac{7 - \sqrt{35}}{2}, \frac{7 + \sqrt{35}}{2}\right\}$.

26) The discriminant is negative. There are no solutions in the real numbers.

27) The discriminant is 41. the two real solutions are $\left\{\frac{-3 - \sqrt{41}}{8}, \frac{-3 + \sqrt{41}}{8}\right\}$.

28) The discriminant is 0, there is one real solution. $\{\tfrac{1}{3}\}$.

29) The discriminant is 7,360,000, the two real solutions are $\{\pm \tfrac{5}{2}\sqrt{46}\}$.

30) The discriminant is 7,744, the two real solutions are $\{0, \tfrac{11}{2}\}$.

31) The discriminant is $-11{,}859.2$, there are no solutions in the real numbers.

32) The discriminant is 3174.88, the two real solutions are approximately $\{-4.02, 100.32\}$.

33) The discriminant is 24,964, the two real solutions are $\{210, 1000\}$.

34) The discriminant is $-41{,}616$, there are no solutions in the real numbers.

35) The discriminant is 7,116, the two real solutions are $\{-43.797, 96.797\}$.

7.3 Applying Quadratic Equations

In this section, you'll solve quadratic equations in the context of using formulas and functions. We'll begin by revisiting the Pythagorean Theorem. Earlier, you didn't have the tools to solve for the length of a leg or the length of the hypotenuse directly, now you do.

7.3.1 The Pythagorean Theorem

Recall that the Pythagorean Theorem describes a relationship between the lengths of the sides of a right triangle.

The Pythagorean Theorem
Given a right triangle, the sum of the squares of the lengths of the legs equals the square of the length of the hypotenuse. If a and b are the lengths of the legs of a right triangle, and c is the length of the hypotenuse, then $a^2 + b^2 = c^2$.

Let's practice solving quadratic equations that arise from the Pythagorean Theorem.

Practice 7.3.1 The Pythagorean Theorem

Find the length of the unknown side. If necessary, find both an exact and an approximate answer. Round decimal answers to the nearest tenth.

a) The two legs of a right triangle have lengths 2 miles and 4 miles. Find the length of the hypotenuse.

$a^2 + b^2 = c^2$

$2^2 + 4^2 = c^2$ $\Rightarrow$ Substituted the known values and simplified.

$20 = c^2$

$c = \pm\sqrt{20}$
$= \pm 2\sqrt{5}$
≈ 4.5

$\Rightarrow$ Used the square root method to isolate c. The length of the hypotenuse is exactly $2\sqrt{5}$ miles which is about four and a half miles. Only the positive root is necessary since distance is positive.

b) A 20-foot ladder is set against the wall of a house with the base of the ladder six feet away from the wall. If the top of the ladder just reaches the bottom of the second story window, how far above the ground is the bottom of the window?

20 feet
Side of house
b feet
6 feet

$\Rightarrow$ Drew a picture that contains the information from the problem.

$a^2 + b^2 = c^2$
$6^2 + b^2 = 20^2$

$\Rightarrow$ Substituted the known values into the Pythagorean Theorem.

$$b^2 = 400 - 36 = 364$$
$$b = \pm\sqrt{364} = 2\sqrt{91}$$
$$b = 2\sqrt{91} \text{ feet}, \ b \approx 19.1 \text{ feet}$$

⇒ Solved using the square root method. The bottom of the window is exactly $2\sqrt{91}$ feet or a little over 19 feet above the ground.

Homework 7.3: Find the length of the unknown side. If necessary, find both an exact and an approximate answer. Round decimal answers to the nearest tenth.

1) 2)

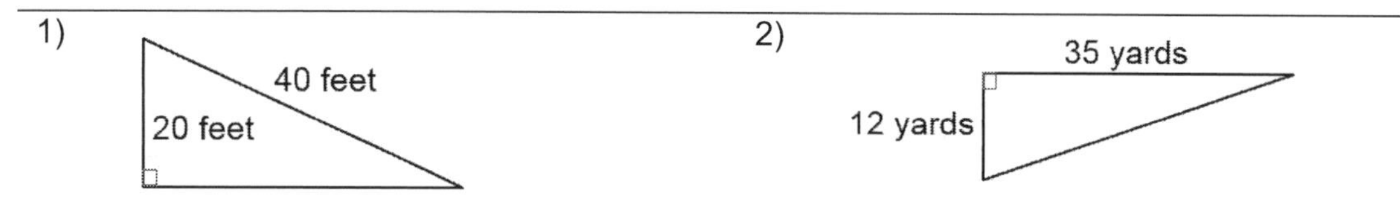

3) The Pythagorean Theorem can help us decide if something we build is "square" (the corners are 90 degrees). You've built a deck that's 12 feet by 16 feet. If the deck is square what should the diagonal distance from corner to corner through the center of the deck be?

4) A daughter is moving out of her parent's home. She wants to take her queen-sized box spring which is 60 inches wide. The dimensions of the back of her parent's SUV is 42 inches wide by 35 inches high. Is there any way to turn the box spring so that it will fit in the SUV? (A box spring is rigid and will not bend.)

5) A student who is 4 feet 8 inches tall wants to estimate the height of a tree to the nearest foot. The tree is supported by a rope that is staked into the ground and tied about a quarter of the way up the trunk from the ground. She paces off the distance from the base of the tree to the stake and found it was about 12 feet. She estimates the length of the rope from the tree to the stake is about 15 feet. What is the estimated height of the tree?

7.3.2 Modeling Profit With Quadratic Equations

Demand, revenue, cost and profit are four common ideas from economics. **Demand** models the relationship between the price of an item and the quantity consumers will purchase. It's often the case that raising an item's price will reduce demand, while lowering the price will increase demand.

Multiplying the price for one item by the number of items demanded gives you the **total revenue**. For instance, if you set the price for one lawn mower at \$250, and consumers demand 10 lawnmowers at that price, then the total revenue would be, \$250(10) = \$2,500.

Of course, there is usually a cost to produce each item and you can find the **total cost** by multiplying the cost to build one item times the number of items demanded. If it costs \$100 to build each lawnmower, then the total cost to build 10 lawnmowers would be \$100(10) = \$1,000.

The **total profit** will be the total revenue minus the total cost. For our lawnmowers, the total profit would be 250(10) – 100(10) = \$1,500. Notice, by the distributive property, that the left expression can be replace with $10(250-100) = 10(150)$. This implies the profit per lawnmower is \$150. Here's some practice with the idea of total profit.

Practice 7.3.2 Modeling Total Profit With Quadratic Equations

Answer the following questions.

A company selling lawn mowers finds their total profit, y, can be predicted using the function $y = -80x^2 + 28{,}200x - 1{,}495{,}000$ where x is the price charged for one lawn mower.

a) Find the profit if the lawn mowers were free (price of \$0) and explain the reason for the value.

At a price of \$0, demand would be high but there wouldn't be any revenue. Costs to build the lawn mowers would remain. The profit (loss) of $-\$1{,}495{,}000$ would account for parts, rent, labor, advertising etc.

⇒ Evaluated the function at a price of \$0.

$$y = -80(0)^2 + 28{,}200(0) - 1{,}495{,}000$$
$$y = -1{,}495{,}000$$

b) The maximum profit the company can earn for selling lawn mowers is \$990,125. (By the end of the course you'll have the tools to find this value for yourself.) Find the price the company should charge per lawn mower to maximize the profit.

To maximize the total profit, the company should charge \$176.25 for a lawn mower.

⇒ Substituted 990,125 for the total profit and solved for the price.

$$990{,}125 = -80x^2 + 28{,}200x - 1{,}495{,}000$$
$$176.25 = x$$

c) If the company earned a total profit of \$750,000, what price were they charging for a lawnmower? Why might there be two prices?

To earn a profit of \$750,000 they were either charging around \$121.46 or \$231.04. At \$121.46 they have a lower price and a higher demand. At \$231.04 they have a higher price but a lower demand.

⇒ Substituted 750,000 for the profit and solved for the price.

$$750{,}000 = -80x^2 + 28{,}200x - 1{,}495{,}000$$
$$x \approx 121.46 \text{ or } x \approx 231.04$$

d) Find the price that will earn the company a total profit of \$1,000,000.

The discriminant is negative so the equation doesn't have a solution in the real numbers. It's not possible to earn a profit of \$1,000,000.

⇒ Substituted 1,000,000 for the profit and solved for the price.

$$1{,}000{,}000 = -80x^2 + 28{,}200x - 1{,}495{,}000$$
$$80x^2 - 28{,}200x + 2{,}495{,}000 = 0$$
$$b^2 - 4ac < 0$$

x is not a real number.

e) What price will lead to a total profit of \$0? Explain the meaning of each answer.

The company will earn a profit of \$0 at a price of \$65 or \$287.50. At \$65 demand will be high but the total revenue will only cover the costs so the profit is \$0. At \$287.50 the price per mower is high but demand will drop due to the high price. Once again, the total revenue will only cover the costs.

⇒ Substituted 0 for the total profit and solved for the price.

$$0 = -80x^2 + 28{,}200x - 1{,}495{,}000$$
$$x = 65 \text{ or } x \approx 287.50$$

Homework 7.3 Answer the following questions.

6) A student starts a lawn care service during the summer. Their monthly profit can be modeled using $y = -10x^2 + 240x - 190$ where x is the price they charge per hour.

   a) Find the student's profit if they charge \$0 and discuss the meaning of the answer.

   b) The maximum monthly profit they can earn is \$1,250. Find the price they should charge to earn the maximum profit.

   c) Under the assumption that raising their hourly price would increase their profit they raise their hourly price to \$13 an hour. What is their monthly profit now? How can it be that raising the price would reduce the profit?

   d) What hourly price would give the student a monthly profit of \$1,000? Discuss both answers.

   e) Find the hourly price that would give a monthly profit of \$1,300.

7) An organization is selling boxes of cookies door to door to raise money. The function $y = -0.75x^2 + 6.3x - 2.4$ predicts the profit to the organization per house given x, the price charged per box.

   a) What price should they charge to reach their maximum profit of \$10.83 per house?

   b) What would be an advantage of dropping the profit per household to \$10.80?

   c) If they lowered their price per box to \$3.20 what would their profit be? What other price would result in this same profit?

   d) Someone in the organization decides that, "To raise the profit we'll just raise the price." Explain what happens to the profit if the price is raised to \$7?

8) A person who makes "vintage" bicycles finds if they charge $\$p$ per bicycle then the function $y = -0.2x^2 + 242x - 42{,}000$ estimates their profit (or loss) for the year.

   a) Find x if the profit is \$30,600 and speculate on the meaning of each answer.

   b) What price should the person charge per bicycle to reach the maximum profit of \$31,205?

   c) If the person is satisfied with an annual profit of \$25,000, and wants to build the least number of bikes, what price should they charge? Round to the nearest dollar.

7.3.3 Some Assorted Functions

Let's finish this section by applying an assortment of functions.

Practice 7.3.3 Some Assorted Functions

Answer the following questions.

a) The number of people attending a wedding reception, y, can be predicted using the number of hours since 6 p.m., x, and the function $y = -10x^2 + 80x + 200$.

   1. What question is $y = 300$ asking and what's the answer to the question?
   2. What question is $x = 2.5$ asking and what's the answer to the question?
   3. What question is $x = 0$ asking and what's the answer to the question?
   4. $y = 360$ is the maximum number of people attending? When will this occur?

1. $y = 300$ is asking when will there be 300 people attending the reception. There will be 300 people at the reception around 7:30 p.m. and 12:30 a.m.	$\Rightarrow$	Substituted 300 for y and solved for x. Added the times to 6 p.m. $300 = -10x^2 + 80x + 200$ $x \approx 1.5$ or $x \approx 6.5$
2. $x = 2.5$ is asking how many people will be attending at 8:30 p.m. At 8:30 p.m. there will be around 338 people at the reception.	$\Rightarrow$	Substituted 2.5 for x and solved for y to find the number of people attending 2.5 hours after 6 p.m. $y = -10(2.5)^2 + 80(2.5) + 200$ $y = 337.5$
3. $x = 0$ is asking for the attendance at 6:00 p.m. There were 200 people at the reception when it began.	$\Rightarrow$	Substituted 0 for x and solved for y to find the number of people attending at exactly 6 p.m.
4. The maximum attendance is expected to happen at 10 p.m.	$\Rightarrow$	Substituted 360 for y and solved $300 = -10x^2 + 80x + 200$, to find the value of x is 4. Added 4 hours to 6 p.m.

Homework 7.3 Answer the following questions.

9) The function $y = 6.7x^2 - 158x + 1372$ predicts y, the cases of Rocky Mountain Spotted Fever in the United States from 1980 until 2009.

a) Find and discuss $y = 441$ (The minimum number of cases for 1980 to 2009).

b) What question is $y = 1{,}000$ asking and what's the answer to the question?

c) What question is $x = 31$ asking and what's the answer to the question?

d) What question is $y = 0$ asking and what's the answer to the question?

10) Since 1970 the function $y = 5.5x^2 - 46.4x + 692$ closely modeled the cost for one year of tuition and fees (in dollars) at a United States four-year public college.

a) Answer the question $x = 0$ is asking.

b) Answer the question $x = 45$ is asking.

c) Answer the question $y = \$6{,}692$ is asking.

d) Find and discuss the meaning of $y = \$12{,}692$. Compare your answer to the answer in terms of the time it took for the price to increase $6,000 when compared to part c above.

11) If an object is thrown into the air with an initial speed of 88 feet per second the function $y = -16x^2 + 88x + 6$ describes its height, in feet, as a function of time in seconds.

a) Find and discuss the meaning of both answers for $y = 118$.

b) Find and discuss the meaning of $x = 0$.

c) What question is $y = 130$ asking? What's the meaning of your answer?

d) What question is $y = 0$ asking? How should we handle the negative solution?

Homework 7.3 Answers

1) $20\sqrt{3}$ ft. ≈ 34.6 ft. 2) 37 yds. 3) 20 ft.

4) No. The opening is only $7\sqrt{61} \approx 54.7$ inches. 5) About 36 feet.

6)
a) Evaluating the function if $x = 0$ gives -190 which implies if they make no profit they're still $190 in debt. (For supplies etc.)

b) Substituting 1,250 for y and solving gives $x = 12$ so they should charge $12 an hour.

c) Evaluating the function when $x = 13$ gives a profit of $1240. Even though they are earning more per hour the hours demanded has fallen due to the higher price.

d) Substituting 1,000 for y and solving gives $x = 7$ or $x = 17$. At an hourly rate of $7 they'll have high demand but earn less per hour. At $17 they'll earn more per hour but have less demand.

e) Substituting 1,300 for y and solving leads to a discriminant of -2,000 so the solution is not a real number. It's not possible for the student to earn a profit $1,300 under the current conditions.

7)
a) Solving $10.83 = -0.75x^2 + 6.3x - 2.4$ shows they should charge $4.20 per box.

b) Solving $10.80 = -0.75x^2 + 6.3x - 2.4$ shows they could charge exactly $4 per box and they wouldn't have to worry about having coins to make change.

c) Evaluating the function when $x = 3.20$ gives a profit of $10.08. Solving $10.08 = -0.75x^2 + 6.3x - 2.4$ shows that setting the price per box to $5.20 will also result in a profit of $3.20.

d) Evaluating the function when $x = 7$ gives a profit of $4.95. It's true they're earning more per box but because of the high price demand has fallen so less boxes are sold.

8)
a) Solving $30{,}600 = -0.2x^2 + 242x - 42{,}000$ gives prices of $550 and $660. $550 will lead to higher demand but a lower price per bike and $660 will lead to reduced demand but a higher price per bike. In either case their profit will be $30,600.

b) They'll achieve their maximum profit of $31,205 by charging $605 per bike.

c) They can achieve a profit of $25,000 by setting their price at either $429 or $782. Since they want to build the least number of bikes they should set their price at $782.

9) a) Solving the function when $y = 441$ shows this happens after around 12 years so the minimum number of cases occurred around 1992.

b) $y = 1{,}000$ is asking to find the year(s) there were 1,000 cases of Rocky Mountain Spotted Fever. Setting y to 1,000 and solving shows there were 1,000 cases around 1983 and 2001.

c) $x = 31$ is asking us to estimate the number of cases in 2011. Evaluating the function for 31 tells us that if the trend continued into 2011 there would be approximately 2,913 cases.

d) $y = 0$ is asking us to find when there won't be any cases. Setting y to 0 and solving leads to a negative discriminant which implies we don't expect the number of cases will drop to 0.

---

10) a) In 1970 the cost for one year of tuition and fees was around $692.

b) In 2015 the expected cost for one year of tuition and fees will be around $9,742.

c) It cost $6,692 in tuition and fees around 2008. (The negative solution isn't useful this time.) So it took 38 years for the tuition to rise $6,000 above the 1970 price.

d) It's expected that one year of college will cost $12,692 in 2021 which is 13 years after 2008. So starting in 1970 and assuming the trend continues, it took about 38 years for the price of tuition and fees to rise $6,000 and only an additional 13 years to add the next $6,000.

---

11) a) The object will be at a height of 118 feet after 2 seconds when it's going up and it will again be at a height of 118 feet after 3.5 seconds on the way down.

b) $x = 0$ describes the height of the object the moment the person throwing the object lets go. The person let go of the object when it was six feet above the ground.

c) $y = 130$ asks how long it will take for the object to be 130 feet off the ground. Setting y to 130 and solving gives a negative discriminant. The object never gets 130 feet off the ground.

d) $y = 0$ is asking, "When is the object on the ground?". Setting y to 0 and solving gives $x \approx -0.07$ and $x \approx 5.6$. The 5.6 tells us that after throwing the object it will take 5.6 seconds for the object to come back down and hit the ground. The -0.07 isn't useful in this context since our time starts when the object is thrown which is at time 0.

7.4 Quadratic Functions in Standard Form

A quadratic function is helpful when the graph of your data looks like this.

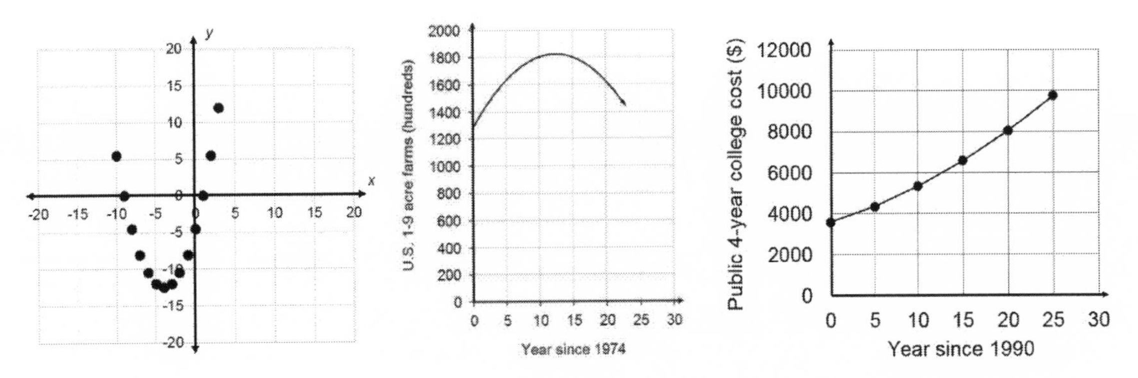

The function $y = ax^2 + bx + c$ is often used to model this kind of parabolic shape. We'll call a function written in the form $y = ax^2 + bx + c$ a quadratic function in standard form. In this section, you'll learn some vocabulary for quadratic functions and practice graphing a quadratic function written in standard form.

7.4.1 Some Vocabulary for Quadratic Functions

It's common to say a parabola like the one to the right "**opens up**" since, as the values of *x* increase left to right, the *y* values decrease to a "turning point" and then increase. The turning point is known as the **vertex** and the *y* coordinate of the vertex is the **minimum** value that *y* takes on. This parabola decreases to the vertex $(0,-4)$ and then increases. The function has a minimum value of -2.

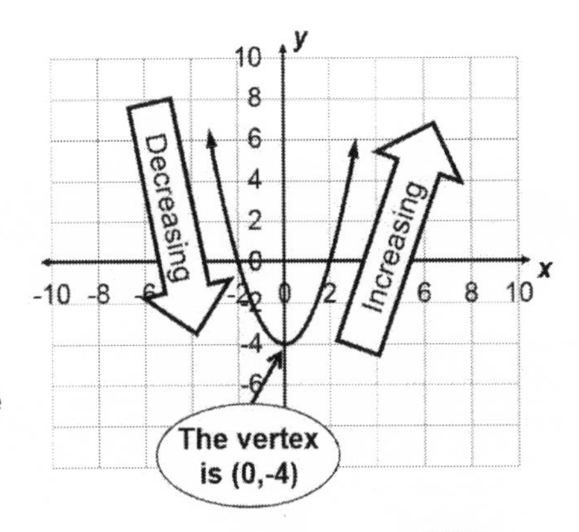

As the values of *x* increase left to right then a parabola that "**opens down**" increases to its vertex and then decreases. The *y* coordinate of the vertex is the **maximum** value the function will take on. The parabola to the right increases to the vertex $(4,6)$ and then decreases. The function has a maximum value of 6.

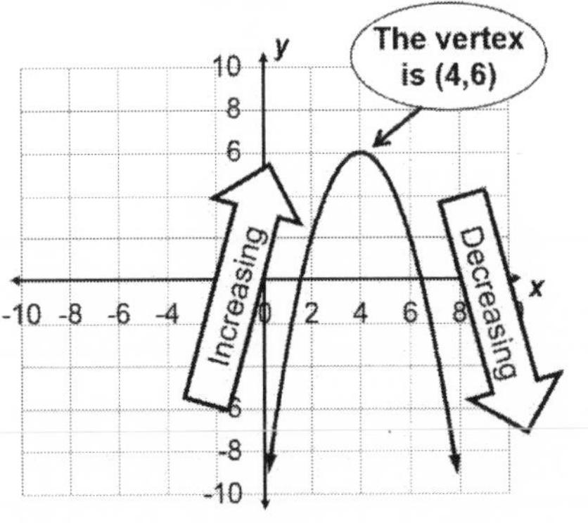

To decide if a parabola opens up or down you only need to check *a*, the second-degree coefficient. If $a < 0$ the parabola opens down. If $a > 0$ the parabola opens up.

If you draw the line through the vertex that's parallel to the *y*-axis you have the axis of symmetry. People often say the axis of symmetry is like a mirror that reflects one branch of the parabola to a second identical looking branch.

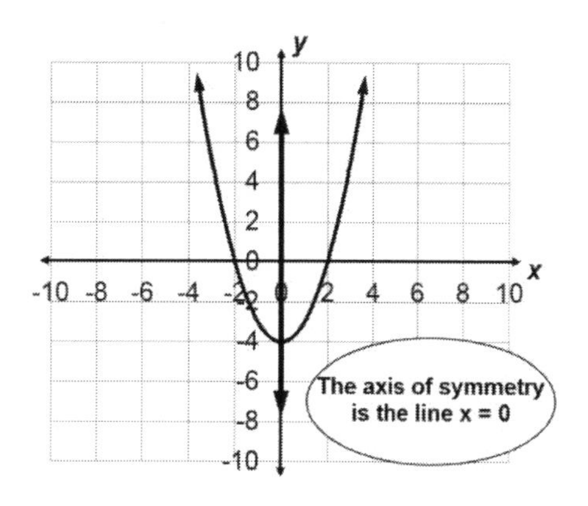

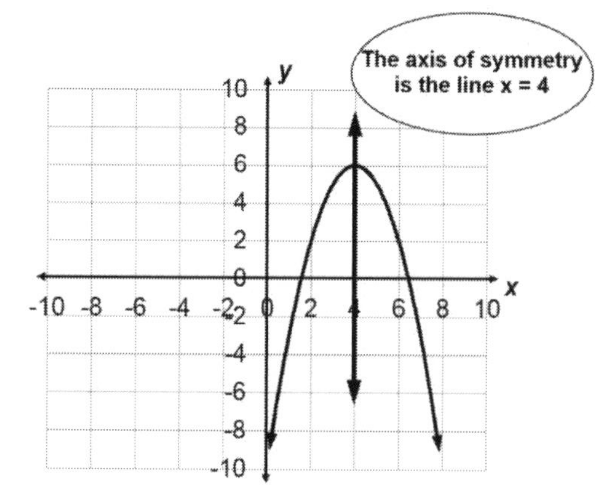

A general quadratic function will have 0, 1 or 2 x-intercepts and 1 y-intercept.

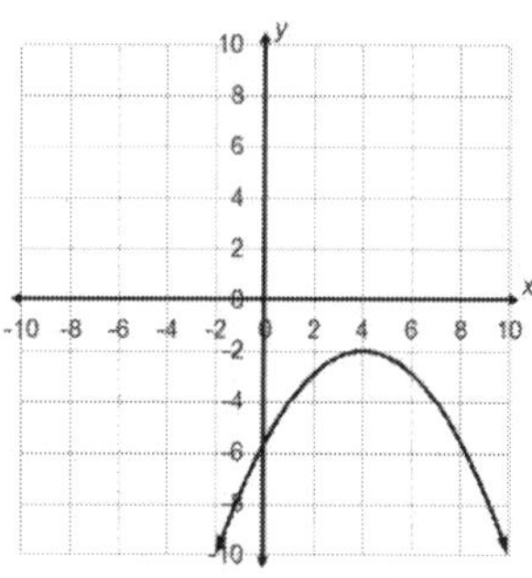

No x-intercepts
One y-intercept

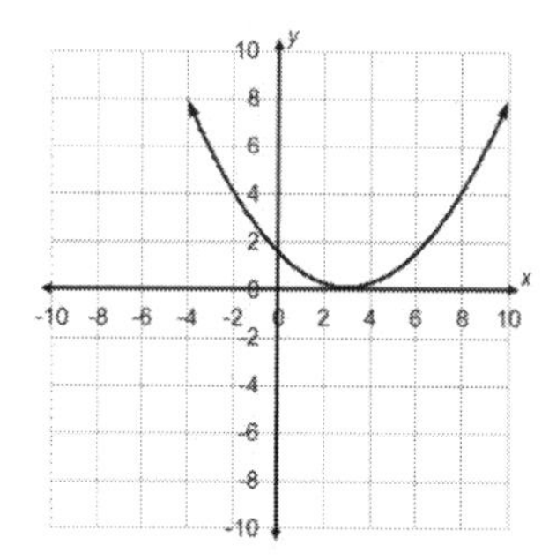

One x-intercept
One y-intercept

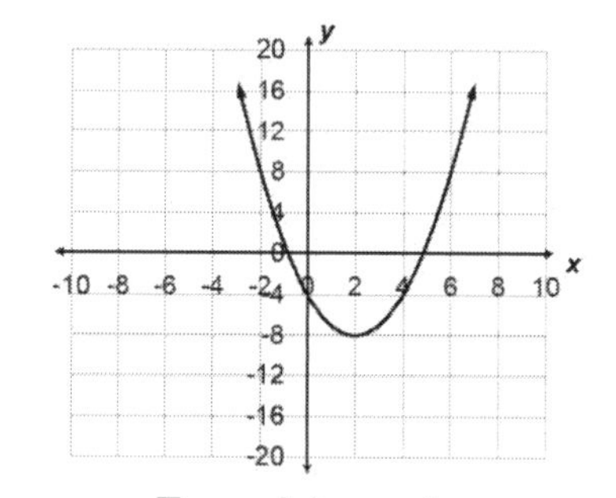

Two x-intercepts
One y-intercept

To find the x-intercepts we'll usually set y to 0 and solve the resulting equation. Notice if we start with the general quadratic function, $y = ax^2 + bx + c$, substitute 0 for x, $y = a(0)^2 + b(0) + c$ and simplify, $y = 0 + 0 + c$, the y value of the y-intercept will always be the same as the value of c.

Practice 7.4.1 Some Vocabulary for Quadratic Functions

Discuss whether the graph opens up or down, the vertex, the maximum or minimum value of the function, the line of symmetry and any intercepts.

---

a)

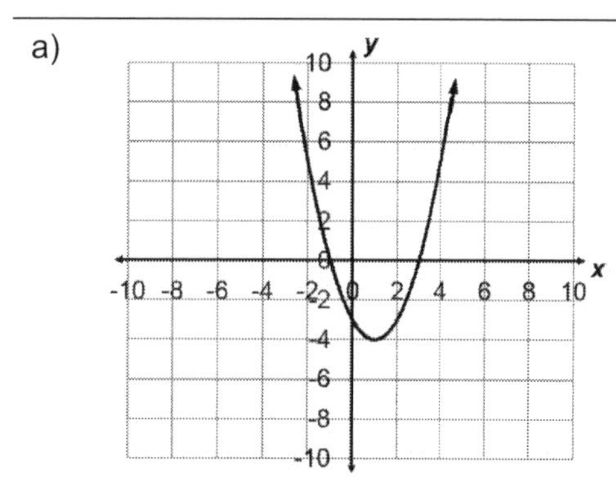

The function opens up. Assuming the vertex is $(1,-4)$ the minimum value is -4 and the axis of symmetry is the line $x = 1$. There are two x-intercepts at about $(-1,0)$ and $(3,0)$ and the y-intercept is around $(0,-3)$.

Homework 7.4 Discuss whether the graph opens up or down, the vertex, the maximum or minimum value of the function, the line of symmetry and any intercepts.

1) 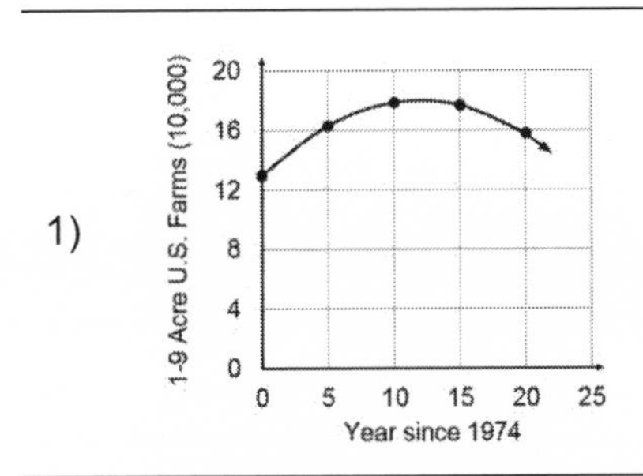

2) 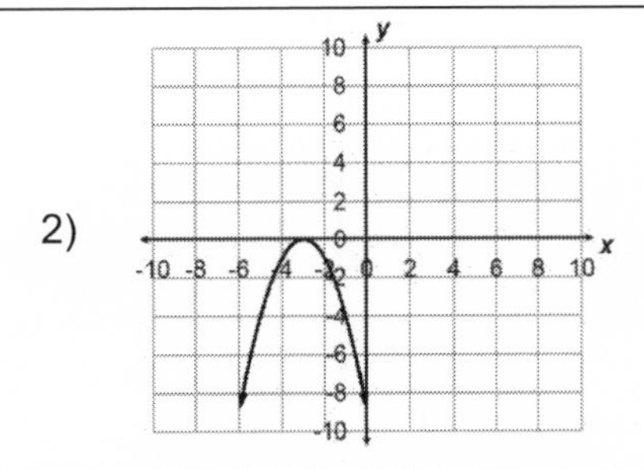

3) 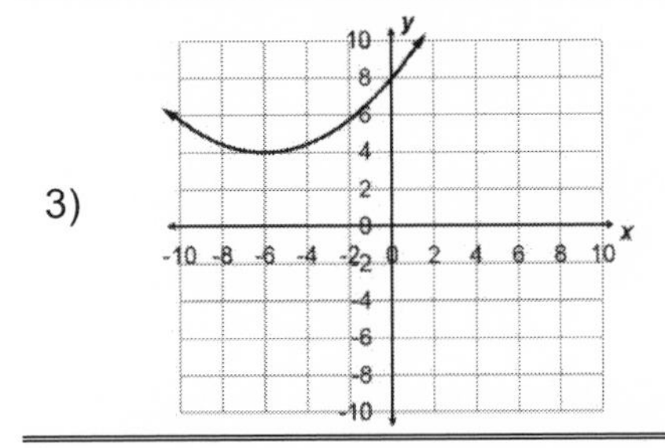

4) 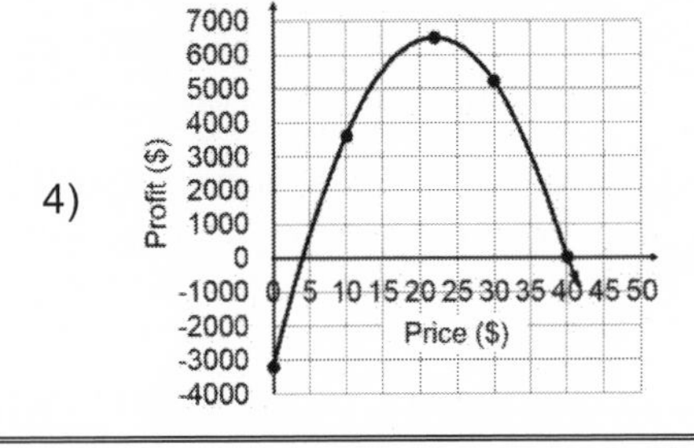

Practice 7.4.1 Some Vocabulary for Quadratic Functions

First, decide how the parabola opens and graph the y-intercept. Next, finish the data table, graph the points and continue the shape. Last, estimate any x-intercepts, the vertex, and the axis of symmetry.

b) $y = x^2 - 2x - 3$

x	y
−3	
−2	
0	
1	
2	
5	

$\Rightarrow$ $y = x^2 - 2x - 3$

x	y
−3	12
−2	5
0	−3
1	−4
2	−3
5	12

Since a is 1 the function opens up. Substituting 0 for *x*, and simplifying, shows the *y*-intercept is $(0,-3)$. There are two *x*-intercepts at about $(-1,0)$ and $(3,0)$ and the vertex is close to $(1,-4)$. The axis of symmetry would be the line $x = -1$.

c) $y = -\frac{1}{2}x^2 + 2x - 5$

x	y
−1	
0	
2	
4	
5	

$\Rightarrow$ $y = -\frac{1}{2}x^2 + 2x - 5$

x	y
−1	−7.5
0	−5
2	−3
4	−5
5	−7.5

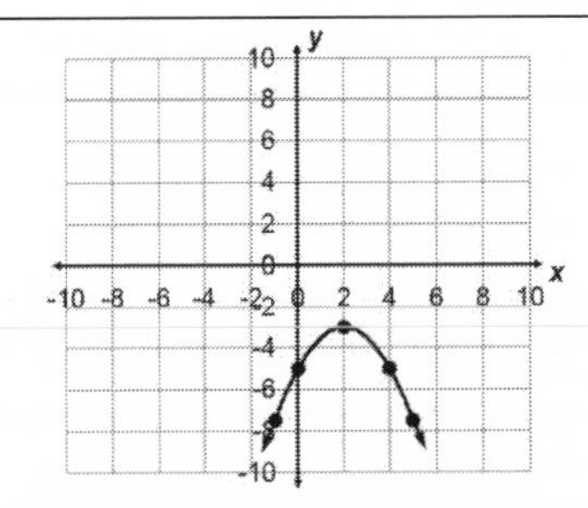

The function opens down and has a *y*-intercept of $(0,-5)$. There are no *x*-intercepts and the vertex is close to $(2,-3)$. The axis of symmetry is the line $x = 2$.

Homework 7.4 *First, decide how the parabola opens and graph the y-intercept. Next, finish the data table, graph the points and continue the shape. Last, estimate any x-intercepts, the vertex, and the axis of symmetry.*

5) $y=-x^2+4x+2$

x	y
−2	
0	
2	
4	
6	

6) $y=1.5x^2+6x+6$

x	y
0	
−1	
−2	
−3	
−4	

7) $y=-\frac{1}{4}x^2-2x-8$

x	y
−8	
−6	
−4	
−2	
0	

8) $y=x^2-6x+5$

x	y
−0.5	
2	
4	
5	
6.5	

7.4.2 Finding the Vertex and the Axis of Symmetry Algebraically

When a function has the form $y=ax^2+bx+c$ the x-coordinate of the vertex is the value $\frac{-b}{2a}$ and the y-coordinate of the vertex is the value $-\frac{b^2}{4a}+c$. Instead of memorizing the expression for the y-coordinate, most people substitute the value they found for the x-coordinate into the function and simplify to find the y-coordinate. Here's some practice.

Practice 7.4.2 Finding the Vertex and the Axis of Symmetry Algebraically

Find the vertex and the axis of symmetry for each function.

a) $y=x^2+4x+7$

The x-coordinate of the vertex is -2. ⇒ Used $\frac{-b}{2a}=\frac{-4}{2(1)}=-2$

The y-coordinate of the vertex is 3 ⇒ Substituted -2 for x and solved for y to find the y-coordinate of the vertex. $y=(-2)^2+4(-2)+7=3$

The vertex is $(-2,3)$. The axis of symmetry is $x=-2$ ⇒ The axis of symmetry is the line parallel to the y-axis which runs through the x axis at 2.

b) $y=5.25x^2-40x+60$

The x-coordinate of the vertex is about 3.8. ⇒ Used $\frac{-b}{2a}=\frac{-(-40)}{2(5.25)}\approx 3.8$ and rounded.

The y coordinate is -16.2. ⇒ Substituted 3.81 for x and solved for y using a calculator to estimate the y coordinate of the vertex. $y=5.25(3.81)^2-40(3.81)+60\approx-16.19$

The vertex is about $(3.8,-16.2)$. The axis of symmetry is $x\approx 3.8$. ⇒ Rounding to another place value would have been fine also.

Homework 7.4 Find the vertex and the axis of symmetry for each function.

9) $y = 2x^2 + 8x - 3$ 10) $y = \frac{1}{4}x^2 - 9$ 11) $y = 1282 - 3.5x^2 + 87x$

12) $y = 0.014x^2 - 0.746x + 16.69$

7.4.3 Finding the x-Intercepts and Uses of the Discriminant

To find the x-intercepts of a quadratic function written in standard form, set the value of y to 0 and solve the resulting equation, $0 = ax^2 + bx + c$, for x. Recall from our earlier work that if the discriminant is less than zero there will be no real x-intercepts. Here's some practice finding x-intercepts.

Practice 7.4.3 Finding the x-Intercepts and Uses of the Discriminant

For each function use the discriminant to decide on the number and kind of x-intercepts. Find any real x-intercepts.

a) $y = 4x + x^2 + 7$

$y = x^2 + 4x + 7$ ⇒ Wrote in the form $y = ax^2 + bx + c$

There are no real x-intercepts ⇒ Since $b^2 - 4ac = 4^2 - 4(1)(7) = -12$ which is less than 0 there are no real x-intercepts.

b) $y = 7x^2 - x$

There are two real x-intercepts. ⇒ Since $b^2 - 4ac = (-1)^2 - 4(7)(0) = 1$ which is positive there are two real x-intercepts.

The x-intercepts are $(0,0)$ and $\left(\frac{1}{7}, 0\right)$. ⇒ Substituted 0 for y and solved for x using the zero-product method.

$$0 = 7x^2 - x$$
$$0 = x(7x - 1)$$
$$x = 0 \text{ or } x = 1/7$$

c) $y = -x^2 - 2x + 4$

There are two real x-intercepts. ⇒ Since $b^2 - 4ac = (-2)^2 - 4(-1)(4) = 20$ there will be two real x-intercepts.

The x-intercepts are $(-1-\sqrt{5}, 0)$ and $(-1+\sqrt{5}, 0)$. ⇒ Substituted 0 for y and solved for x.

$$0 = -x^2 - 2x + 4$$
$$x = -1 - \sqrt{5} \text{ or } x = -1 + \sqrt{5}$$

Homework 7.4 For each function use the discriminant to decide on the number and kind of x-intercepts. Find any real x-intercepts.

13) $y = x^2 - 2x + 5$ 14) $y = 2 - 3x^2 + x$ 15) $y = -5x^2 + 3$

16) $y = -7.8x + 1.3x^2 + 11.7$ 17) $y = 5.25x^2 - 40x + 660$ 18) $y = 9x^2 - 24x - 34$

7.4.4 Graphing a Quadratic Function Written in Standard Form

You now have a set of ideas which are helpful when graphing a quadratic function in standard form.

Procedure **– Graphing a Quadratic Function in Standard Form**
1. Write the function in the form $y = ax^2 + bx + c$ 2. Decide if the graph opens up $(a > 0)$ or opens down $(a < 0)$. 3. Substitute 0 for x and simplify to find the y-intercept. 4. Find the x-coordinate of the vertex using $\frac{-b}{2a}$. Then substitute to find the y-coordinate. 5. Find the axis of symmetry using $x = -\frac{b}{2a}$ 6. If the discriminant shows x-intercepts exist, find them.

Practice 7.4.4 Graphing a Quadratic Function Written in Standard Form

Use the procedure to graph the quadratic function.

a) $y = x^2 + 6x + 5$

The graph opens up	$\Rightarrow$	$a = 1$ which is greater than 0.
The y-intercept is $(0,5)$	$\Rightarrow$	Substituted 0 for x and solved to find $y = 5$.
The vertex is $(-3,-4)$ The axis of symmetry is $x = -3$	$\Rightarrow$	The x-coordinate of the vertex is $\frac{-b}{2a} = \frac{-6}{2(1)} = -3$. When $x = -3$ then $y = (-3)^2 + 6(-3) + 5 = -4$
There are two x-intercepts.	$\Rightarrow$	$b^2 - 4ac = 36 - 20 = 16$ which is greater than zero.
The x-intercepts are $(-5,0)$ and $(-1,0)$.	$\Rightarrow$	Decided to try the zero-product method. $0 = x^2 + 6x + 5 \Rightarrow 0 = (x+5)(x+1)$ so $x = -5$ and $x = -1$.

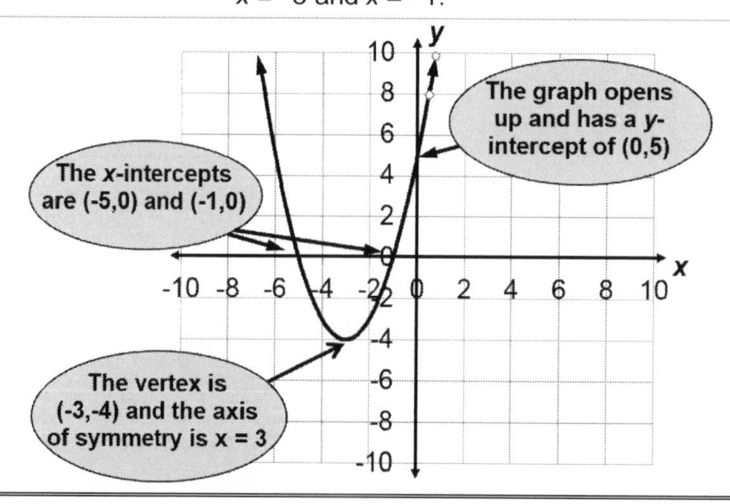

Homework 7.4 Use the procedure to graph the quadratic function.

19) $y = x^2 - 1$

20) $y = x^2 - 2x + 1$

21) $y = -2x^2 + x$

22) $y = x^2 - 2x - 8$

23) $y = 5x^2 + 10x - 2$

24) $y = 4x^2 - 4x - 5$

25) $y = 5.25x^2 - 40x + 660$ (Assume $x \geq 0$)

26) $y = -\frac{1}{2}x^2 - x + 2$

27) $y = -2x^2 - 8x - 2$

28) $y = 2x^2 + 17x + 8$

29) $y = -9x^2 + 6x - 1$

30) $y = -x^2 + 12x - 40$

31) $y = -16x^2 + 88x$ (Assume $y \geq 0$)

32) $y = 0.6t^2 - 1.8t - 2.7$

Homework 7.4 Answers

1) The function opens down, the vertex is around $(12,18)$ so the maximum value is about 18 and the axis of symmetry close to $x = 12$. The *y*-intercept is about $(0,13)$.

2) The function opens down, if the vertex is at $(-3,0)$ then the maximum value is 0 and the axis of symmetry is $x = -3$. The *y*-intercept is about $(0,-9)$ and the *x*-intercept is $(-3,0)$.

3) The function opens up, if the vertex is at $(-6,4)$ then the minimum value is 4 and the axis of symmetry is $x = -6$. The *y*-intercept is about $(0,8)$.

4) The function opens down, if the vertex is at $(22,6500)$ then the maximum value is 6,500 and the axis of symmetry is $x = 22$. The *y*-intercept is about $(0,-3200)$ and the *x*-intercepts are about $(4,0)$ and $(40,0)$.

5) $y = -x^2 + 4x + 2$

x	y
−2	−10
0	2
2	6
4	2
6	−10

$\Rightarrow$

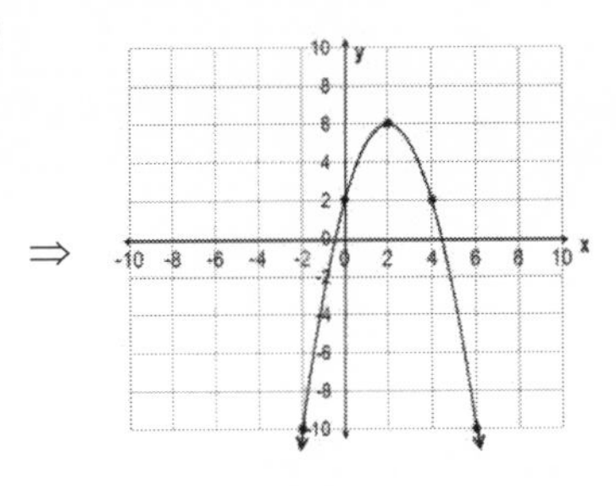

The function opens down, has a *y*-intercept of $(0,2)$ *x*-intercepts at about $(4.5,0)$ and $(-0.5,0)$ and a vertex at $(2,6)$. The axis of symmetry is the line $x = 2$ and the maximum *y* value is 6.

6) $y = 1.5x^2 + 6x + 6$

x	y
0	6
−1	1.5
−2	0
−3	1.5
−4	6

$\Rightarrow$

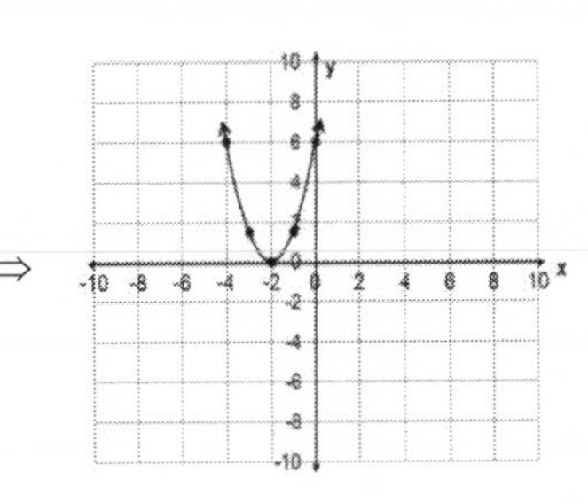

The function opens up, has a *y*-intercept of $(0,6)$ an *x*-intercept of $(-2,0)$ and a vertex at $(-2,0)$. The axis of symmetry is the line $x = -2$ and the minimum *y* value is 0.

7) $y = -\frac{1}{4}x^2 - 2x - 8$

x	y
−8	−8
−6	−5
−4	−4
−2	−5
0	−8

$\Rightarrow$

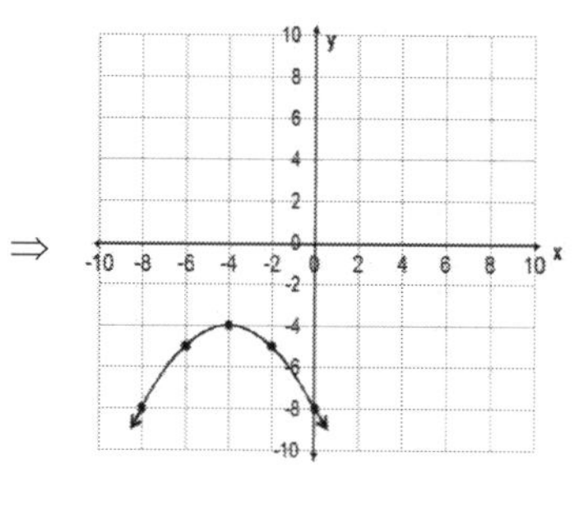

The function opens down, has a *y*-intercept of $(0,-8)$ no *x*-intercepts and a vertex at $(-4,-4)$. The axis of symmetry is the line $x = -4$ and the maximum *y* value is -4.

8) $y = x^2 - 6x + 5$

x	y
−0.5	8.25
2	−3
4	−3
5	0
6.5	8.25

$\Rightarrow$

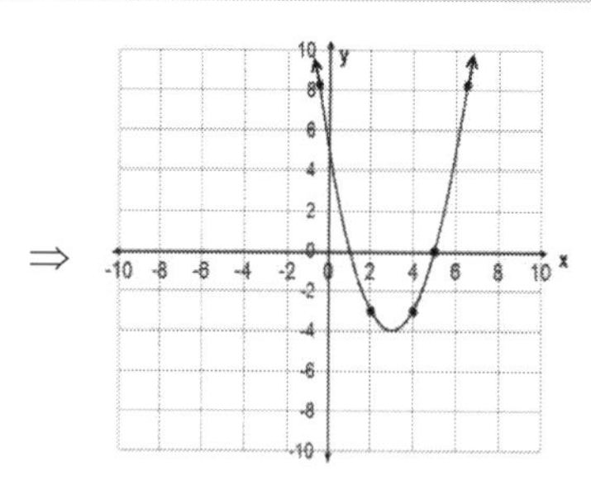

The function opens up, has a *y*-intercept of $(0,5)$ *x*-intercepts at $(1,0)$ and $(5,0)$ and a vertex at $(3,-4)$. The axis of symmetry is the line $x = 3$ and the minimum *y* value is -4.

9) The *x* coordinate of the vertex is $\frac{-b}{2a} = \frac{-(8)}{2(2)} = \frac{-8}{4} = -2$. The *y* coordinate is $2(-2)^2 + 8(-2) - 3 = -11$. The vertex is $(-2,-11)$ and the axis of symmetry is $x = -2$.

10) The *x* coordinate is $\frac{-b}{2a} = \frac{-(0)}{2(1/4)} = \frac{0}{1/2} = 0$. The *y* coordinate is $\frac{1}{4}(0)^2 - 9 = -9$. The vertex is $(0,-9)$ and the axis of symmetry is $x = 0$, which is the *y* axis.

11) The *x* coordinate is $\frac{-b}{2a} = \frac{-(87)}{2(-3.5)} = \frac{-87}{-7} = 12.4$. The *y* coordinate is $-3.5(12.4)^2 + 87(12.4) + 1282 = 1{,}822.6$. The vertex is $(12.4, 1822.6)$ and the axis of symmetry is $x = 12.4$.

12) The *x* coordinate is $\frac{-b}{2a} = \frac{-(-0.746)}{2(0.014)} = \frac{0.746}{0.028} = 26.64$. The *y* coordinate is $0.014(26.64)^2 - 0.746(26.64) + 16.69 = 6.75$. The vertex is $(26.64, 6.75)$ and the axis of symmetry is $x = 26.64$.

13) The discriminant is $b^2 - 4ac = (-2)^2 - 4(1)(5) = -16$ which is less than 0. There are no real *x*-intercepts.

14) The discriminant is $b^2 - 4ac = 1^2 - 4(-3)(2) = 25$ which is greater than 0 so there are 2 real *x*-intercepts. The *x*-intercepts can be found using $x = \frac{-1 \pm \sqrt{25}}{2(-3)}$ and are $\left(-\frac{2}{3}, 0\right)$ and $(1,0)$.

15) The discriminant is $b^2 - 4ac = (0)^2 - 4(-5)(3) = 60$ which is greater than 0 so there are 2 real x-intercepts. The x-intercepts can be found using, for example, the square root property $x = \pm\sqrt{\frac{3}{5}}$. The x-intercepts are $\left(-\frac{\sqrt{15}}{5}, 0\right)$ and $\left(\frac{\sqrt{15}}{5}, 0\right)$ after rationalizing the denominator.

16) The discriminant is $b^2 - 4ac = (-7.8)^2 - 4(1.3)(11.7) = 0$ so there is one real x-intercept. The x-intercept can be found using $\frac{-(-7.8) \pm \sqrt{0}}{2(1.3)}$. The x-intercept is $(3,0)$.

17) The discriminant is $b^2 - 4ac = (-40)^2 - 4(5.25)(660) = -12{,}260$ which is less than 0, so there are no real x-intercepts.

18) The discriminant is $b^2 - 4ac = (-24)^2 - 4(9)(-34) = 1800$ so there are two real x-intercepts. The quadratic formula gives

$\frac{24 \pm \sqrt{1800}}{18} = \frac{24 \pm 30\sqrt{2}}{18} = \frac{24}{18} \pm \frac{30\sqrt{2}}{18} = \frac{4}{3} \pm \frac{5\sqrt{2}}{3}$. The x-intercepts are $\left(\frac{4}{3} \pm \frac{5\sqrt{2}}{3}, 0\right)$

which are approximately $(-1,0)$ and $(3.7,0)$.

19) Opens up, y-intercept $(0,-1)$, vertex $(0,-1)$, axis of symmetry $x = 0$, the discriminant is 4 so there are two real x-intercepts $(-1,0)$ and $(1,0)$.

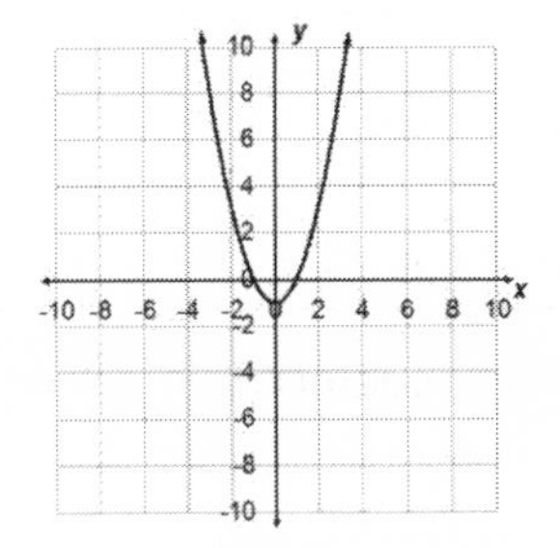

20) Opens up, y-intercept $(0,1)$, vertex $(1,0)$, axis of symmetry $x = 1$, the discriminant is 0 so there is one real x-intercept is $(1,0)$.

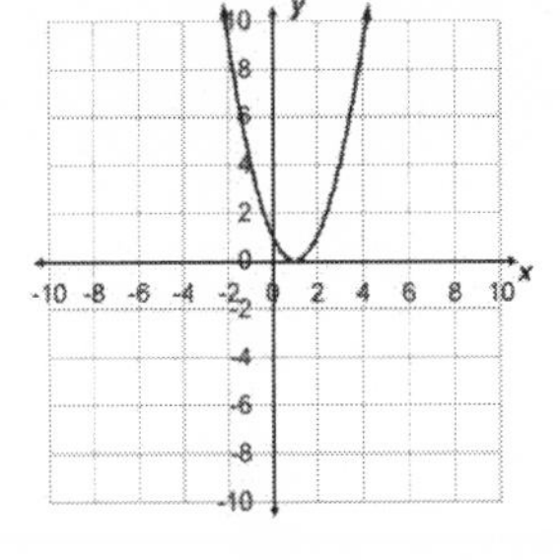

21) Opens down, vertex $\left(\frac{1}{4}, \frac{1}{8}\right)$, axis of symmetry $x = \frac{1}{4}$, y-intercept $(0,0)$, the discriminant is 1 so there are two real x-intercepts $(0,0)$ and $\left(\frac{1}{2}, 0\right)$.

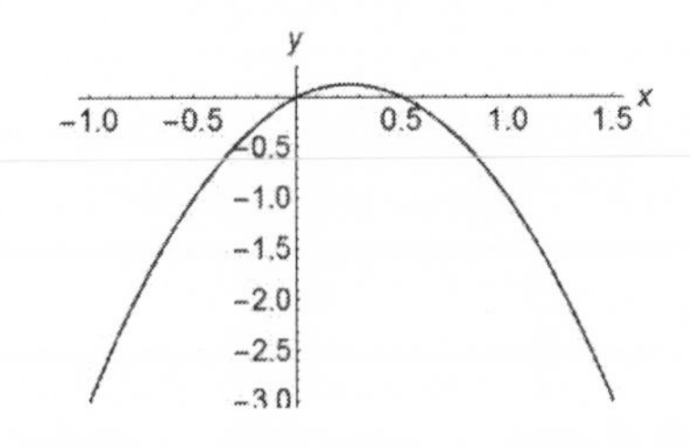

22) Opens up, y-intercept $(0,-8)$, vertex $(1,-9)$, axis of symmetry $x=1$, the discriminant is 36 so there are two real x-intercepts $(-2,0)$ and $(4,0)$.

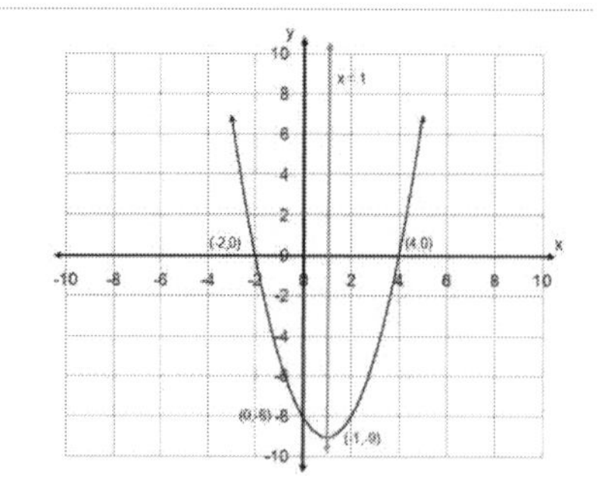

23) Opens up, y-intercept $(0,-2)$, vertex $(-1,-7)$, axis of symmetry $x=-1$, The radicand is 140 so there are two real x-intercepts. The x-intercepts $\left(-1-\frac{\sqrt{35}}{5},0\right)$ and $\left(-1+\frac{\sqrt{35}}{5},0\right)$.

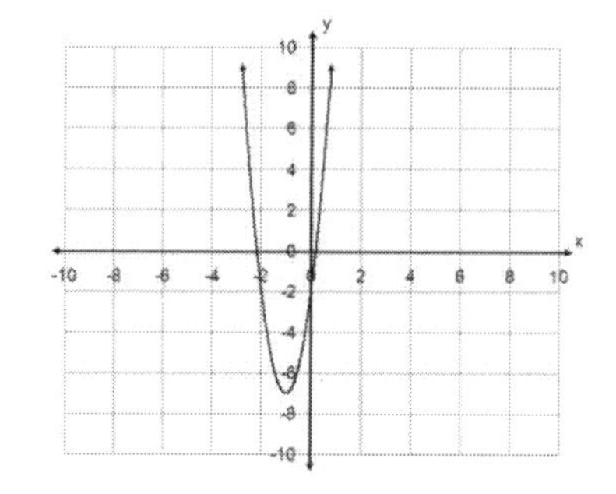

24) Opens up, y-intercept $(0,-5)$, vertex $\left(\frac{1}{2},-6\right)$, axis of symmetry $x=\frac{1}{2}$, the discriminant is 96 and the 2 real x-intercepts are $\left(\frac{1}{2}-\frac{\sqrt{6}}{2},0\right)$ and $\left(\frac{1}{2}+\frac{\sqrt{6}}{2},0\right)$.

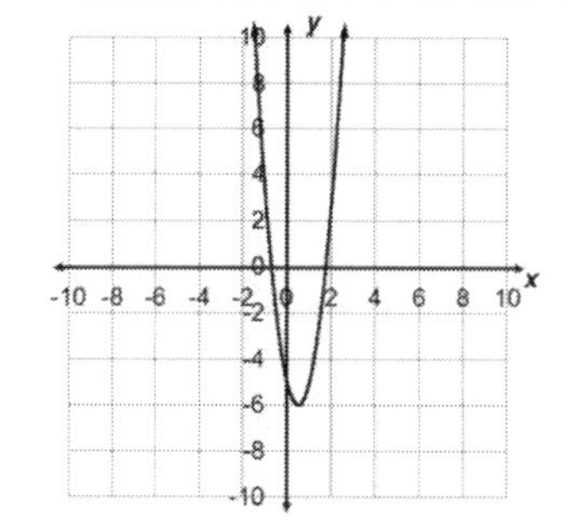

25) Opens up, y-intercept $(0,660)$, vertex $(3.8,583.81)$, axis of symmetry $x=3.8$, The discriminant is $-12,260$ so there are no x-intercepts.

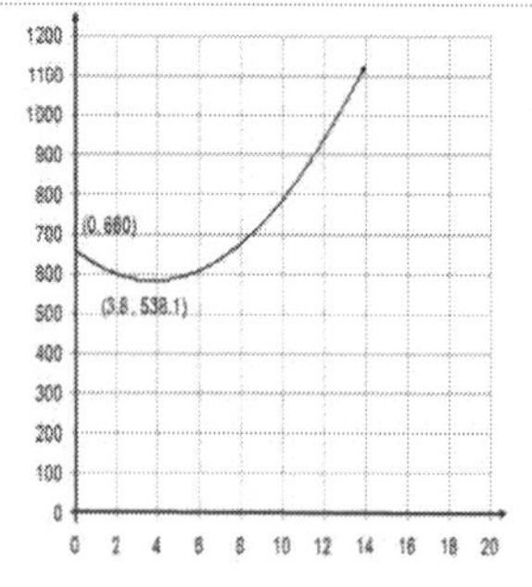

26) Opens down, y-intercept $(0,2)$, vertex $\left(-1,\frac{5}{2}\right)$, axis of symmetry $x=-1$, the discriminant is 5 so there are two real x-intercepts $\left(-1-\sqrt{5},0\right)$ and $\left(-1+\sqrt{5},0\right)$ which are approximately $(-3.2,0)$ and $(1.2,0)$.

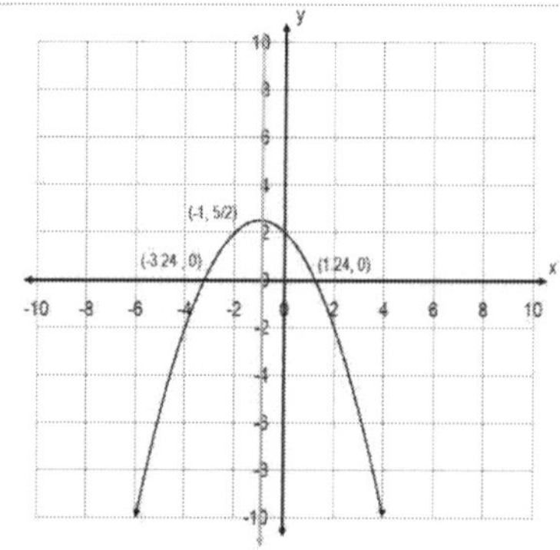

27) Opens down, vertex $(-2,6)$, axis of symmetry $x=-2$,
y-intercept $(0,-2)$, the discriminant is 48 so there are two real x-intercepts $\left(-2-\sqrt{3},0\right)$ and $\left(-2+\sqrt{3},0\right)$ which are approximately $(-3.7,0)$ and $(-0.3,0)$.

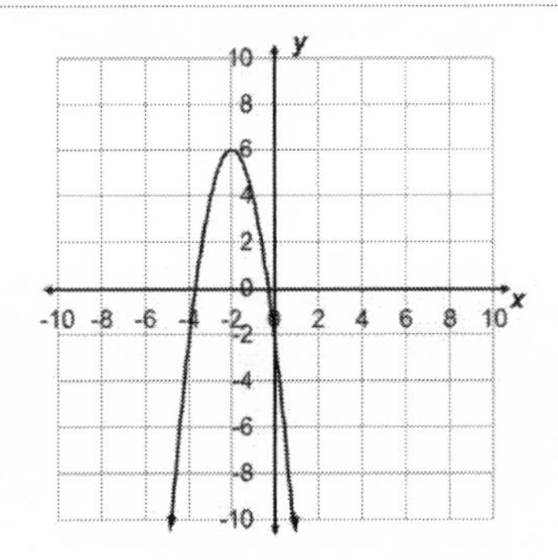

28) Opens up, vertex $\left(-\frac{17}{4},-\frac{225}{8}\right)$, axis of symmetry
$x=-4.25$,
y-intercept $(0,17)$, the discriminant is 225 so there are two real x-intercepts $(-8,0)$ and $\left(-\frac{1}{2},0\right)$.

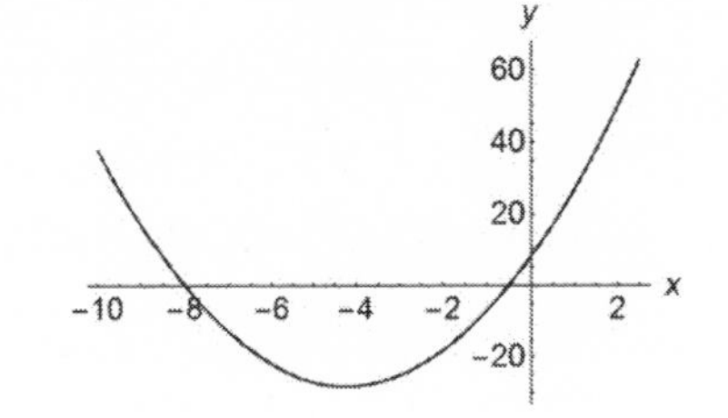

29) Opens down, vertex $\left(\frac{1}{3},0\right)$, axis of symmetry $x=\frac{1}{3}$,
y-intercept $(0,-1)$, the discriminant is 0 so there is one real x-intercept $\left(\frac{1}{3},0\right)$.

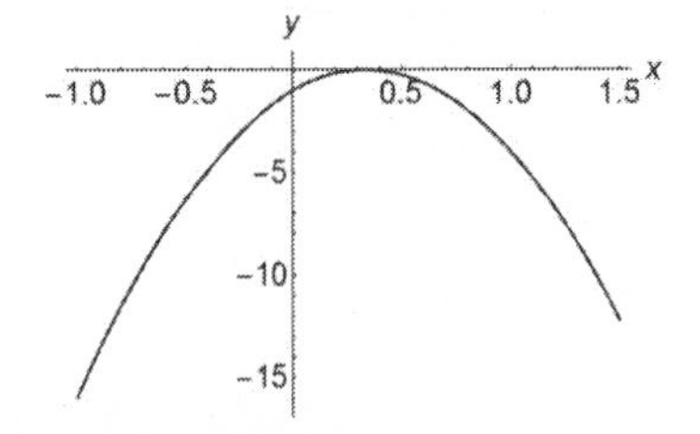

30) Opens down, vertex $(6,-4)$, axis of symmetry $x=6$,
y-intercept $(0,-40)$, the discriminant is -16 so there are no real x-intercepts.

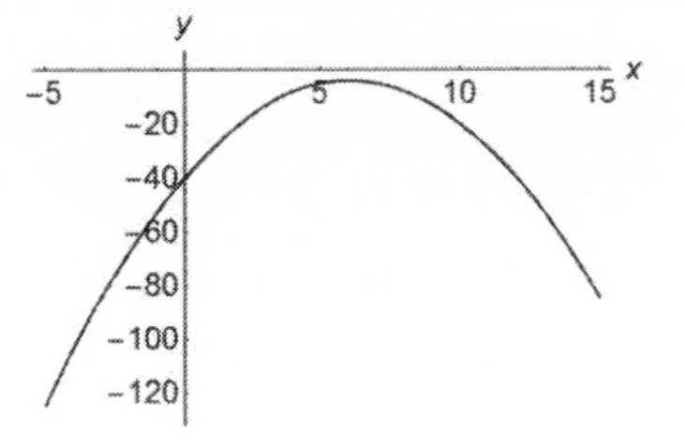

31) Opens down, vertex $(2.75,121)$, axis of symmetry
$x=2.75$,
y-intercept $(0,0)$, the discriminant is 7,744 so there are two real x-intercepts $(0,0)$ and $(5.5,0)$.

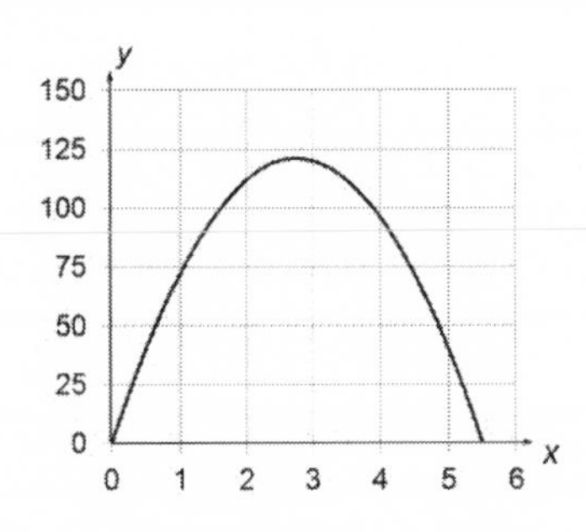

32) Opens up, vertex $(1.5, -4.05)$, axis of symmetry $x = 1.5$, y-intercept $(0, -2.7)$, the discriminant is 9.72 so there are two real x-intercepts which are approximately $(-1.1, 0)$ and $(4.1, 0)$.

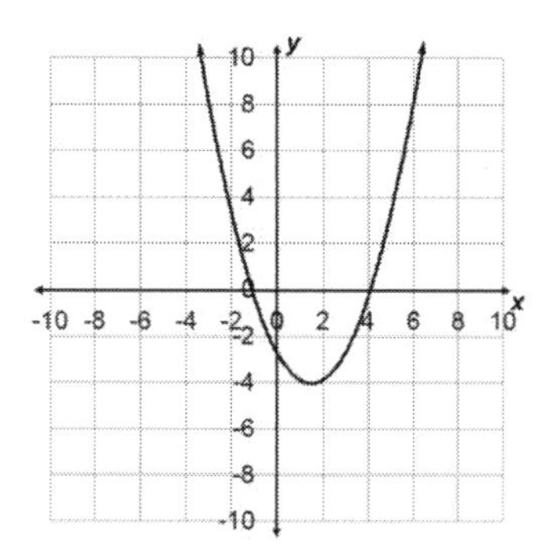

7.5 Applying Quadratic Functions

Usually students who plan on majoring in subjects like engineering or economics accept the importance of being literate with functions. Unfortunately, students majoring in subjects like psychology or journalism often tell themselves functions aren't that important. That is, until their senior year psychology professor presents information using mathematical models, or, after a press conference, they can't honestly evaluate the scientific information they've just seen.

You on the other hand, now have lots of practice with functional literacy. Let's apply those skills to quadratic functions.

7.5.1 Applying a Quadratic Function Written in Standard Form

Here's a couple of worked examples to help you get used to the ideas we'll need.

Practice 7.5.1 Applying a Quadratic Function Written in Standard Form

Use the given function to answer the following questions.

---

a) The per capita (per person) spending on prescription drugs (in dollars) between 1995 and 2008 can be approximated by the quadratic model $y=-0.54x^2+51.84x+217.84$ where x is the number of years after 1995.

---

1. Determine whether the parabola opens up or down and the location of the vertex. Discuss the implications of the information.

   Since $-0.54<0$ the parabola opens down so there could be a maximum value for the per capita cost. The vertex is at $(48,1462)$ so if the current trend continues, spending will peak at \$1,462 per person in 2043 $(1995+48=2043)$. Also, although spending is rising each year the rate of increase from year to year is getting smaller.

---

2. Answer the question that $y=1{,}000$ is asking.

   $y=1{,}000$ is asking us to find the year that per capita spending will be \$1,000.

   Solving the equation $1{,}000=-0.54x^2+51.84x+217.84$ tells us $x\approx 19$ and $x\approx 77$ which implies spending will be \$1,000 in the years 2014 and 2072. The year 2072 is probably too far into the future to be useful.

---

3. Ask the question, "What's the expected spending in 2009?" using x or y and then find and discuss the answer.

   The question is about the year 2009 which implies It's a question dealing with x values. $x=14$ will ask the question. Solving $y=-0.54(14)^2+51.84(14)+217.84$ shows $y\approx 838$ so in 2009, (year 14) the expected per capita spending should be about \$838.

---

b) *The number of farms in the United States between 1975 and 2009 (in thousands) can be modeled using* $y = 0.6x^2 - 32x + 2{,}544$ where x is the number of years after 1975.

1. Verify that the graph should be opening up.

   Since $a = 0.6$, and $0.6 > 0$, the parabola should open up.

2. Find the *y*-intercept and discuss its meaning.

   Setting $x = 0$ and solving shows the *y*-intercept is $(0, 2544)$ so in 1975 there were about 2,544,000 farms in the United States.

3. Find the year when the number of farms was at a minimum.

   Questions of a minimum or maximum value has to do with the vertex. Simplifying $\frac{-b}{2a} = \frac{-(-32)}{2(0.6)} \approx 27$ shows the number of farms will reach a minimum in 2002 (year 27).

4. Find the minimum number of farms.

   Again, questions of a minimum or maximum value has to do with the vertex.

   Substituting 27 for *x* and solving, $y = 0.6(27)^2 - 32(27) + 2544 \approx 2117.4$, shows that in 2002 the number of farms in the U.S. was around 2,117,400 which was the minimum number of farms.

5. Scale the *y*-axis from 0 to 3,000 and the *x*-axis from 0 to 60. Using the *y*-intercept, the vertex and the axis of symmetry graph the function.

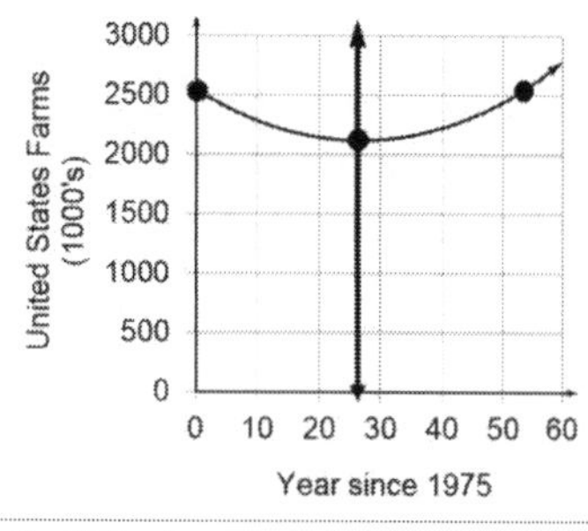

$\Rightarrow$ Started at the *y*-intercept (0,2544) and the vertex (27,2117). Next, included the axis of symmetry which is the line $x = 27$. Last, since 27 units to the left of the vertex the *y* value is around 2,544 then, by symmetry, moving 27 units to the right of vertex the *y* value should again be 2,544. Drew a third point at around (54, 2544) and connected the points with a curve.

6. What question is $y = 2{,}250$ asking? Use the graph, and then the algebraic function, to answer the question.

   $y = 2{,}250$ predicts the year(s) there will be 2,250,000 farms. Since 2,250 is half way between 2,000 and 2,500, I drew a horizontal line from $y = 2{,}250$, and then a vertical line where the horizontal line met the graph, to estimate that the values of *x* would be around 10 and 42. This implies there will be 2,250,000 farms in the United States around 1985 (year 10) and 2017 (year 42).Substituting 2,250 for y and solving gives

   $2250 = 0.6x^2 - 32x + 2544 \Rightarrow x \approx 9$ or $x \approx 44$. This implies there will be 2,250,000 farms in the United States around 1984 and 2019 which is close to the previous values found using the graph.

7. Assuming the pattern continues use the graph, and then the algebraic function, to predict the number of farms in 2025.

Drawing a vertical line from 50 and then a horizontal line from where the vertical line touches the graph gives an approximate value of 2,450,000 farms. Substituting 50 for *x* and solving gives $y = 0.6(50)^2 - 32(50) + 2544 = 2,444$. This again implies there may be around 2,444,000 farms in the United States in 2025 if the trend continues.

8. Ask the question, "When will there be 2,000,000 farms?", using *x* or *y* and then answer the question using the algebraic function. Verify your conclusion using the graph.

$y = 2,000$ is asking when there will be 2,000,000 farms in the United States.

Substituting 2,000 for *y* and solving does not return a real number (the discriminant is negative), which implies that there isn't a year (an *x* value) that will return a value of 2,000. The graph verifies this is true since 2,000 is below the minimum value of 2,117,400 farms.

Homework 7.5 Use the given function to answer the following questions.

1) The function $y = -0.1x^2 + 4.3x + 16.775$, where *x* is the years since 1980, predicts the number of cable subscribers in millions between the year 1980 and 2010.

a) Scale your *y*-axis from 0 to 80 and your *x*-axis from 0 to 40. Graph the parabola.

b) Say you were an investor trying to decide in 1995 if you should purchase stock in a cable company. Your hope is that the price of the stock will rise over the next ten years. To help make your decision you've built the above function from the data for 1980 to 1995. Discuss the implications of the vertex for your decision to purchase the stock.

c) Assuming the trend continues use the graph and the algebraic function to predict when the number of subscribers will return to 1990 levels.

d) Answer the question $x = 35$ is asking using both the graph and the algebraic function.

e) When do you expect the number of subscribers will fall to ten million less than the peak number of subscribers?

2) From 1970 until 1995 the function $y = 4x^2 - 3x + 494$ gave a good approximation of the cost for one year of tuition and fees (in dollars) at a United States four-year public university.

a) Answer the question $x = 0$ is asking.

b) Using the value of *a* (from $ax^2 + bx + c$), and the vertex, discuss the implications for the cost of college from 1970 through 1995.

c) Scale the *x*- axis from 0 through 40 and the *y*-axis from 0 through 6,000. Since the *y*-intercept and the vertex are so close you'll have to graph some points. Use the values 0, 10, 20 and 30 for *x*.

d) What question is $x = 35$ asking? Use the graph, and then the function, to answer the question.

e) Discuss the question $y = \$3,000$ is asking and answer the question using the algebraic function. Check your answer using the graph.

f) Ask the question, "What is the expected cost in 2015?" using either *x* or *y* and then answer the question using the algebraic function. (Assume the trend continues through 2015.)

3) Although modeling a disease like Rocky Mountain Spotted Fever, which has a lot of variation year to year, is not as accurate as some other models, it can give us information about general trends. The number of cases of Rocky Mountain Spotted Fever between 1983 and 2011 can be modeled with the function $y = 8.8x^2 - 167x + 1175$.

a) Discuss the implication of "*a*", then find the vertex and discuss its meaning.

b) Translate the idea, "I wonder when the number of cases will return to the 1983 level?" into using either *x* or *y*, and then use the algebraic function to answer the question.

c) Answer the question $y = 2,000$ is asking.

d) Translate, "Approximately how many cases can we expect in 2010?" using either *x* or *y* and answer the question using English.

4) A person who makes "vintage" bicycles finds if they charge $\$x$ per bicycle, then the function $y = -0.2x^2 + 242x - 42,000$ estimates, *y*, their profit (or loss) for the year.

a) Find the price that will maximize the profit and find this maximum profit.

b) Find $x = 0$ and suggest a reason for the answer.

c) Find *x* if $y = 30,600$ and speculate on the meaning of each answer.

d) Find the change in profit/loss if prices are raised from \$210 to \$310. Find the change in profit if prices are raised from \$500 to \$600. Discuss this difference in profit for the same \$100 increase in price.

5) The number of people attending a wedding reception, *y*, can be predicted using the number of hours after 6 p.m., *x*, by the function $y = -10x^2 + 80x + 200$.

a) Scale your *x*-axis from 0 to 10 and your *y*-axis from 0 to 400. Graph the parabola using the *y*-intercept, the vertex and the axis of symmetry.

b) Use the graph and algebraic function to answer the question $y = 300$ is asking.

c) Use the graph and algebraic function to answer the question $x = 2.5$ is asking.

d) Write the question, "How many people are expected when the reception starts?" using *x* or *y*. What's the answer to the question?

e) Find the maximum number of people attending the reception and when this will occur?

f) Ask the question, "When will the last person theoretically be leaving the reception?" using *x* or *y* and use the algebraic function to estimate the answer.

6) Between 1972 and 2000 the function $y = -0.5x^2 + 10x + 145$ modeled the number of cases of Chickenpox (in thousands) in the United States.

a) Discuss why you suspect this function may have a maximum value.

b) Scale your *y*-axis from 0 to 250 and your *x*-axis from 0 to 30. Graph the parabola using the *y*-intercept, the vertex and the axis of symmetry.

c) Using the algebraic function approximate the year the number of cases of Chickenpox peaked and find the number of cases that year. Verify the answer using the graph.

d) Ask the question, "How many cases of chickenpox were there in the United States in 1994?" using *x* or *y* and then use the graph and equation to answer the question.

e) Answer the question $y = 175$ is asking using both the graph and the algebraic function.

f) Find the year the number of cases returned to the 1977 level.

g) Theoretically what's the first year there will be no cases of Chickenpox in the United States?

7) If an object is thrown into the air from the top of a building with an initial speed of 88 feet per second, the function $y = -16x^2 + 88x + 84$ describes its height in feet as a function of time.

a) Why do suspect the object might have a maximum height?

b) Scale the *y* axis from 0 to 250 and the *x* axis from 0 to 7. Using the *y*-intercept, vertex and axis of symmetry, graph the function.

c) Find the maximum height the object will reach and how long it takes to get there.

d) If the person is reaching about 7 feet high when the object leaves their hand, approximately how tall is the building?

e) Using the graph estimate when the object will have a height that's 50 feet less than its maximum height. Then use the algebraic function to find the time(s).

f) Answer the question $x = 1$ is asking.

g) Find and discuss the meaning of the *x*-intercepts.

h) Use the algebraic function to find when the object will return to the height it was at after half a second. Use the graph to verify your answer(s).

8) The function $y = 0.017x^2 + 0.78x + 21.1$ uses the year since 1980 to predict the millions of tons of paper and paperboard recovered for reuse (recycled).

a) Discuss whether the parabola opens up or down and what that implies.

b) Ask the questions, "How many millions of tons were recycled in 1980." using *x* or *y* and then answer the question.

c) Find and discuss the meaning of the vertex.

d) In Quadrant 1 scale your y-axis from 0 to 100 and your x-axis from 0 to 50. Since the vertex is far outside our data set we should plot a few points to draw our graph. Make a data table using 0, 10, 30 and 50 for the x values and graph the curve.

e) Use the graph to estimate the tons recovered in 1999 and then use the algebraic function to estimate the tons recovered in 1999.

f) Use the algebraic function to find $y = 10$ and discuss the meaning of what you find. Verify your answer using the graph.

g) Use the graph to predict the first year eighty-million tons of paper and paperboard will be recovered in the United States. Then ask the question using either x or y and answer the question using the algebraic function.

9) A college student starts a lawn care service during the summer. Their monthly profit can be modeled using $y = -10x^2 + 240x - 190$ where x is the price they charge per hour.

a) Scale your y-axis from -400 to 1,600 and your x-axis from 0 to 20. Graph the parabola using the y-intercept, the vertex and the axis of symmetry.

b) Find the student's profit if they charge \$0 and discuss the meaning of the answer.

c) Find the price they should charge to maximize their monthly profit and what the monthly profit would be.

d) Under the assumption that raising the hourly price would increase the profit they raise their hourly price to \$16 an hour. Use the graph and the algebraic function to estimate their monthly profit now? Why does raising the price reduce the profit?

e) Use the graph and the algebraic function to predict the hourly price that would give the student a monthly profit of \$1,000? Discuss both answers in terms of price, demand and which you'd consider "better".

f) Find the hourly price that would give a monthly profit of \$1,300. Discuss what you find in terms of the vertex.

---

10) From 1990 through 2016 the function $y = 4.8x^2 + 126x + 3{,}591$ gave a good approximation of the cost for one year of tuition and fees (in current dollars) at a United States four-year public university.

a) Answer the question $x = 0$ is asking.

b) Using the value of a, and the vertex, discuss the implications for the cost of college from 1990 through 2016.

c) Scale the x axis from 0 through 40 and the y axis from 0 through 12,000 then graph the function.

d) Ask the question, "What's the projected cost in 2020?" using x or y. Use the graph and the algebraic function to answer the question.

e) Answer $y = \$6{,}000$ using the algebraic function. Check your answer using the graph.

f) Assuming the trend continues, predict the cost in 2025.

Homework 7.5 Answers

1)

a) Used the vertex, y-intercept and axis of symmetry to graph the function.

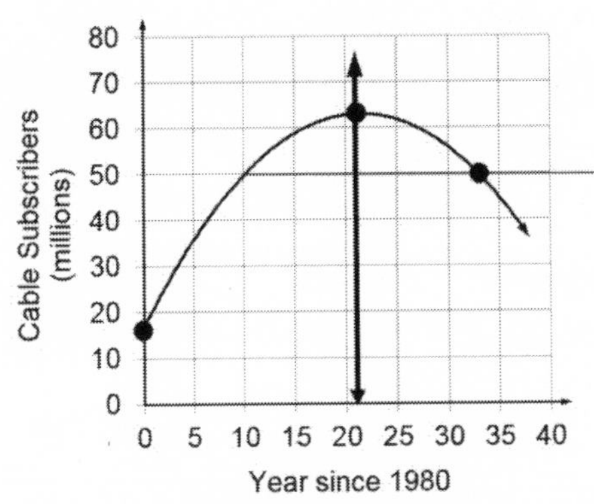

b) The vertex will be around $(21.5, 63)$ which implies the number of subscribers will peak in 2001 or 2002 at 63,000,000 subscribers and then start to decline. If the company has not begun discussing how to increase revenue during a time of decreasing subscriptions, they may not be a good investment.

c) In 1990 there were around 50 million subscribers $y = -0.1(10)^2 + 4.3(10) + 16.8 \Rightarrow y = 49.8$. The graph predicts the number of subscribers will return to 50 million around year 33 which is 2013. The algebraic function also predicts this will happen in 2013, $50 = -0.1x^2 + 4.3x + 16.8 \Rightarrow x \approx 33$.

d) $x = 35$ asks us to predict the number of subscribers in 2015. The graph predicts around 45,000,000 subscribers. The algebraic function predicts around 44,800,000 subscribers.

e) As the vertex shows, 63,000,000 is the peak number of subscribers. Solving $53 = -0.1x^2 + 4.3x + 16.8 \Rightarrow x \approx 31.5$ shows the number of subscribers will be ten million less than the peak number of subscribers around 2011-12.

2)

a) The cost in 1970 was about \$494.

b) The value of a is 4 which implies the graph is opening up. The vertex is about $(0.38, 494)$ which implies the cost of college is continuously rising.

c)

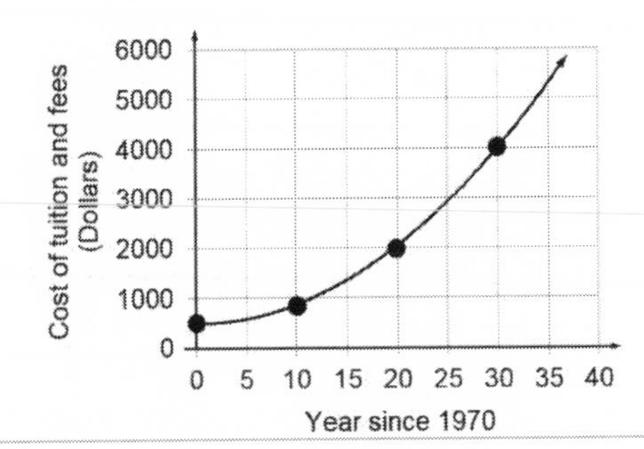

d) $x = 35$ is asking for the cost in 2005. An estimated answer from the graph would be around $5,400. After substituting 35 for *x* the function returns the cost $5,299.

e) Solving $3{,}000 = 4x^2 - 3x + 494$ gives *x* values of about -25 and 25. Disregarding the negative value, 25 tells us the cost will be $3,000 around 1995 (year 25). The graph verifies the answer.

f) $x = 45$, Solving $y = 4(45)^2 - 3(45) + 494$ predicts the cost will be around $8,459.

3)

a) Since $a > 0$ the parabola opens up. This implies that the number of cases may reach a minimum and then start increasing. The vertex is $(9.5, 383)$ which implies the number of cases reached a minimum around 1992-1993 at 383 cases.

b) Since there were about 1,175 cases in 1983 the question would be $y = 1{,}175$. Solving gives us two values for *x*, 0 and 19. The 0 represents 1983 so the answer would be, "The number of cases will return to the 1983 level around 2002."

c) Solving $8.8(t - 9.5)^2 + 381 = 2{,}000$ returns -4 and 23. I'm comfortable saying, "There might be around 2,000 cases in 2006." The -4 implies there may have also been 2,000 cases in 1979, but you should always be wary of extrapolating that far into the past.

d) $x = 27$. In 2010 the function predicts there will be approximately 3,081 cases.

4)

a) Questions about minimum or maximum values have to do with the vertex. For this function the vertex, $(605, 31205)$, tells us that setting the price at $605 per bicycle will maximize profit at $31,205 per year.

b) If $x = 0$ then $y = -42{,}000$ which implies setting the price charged for a bicycle at $0 will lead to a $42,000 loss. Probably for initial expenses such as parts.

c) Solving for *x* gives the answers $550 and $660. This implies if we set the price at $55 less than the maximum price (of $605) we will lose profit due to charging too little (the $550) or we lose profit due to decreasing demand (at $650).

d) The profit at $210 is $0, while the profit at $310 is $13,800 for a gain of $13,800. The profit at $500 is $29,000, while the profit at $600 is $31,200 for a gain of only $2,200. In both cases profit is increasing but from $210 to $310 the price is low enough that we still have high demand. Once we're at $500 to $600, although we're making more per bike, demand has decreased quite a bit. So in both cases, profit continues to increase, but at the higher price, demand is falling off.

5)

a)

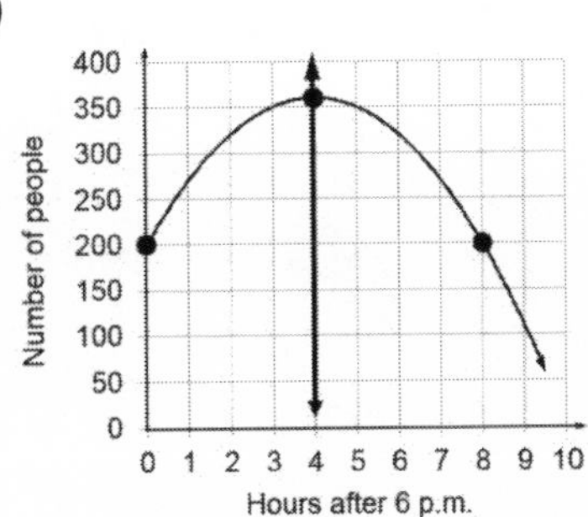

b) $y = 300$ is asking when will 300 people be attending the reception? Using the graph, a good estimate would be around 7:30 p.m. and 12:30 a.m. The algebraic function backs this up with times of ≈ 1.5 hours (7:30 p.m.) and ≈ 6.5 hours (12:30 a.m.)

c) $x = 2.5$ is asking how many people will be attending the reception at 8:30 p.m. Using the graph a good estimate would be around 330 people. The algebraic function approximates there will be 338 people.

d) $x = 0$ asks, "How many people are expected when the reception starts?" The answer is approximately 200 people.

e) Questions about a maximum have to do with the vertex. The vertex tells us that at 10 p.m. we can expect around 360 people which will be the maximum attendance.

f) $y = 0$ predicts when the last person will leave the reception. Solving the equation $0 = -10x^2 + 80x + 200$ estimates the last person will leave at 4 a.m.

6)

a) Since *a* is negative the parabola opens down so the vertex will have a maximum.

b)

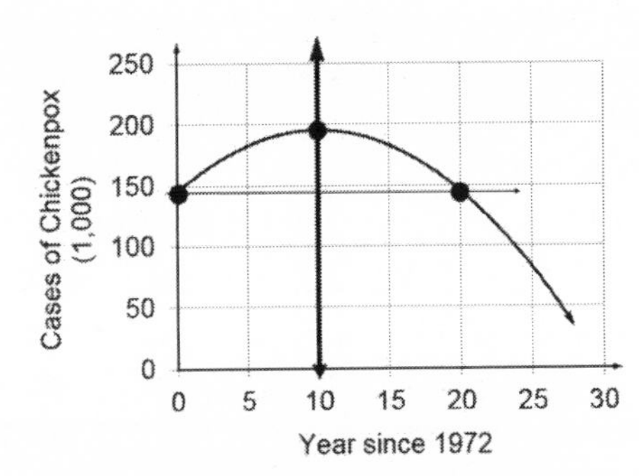

c) The vertex tells us that around 1982 the number of cases of Chickenpox peaked at 195,000 cases.

d) $x = 22$ asks about the number of cases in 1992. Using the graph the estimate is around 125,000 cases. Using the algebraic function the number of cases is 123,000 cases.

e) $y = 175$ is asking us to predict the year(s) the number of cases will reach 175,000. Looking at the graph a good prediction would be years 4 and 17 which would be 1976 and 1989. Solving $175 = -0.5t^2 + 10t + 145$ predicts years 3.7 and 16.3 which would be 1976 and 1988.

f) Solving $y=-0.5(5)^2+10(5)+145$ tells us that in 1977 there was around 182,500 cases.

Solving $182.5=-0.5x^2+10x+145$ gives us years 5 and 15. We already knew 5, it's 1977. The 15 tells us the number of cases will first return to 1977 levels around 1987.

g) Solving $0=-0.5t^2+10t+145$ tells us that in 2002 the number of cases will first go to 0. This didn't happen. Instead, the reason for the decline in Chickenpox was that a vaccine for Chickenpox was introduced in 1995 and since then a rational function is a much better model for cases of Chickenpox. Rational functions are usually first discussed in an intermediate or college algebra course.

7)

a) Since $-16<0$ the parabola is opening down which implies there might be a maximum height.

b)

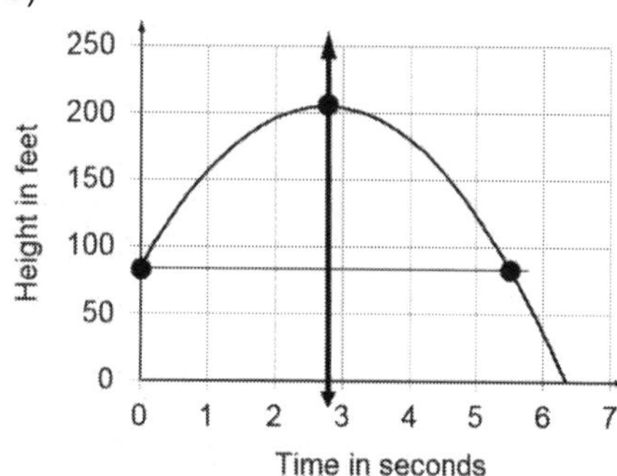

c) The vertex answers both these questions. The maximum height will be 205 feet and it will happen after 2.75 seconds.

d) The *y*-intercept is at 84 feet, subtracting 7 feet for the person tells us the building is about 77 feet high.

e) Drawing a horizontal line from $y=155$ gives approximate times of 1 second and 4.5 seconds. Using the algebraic function gives $x\approx 0.98$ and $x\approx 4.52$ which is very close to the estimated answers from the graph.

f) $x=1$ is asking us to find the height after 1 second. From our work in the previous problem we know the height should be around 155 feet. The algebraic function returns 156 feet.

g) The *x*-intercepts tell us when the object is on the ground. Setting $y=0$, and solving, returns the values $x\approx -0.8$ and $x\approx 6.3$. The negative value isn't useful in this context but the positive value tells us to first expect the object will be on the ground after about 6.3 seconds.

h) $x=0.5$ gives 124. Replacing y with 124, and solving for *x*, gives back 0.5 as expected but also gives 5. On the graph, after 5 seconds, the object should be close to 124 feet. This turns out to be correct.

8)

a) Since $0.017 > 0$ the parabola opens up. This implies that, in time, the tons of paper and paperboard recovered will continue to increase from a minimum value.

b) $x = 0$ asks this question. In 1980 (year 0). 21.1 million tons were recovered.

c) The x-coordinate of the vertex, approximately -23, gives a minimum value for R in the year 1957. We don't have data for the paper and paperboard recovered in 1957 because recycling hadn't become a common topic of interest. This is a case where the vertex doesn't have a relevant meaning for our data. Because the vertex is less than 1980 though, we do know that recycling is increasing from 1980 on.

d)

x	y
0	21.1
10	30.5
30	59.8
50	102.6

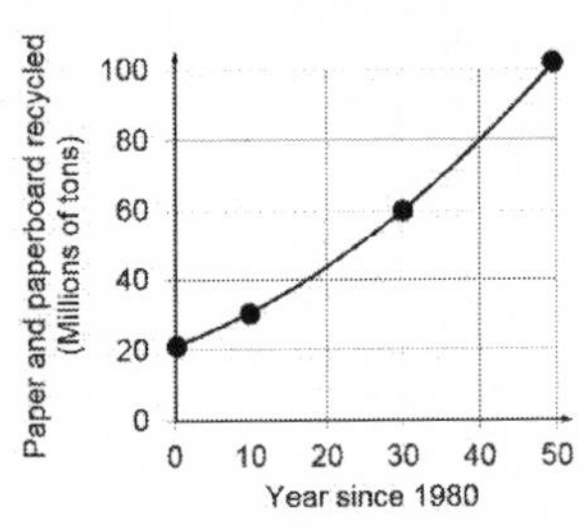

e) 1999 is a little to the left of year 20. The graph shows that in 1999 about 42 million tons were recovered. Substituting 19 for t and solving, predicts that in 1999 about 42 million tons were recycled.

f) This is asking when 10 million tons of paper and paperboard were recovered. Setting y to 10 and solving leads to a negative discriminant so the equation has no solution in the real numbers. This implies the function can't discuss this value. Looking at the graph you can see that 10 is below our starting value of 21.1.

g) Moving horizontally from $R = 80$ and then vertically to the year it seems this should occur around 2020. $y = 80$ asks the question and solving $0.017t^2 + 0.78t + 21.1 = 80$ finds again that this will first happen in 2020. The other answer, -86, can be disregarded.

9) a)

b) Charging \$0 implies we're interested in the y-intercept. Charging \$0 results in a loss of \$190 probably for initial expenses.

c) The vertex tells us they'll maximize their profit at \$12 an hour for a maximum profit of \$1,250 per month.

d) At \$16 an hour the graph shows their monthly profit will fall to around \$1,100. The algebraic function predicts their profit will fall to \$1,090. Raising the price has lowered the demand.

e) The graph predicts that setting the price at either \$7 an hour or \$17 an hour will result in a monthly profit of \$1,000. The algebraic function predicts prices of \$7 and \$17. At \$7 an hour demand will be high but the price will be low. At \$17 the price will be high but the higher price will result in reduced demand. Personally, I'd rather charge \$17 since I'll make the same amount as charging \$7 with less time spent on lawns.

f) Trying to solve $1{,}300 = -10p^2 + 240p - 190$ results in a negative discriminant. This implies that under the current conditions the student can't earn \$1,300 a month. This makes sense since, as the vertex shows, the maximum profit is \$1,250.

10)

a) The cost in 1990 was about \$3,591.

b) The value of a is 4.8 which implies the graph is opening up. The vertex is about (–13, 2764) which implies the cost of college will be continually increasing from \$3,591.

c)

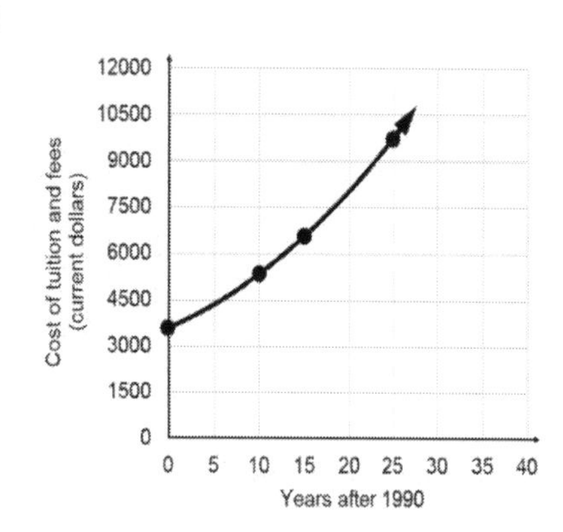

d) $x = 30$ is asking for the cost in 2020. An estimated answer from the graph would be around \$12,000. The algebraic function returns \$11,691.

e) Solving $6{,}000 = 4.8x^2 + 126x + 3{,}591$ gives x values of about -39 and 13. Disregarding the negative value the 13 tells us the cost will be \$6,000 around 2003. The graph verifies the answer.

f) Setting $x = 35$ predicts the cost will be around \$13,881 for one year of tuition and fees in 2025.

7.6 Completing the Square

You've solved quadratic equations using the quadratic formula, the zero-product method and the square-root method. In this section, you'll practice completing the square which is a fourth way to solve quadratic equations. Completing the square transforms a quadratic equation from standard form to "vertex" form so you can use the square-root method to solve the equation.

7.6.1 The Idea of Completing the Square

To "complete the square" we add a term to a binomial which creates a perfect square trinomial. For instance, if we start with the binomial x^2+6x, and add 9, x^2+6x+9, we create a trinomial that factors to a perfect square, $x^2+6x+9=(x+3)(x+3)=(x+3)^2$.

Visually you can see how the constant 9, "completes the square." If we had a square with side lengths of $x+3$, then the area of the square would be $(x+3)(x+3)$ (length times width) which, after multiplying, would be x^2+6x+9. If you look to the right though you'll see that currently we haven't completed the area of the square. We have an area of x^2 plus two areas of 3*x* which is x^2+6x but to "complete" the area of the square we need to fill in the little square at the bottom right corner which has area 3×3 or 9. Adding 9, x^2+6x+9

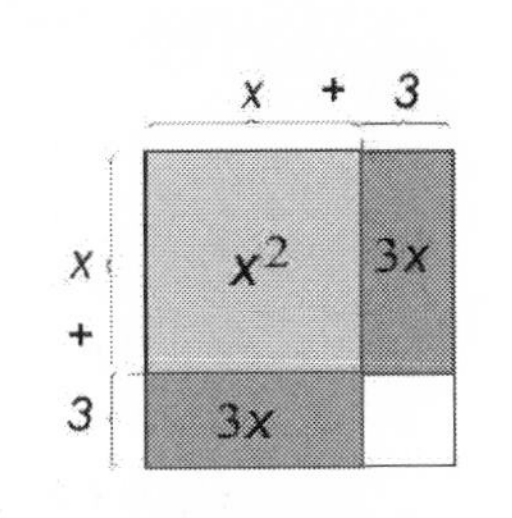

completes the area of the square and also implies that the four separate <u>terms</u> $x^2+3x+3x+9$ (or x^2+6x+9), is equivalent to 2 <u>factors</u> of $x+3$ (length times width) or $(x+3)^2$.

To find the term that completes the square, identity *b* (the first-degree coefficient), and find $\left(b/2\right)^2$. For example, with x^2+6x, *b* is 6, and $\left(6/2\right)^2=3^2=9$. If *b* is a fraction, you may find it easier to work with $\left(\frac{1}{2}\times b\right)^2$.

Practice 7.6.1 The Idea of Completing the Square

Add a constant to the binomial to "complete" the square.

---

a) x^2-10x

$b=-10 \quad \Rightarrow$ Remembered to think of polynomials as a **sum** of terms.

$\left(-10/2\right)^2=(-5)^2=25 \quad \Rightarrow$ Squared half of negative 10.

$x^2-10x+25 \quad \Rightarrow$ Added my value to the binomial.

b) $y^2 + \frac{2}{3}y$

$b = 2/3$	$\Rightarrow$	b is the first degree coefficient.
$\frac{1}{2} \times \frac{2}{3} = \frac{1}{3}$	$\Rightarrow$	With fractions, it's often easier to think of dividing by 2 as multiplication by one-half.
$\left(\frac{1}{3}\right)^2 = \frac{1}{9}$	$\Rightarrow$	Squared half of two-thirds.
$y^2 + \frac{2}{3}y + \frac{1}{9}$	$\Rightarrow$	Added my value to the binomial.

Homework 7.6 Add a constant to the binomial to "complete" the square.

1) $q^2 - 2q$ 2) $b^2 + 4b$ 3) $w^2 - w$ 4) $p^2 - 3p$

5) $y^2 - 5y$ 6) $m^2 - \frac{3}{2}m$ 7) $h^2 + \frac{5}{3}h$

7.6.2 Factoring to a Perfect Square

In practice, it's sometimes difficult to figure out how to factor the trinomial to the square of a binomial. One idea is to use the form $\left(x + \left(\frac{b}{2}\right)\right)^2$. For instance, if you start with $x^2 + \frac{4}{3}x$, find $\left(\frac{b}{2}\right)^2$, $\left(\frac{1}{2} \times \frac{4}{3}\right)^2 = \left(\frac{2}{3}\right)^2 = \frac{4}{9}$ and build the trinomial, $x^2 + \frac{4}{3}x + \frac{4}{9}$, your factored form will be $\left(x + \left(\frac{b}{2}\right)\right)^2$ or in this case $\left(x + \left(\frac{1}{2} \times \frac{4}{3}\right)\right)^2 = \left(x + \frac{2}{3}\right)^2$.

Practice 7.6.2 Factoring to a Perfect Square

Build the factored form using $\left(x + \left(\frac{b}{2}\right)\right)^2$.

a) $x^2 - 10x + 25$

$x^2 - 10x + 25$	$\Rightarrow$	Added $\left(\frac{-10}{2}\right)^2 = 25$ to build the trinomial.
$\left(x + \left(\frac{-10}{2}\right)\right)^2 = (x + -5)^2$ $(x - 5)^2$	$\Rightarrow$	Went immediately to the square of the binomial and wrote adding a negative as subtraction.

b) $y^2 + \frac{2}{3}y + \frac{1}{9}$

$\left(y + \left(\frac{1}{2} \times \frac{2}{3}\right)\right)^2$ $\Rightarrow$ Went immediately to the square of the binomial.

$\left(y + \frac{1}{3}\right)^2$ $\Rightarrow$ Simplified.

Homework 7.6 Build the factored form using $\left(x + \left(\frac{b}{2}\right)\right)^2$.

8) $q^2 - 2q + 1$ 9) $b^2 + 4b + 4$ 10) $w^2 - w + \frac{1}{4}$ 11) $p^2 - 3p + \frac{9}{4}$

12) $y^2 - 5y + \frac{25}{4}$ 13) $m^2 - \frac{3}{2}m + \frac{9}{16}$ 14) $h^2 + \frac{5}{3}h + \frac{25}{36}$

7.6.3 Solving by Completing the Square When $a = 1$

Completing the square helps us rewrite $ax^2 + bx + c = 0$ into a form where we can use the square-root method to solve the equation. Here are the steps when $a = 1$.

Procedure –Solving by Completing the Square when $a = 1$
1. If it's necessary, use addition or subtraction so the constant term is isolated. 2. Identify the value of b and add $\left(\frac{b}{2}\right)^2$ to both sides. 3. Factor the left side to the square of a binomial and simplify the right side. 4. Solve using the square-root method.

Practice 7.6.3 Solving by Completing the Square When $a = 1$

Solve by completing the square.

a) $y^2 - 8y + 13 = -2$

$y^2 - 8y = -15$ $\Rightarrow$ Isolated the constant term by subtracting 13 from both sides.

$y^2 - 8y + 16 = -15 + 16$ $\Rightarrow$ $\left(\frac{b}{2}\right)^2 = \left(\frac{-8}{2}\right)^2 = (-4)^2 = 16$. Added 16 to both sides.

$(y - 4)^2 = 1$ $\Rightarrow$ Factored the left side and simplified the right side.

$$y - 4 = \pm\sqrt{1}$$
$$y = 4 \pm 1 \quad \Rightarrow \quad \text{Solved using the square-root method.}$$
$$y = 5 \quad \text{or} \quad y = 3$$

---

b) $y^2 + \frac{2}{3}y - \frac{5}{9} = 0$

$y^2 + \frac{2}{3}y = \frac{5}{9}$ $\Rightarrow$ Isolated the constant term.

$y^2 + \frac{2}{3}y + \frac{1}{9} = \frac{5}{9} + \frac{1}{9}$ $\Rightarrow$ $\left(\frac{1}{2} \times b\right)^2 = \left(\frac{1}{2} \times \frac{2}{3}\right)^2 = \left(\frac{1}{3}\right)^2 = \frac{1}{9}$. Added $\frac{1}{9}$ to both sides.

$\left(y + \frac{1}{3}\right)^2 = \frac{2}{3}$ $\Rightarrow$ Factored the left side and simplified the right side.

$y + \frac{1}{3} = \pm\sqrt{\frac{2}{3}}$

$y = -\frac{1}{3} \pm \sqrt{\frac{2}{3}}$ $\Rightarrow$ Solved using the square-root method. Made sure to rationalize the denominator at the end.

$y = -\frac{1}{3} \pm \frac{\sqrt{6}}{3}$

Homework 7.6 Solve by completing the square.

15) $q^2 - 2q - 3 = 0$

16) $x^2 - 6x + 7 = 0$

17) $v^2 - 10v + 13 = 0$

18) $w^2 - w - 1 = 0$

19) $y^2 + 2y - 4 = 4$

20) $a^2 + 3a + 6 = 10$

21) $x^2 - x + 1 = \frac{23}{4}$

22) $y^2 + \frac{1}{3}y - \frac{2}{9} = 0$

23) $m^2 - \frac{4}{3}m + 2 = \frac{8}{3}$

24) $Q^2 + \frac{8}{3}Q + \frac{2}{3} = \frac{7}{18}$

25) $r^2 - 4r + \frac{1}{3} = -\frac{7}{3}$

7.6.4 *Solving by Completing the Square*

Now let's assume the value of *a* isn't 1 and practice the full procedure.

Procedure –Solving by Completing the Square
1. If $a \neq 1$ divide each term by *a* and simplify (if it's possible to). 2. If it's necessary, use addition or subtraction so the constant term is isolated. 3. Identify the value of *b* and add $\left(\frac{b}{2}\right)^2$ to both sides. 4. Factor the left side to the square of a binomial and simplify the right side. 5. Solve using the square root method.

Practice 7.6.4 Solving by Completing the Square

Solve by completing the square.

a) $2x^2 - 12x = -17$

$x^2 - 6x = -\frac{17}{2}$	$\Rightarrow$	Divide each term by *a* (which in this case was 2).
$x^2 - 6x + 9 = -\frac{17}{2} + 9$	$\Rightarrow$	$\left(\frac{-6}{2}\right)^2 = (-3)^2 = 9$. Added 9 to both sides.
$(x-3)^2 = \frac{1}{2}$	$\Rightarrow$	Factored the left side and simplified the right side.
$x - 3 = \pm\sqrt{\frac{1}{2}} = \pm\frac{\sqrt{2}}{2}$	$\Rightarrow$	Used the square-root method and rationalized the denominator on the right side.
$x = 3 \pm \frac{\sqrt{2}}{2}$	$\Rightarrow$	Added 3 to both sides.

b) $3w^2 + 2w - 6 = 0$

$w^2 + \frac{2}{3}w - 2 = 0$	$\Rightarrow$	Divide each term by *a*. On the right side $\frac{0}{2}$ is still 0.
$w^2 + \frac{2}{3}w = 2$	$\Rightarrow$	Added 2 to both sides.
$w^2 + \frac{2}{3}w + \frac{1}{9} = 2 + \frac{1}{9}$	$\Rightarrow$	$\left(\frac{1}{2} \times \frac{2}{3}\right)^2 = \left(\frac{1}{3}\right)^2 = \frac{1}{9}$. Add one-ninth to both sides.
$\left(w + \frac{1}{3}\right)^2 = \frac{19}{9}$	$\Rightarrow$	Factored the left side and simplified the right side.
$w + \frac{1}{3} = \pm\sqrt{\frac{19}{9}} = \pm\frac{\sqrt{19}}{3}$	$\Rightarrow$	Used the square-root method and on the right used the quotient property for roots and simplified.
$w = -\frac{1}{3} \pm \frac{\sqrt{19}}{3}$	$\Rightarrow$	Subtracted $\frac{1}{3}$ from both sides.

Homework 7.6 Solve by completing the square

26) $4x^2+12x-7=0$	27) $3x^2+2x-\frac{8}{3}=0$	28) $2t^2-6t+5=0$
29) $4x^2+2x-3=0$	30) $9k^2-12k-23=0$	31) $2x^2+\frac{2}{3}x-\frac{1}{2}=0$
32) $20x^2-60x=-37$	33) $3p^2-\frac{2}{3}p-\frac{5}{3}=0$	34) $4w^2-10w+11=0$
35) $\frac{1}{2}y^2-\frac{1}{4}y-1=0$	36) $\frac{3}{4}x^2+\frac{9}{8}x+\frac{3}{8}=0$	

Homework 7.6 Answers

1) q^2-2q+1	2) b^2+4b+4	3) $w^2-w+\frac{1}{4}$	4) $p^2-3p+\frac{9}{4}$	5) $y^2-5y+\frac{25}{4}$
6) $m^2-\frac{3}{2}m+\frac{9}{16}$	7) $h^2+\frac{5}{3}h+\frac{25}{36}$	8) $(q-1)^2$	9) $(b+2)^2$	10) $\left(w-\frac{1}{2}\right)^2$
11) $\left(p-\frac{3}{2}\right)^2$	12) $\left(y-\frac{5}{2}\right)^2$	13) $\left(m-\frac{3}{4}\right)^2$	14) $\left(h+\frac{5}{6}\right)^2$	15) $\{-1,3\}$
16) $\{3\pm\sqrt{2}\}$	17) $\{5\pm2\sqrt{3}\}$	18) $\left\{\frac{1}{2}\pm\frac{\sqrt{5}}{2}\right\}$	19) $\{-4,2\}$	20) $\{-4,1\}$
21) $\left\{\frac{1}{2}\pm\sqrt{5}\right\}$	22) $\left\{-\frac{2}{3},\frac{1}{3}\right\}$	23) $\left\{\frac{2}{3}\pm\frac{\sqrt{10}}{3}\right\}$	24) $\left\{-\frac{4}{3}\pm\frac{\sqrt{6}}{2}\right\}$	25) $\left\{2\pm\frac{2\sqrt{3}}{3}\right\}$
26) $\left\{\frac{-7}{2},\frac{1}{2}\right\}$	27) $\left\{-\frac{4}{3},\frac{2}{3}\right\}$	28) No solution	29) $\left\{-\frac{1}{4}\pm\frac{\sqrt{13}}{4}\right\}$	30) $\left\{\frac{2}{3}\pm\sqrt{3}\right\}$
31) $\left\{-\frac{1}{6}\pm\frac{\sqrt{10}}{6}\right\}$	32) $\left\{\frac{3}{2}\pm\frac{\sqrt{10}}{5}\right\}$	33) $\left\{\frac{1}{9}\pm\frac{\sqrt{46}}{9}\right\}$	34) No solution	35) $\left\{\frac{1}{4}\pm\frac{\sqrt{33}}{4}\right\}$
36) $\left\{-1,-\frac{1}{2}\right\}$				

7.7 Literal Equations

A **literal equation** is an equation with more than one type of variable. Since literal equations are important for fields as diverse as culinary arts and nuclear physics, your future will probably include some literal equations. An important skill is being able to rewrite a literal equation to change your point of view. For example, the formula for finding the length of the hypotenuse of a right triangle $c=\sqrt{a^2+b^2}$ is a literal equation. As you saw in an earlier assignment, instead of finding the length of the hypotenuse we might want to find the length of one of the legs. That means we might want to rewrite $c=\sqrt{a^2+b^2}$ as $a=\sqrt{c^2-b^2}$. In this section you'll practice rewriting literal equations.

Before we start it's important to discuss some subtleties about notation. Lowercase and uppercase letters in the same literal equation represent different items. For instance, the formula $P=-2p^2+40p-150$ might show that *P*, our profit, depends on *p*, the price we charge. A letter with a subscript represents a different item than a letter with no subscript so A_0 would be different from A. Also, as you saw with the slope formula, letters with different subscripts are different, so x_1 is different from x_2.

7.7.1 Equivalent Ways to Write a Literal Expression

The solution to a literal equation is often an expression with more than one operation. Usually there are equivalent ways to write this expression. As a rule of thumb, an expression with fewer operators is considered simpler but sometimes the final form is more a matter of tradition than mathematics. Let's practice rewriting some literal expressions so your answer and my answer will look the same.

Practice 7.7.1 Equivalent Ways to Write a Literal Expression

Rewrite the following literal expressions.

a) Show that $\frac{P-2L}{2}$ can be written as $\frac{1}{2}P-L$.

$\frac{P}{2}-\frac{2L}{2}$	$\Rightarrow$	Wrote the fraction as two terms with the common denominator.
$\frac{P}{2}-\frac{\cancel{2}L}{\cancel{2}}=\frac{P}{2}-L$	$\Rightarrow$	Now that I had factors I reduced the common factor of 2.
$\frac{1}{2}P-L$	$\Rightarrow$	Made the coefficient explicit. $\frac{P}{2}=\frac{1}{2}\left(\frac{P}{1}\right)=\frac{1}{2}P$

b) Show that $\dfrac{A}{\frac{1}{2}b}$ can be written as $\dfrac{2A}{b}$.

$$\frac{A}{\frac{1}{2}b} = \frac{2}{2} \times \frac{A}{\frac{1}{2}b} = \frac{2A}{\frac{2}{1}\left(\frac{1}{2}b\right)} = \frac{2A}{b} \Rightarrow$$

After multiplying both the numerator and denominator by 2 the 1/2 in the denominator can be reduced. This leaves $2A$ in the numerator and b in the denominator.

$$\frac{A}{\frac{1}{2}b} = \frac{\frac{A}{1}}{\frac{b}{2}} = \frac{A}{1} \times \frac{2}{b} = \frac{2A}{b} \Rightarrow$$

A second approach is to write $\frac{1}{2}b$ as $\frac{b}{2}$ and then use the procedure for dividing fractions.

c) Show that $\dfrac{1-m}{-m}$ can be written as $\dfrac{m-1}{m}$.

$$\frac{1-m}{-m} = \frac{-1(1-m)}{-1(-m)} = \frac{-1+m}{m} \Rightarrow$$

Used the multiplicative identity and multiplied both the numerator and denominator by negative one. Distributed in the numerator.

$$\frac{m+-1}{m} = \frac{m-1}{m} \Rightarrow$$

Used the commutative property to rewrite the numerator. Noticed the m's can't be reduced because the m in the numerator is a term.

$$\frac{-1(-1+m)}{-1(m)} = \frac{\cancel{-1}(-1+m)}{\cancel{-1}(m)} = \frac{-1+m}{m} = \frac{m-1}{m} \Rightarrow$$

A second approach would be to factor out a factor of -1 and reduce.

Homework 7.7 Rewrite the following literal expressions.

1) Show $\dfrac{S-24}{3}$ can be written as $\dfrac{S}{3}-8$.

2) Show $\dfrac{z-8x}{-4}$ can be written as $2x-\dfrac{1}{4}z$.

3) Show $\dfrac{A}{\frac{b}{2}}$ can be written as $\dfrac{2A}{b}$.

4) Show $\dfrac{A-\frac{b}{2}}{\frac{1}{2}}$ can be written as $2A-b$.

5) Show $\dfrac{mx-nx}{nx}$ can be written as $\dfrac{m}{n}-1$.

6) Show $-\dfrac{x}{x-1}$ can be written as $\dfrac{x}{1-x}$.

7) Show $\dfrac{\frac{yz}{1-y}}{z}$ can be written as $\dfrac{y}{1-y}$.

8) Show $\dfrac{\frac{2A-ah}{2}}{h}$ can be written as $\dfrac{A}{h}-\dfrac{1}{2}a$.

7.7.2 Solving Literal Equations Using a Two-Column Table

We say $c=\sqrt{a^2+b^2}$ is "solved for" c because c is isolated. I've solved $c=\sqrt{a^2+b^2}$ for a, when I've isolated the letter a instead. To solve $c=\sqrt{a^2+b^2}$ for a, I'll have to use inverse operations to "undo" the three operations "acting on" a (squaring, adding and square-root). At first, you may find it difficult to "see" how to order the inverse operations. The two-column table is a systematic way to help you decide on the right order.

Procedure – Solving Literal Equations Using a Two-Column Table

1. Build a two-column table. Label the left column for operations and the right column for inverses.
2. Imagine any letter that you're not solving for is a number and make a row for each operation that "acts on" the variable.
3. In the left column, list the operations, in order, that you would use to simplify the expression (remember to think of each letter you're not solving for as a number).
4. In the right column list the inverse operations for the operations found in step 3.
5. Isolate the variable by starting at the bottom of the right column and working your way to the top.

As an example, I'll build a two-column table to solve the equation $y = mx + b$ for *x*.

I began by making a two-column table. The left column is for operations and the right column is for inverses.

Operations	Inverses

To decide on the number of additional rows, I pretend every letter on the right side that isn't *x* (my "solved for" variable) is a number. For example, I might imagine that *m* is 5 and *b* is 4, $y = 5x + 4$. Since there are two operations acting on *x*, multiplying and adding, I need to make two additional rows.

Operations	Inverses

Still imagining the equation is $y = 5x + 4$, the order of operations tells me that multiplication would be first and addition second. Shifting my attention back to $y = mx + b$ I record that order (multiplication first, adding second), top to bottom, in the left column.

Operations	Inverses
$\times m$	
$+b$	

Moving to the right column I record the inverse operations. The inverse for multiplying by *m* is dividing by *m* and the inverse for adding *b* is subtracting *b*.

Operations	Inverses
$\times m$	$\div m$
$+b$	$-b$

Since, to isolate *x*, I am "undoing" operators, I need to start at the bottom of the right column and work to the top. First, I'll subtract *b* from both sides and simplify on the right side.

$$y = mx + b$$
$$y - b = mx + b - b$$
$$y - b = mx$$

Now, moving up a row, I see I need to divide both expressions by *m*. After dividing I reduce the common factor of *m* on the right side.

$$\frac{y-b}{m} = \frac{\cancel{m}x}{\cancel{m}}$$
$$\frac{y-b}{m} = x$$

I've now solved for *x* since *x* is isolated. It's common, but not necessary, to write the "solved for" letter on the left. Here's some practice building a two-column table to help solve literal equations.

$$x = \frac{y-b}{m}$$

Practice 7.7.2 Solving Literal Equations Using a Two-Column Table

Use a two-column table to solve for the indicated variable.

a) Solve $P = 2L + 2W$ for W.

Operations	Inverses
$\times W$	
$+2L$	

$\Rightarrow$ Realized the operations acting on W are multiplying and adding so made two rows. Recorded the operations, in order, top to bottom.

Operations	Inverses
$\times W$	$\div W$
$+2L$	$-2L$

$\Rightarrow$ Recorded the inverse operations.

$$P = 2L + 2W$$
$$P - 2L = 2L + 2W - 2L$$
$$P - 2L = 2W$$

$\Rightarrow$ Started at the bottom of the right column, subtracted $2L$ from both expression and simplified.

$$\frac{P-2L}{2} = \frac{\cancel{2}W}{\cancel{2}}$$
$$\frac{P-2L}{2} = W$$

$\Rightarrow$ Moved up a row, divided both expressions by 2 and simplified.

$$W = \frac{P-2L}{2} = \frac{1}{2}P - L$$

$\Rightarrow$ Kept in mind the answer can take different forms. (See if you're able to build the second form.)

b) Solve $c = \sqrt{a^2 + b^2}$ for a.

Operations	Inverses
^2	
$+b^2$	
$\sqrt{\ }$	

$\Rightarrow$ Since there are three operations happening to a, (squaring, adding and square-root) made three rows. Recalled the radicand has implicit grouping so by the order of operations a is squared first, then b^2 is added and taking the square root is last. Recorded the operations in the left column.

Operations	Inverses
^2	$\pm\sqrt{\ }$
$+b^2$	$-b^2$
$\sqrt{\ }$	^2

$\Rightarrow$ Recorded the inverse operations in the right column. Recalled that the square-root method needed to include the $\pm$ symbol when finding the inverse of squaring.

$$c^2 = \left(\sqrt{a^2 + b^2}\right)^2$$
$$c^2 = a^2 + b^2$$

$\Rightarrow$ Started at the bottom of the right column and squared both expressions. On the right side, remembered that squaring a square-root leaves just the radicand.

$$c^2 - b^2 = a^2 + b^2 - b^2$$
$$c^2 - b^2 = a^2$$

$\Rightarrow$ Moved up a row and subtracted b^2 from both expressions. Kept in mind I'm trying to isolate a.

$\pm\sqrt{c^2-b^2}=\sqrt{a^2}$ $\pm\sqrt{c^2-b^2}=a$ $a=\pm\sqrt{c^2-b^2}$	$\Rightarrow$	Moved up a row and took the square root of both expressions. Made sure to include the $\pm$ symbol. Wrote the solved for variable on the left.

c) Solve $Q_2=\dfrac{M+AT}{Q_1}$ for A

Operations	Inverses
$\times T$	$\div T$
$+M$	$-M$
$\div Q_1$	$\times Q_1$

$\Rightarrow$ The table needed three rows since multiplication, addition and division are acting on A. Kept in mind the numerator has implicit grouping so by the order of operations multiplication would be first, addition next and division last. Wrote the inverse operations in the right column.

$Q_1Q_2=\dfrac{\cancel{Q_1}}{1}\left(\dfrac{M+AT}{\cancel{Q_1}}\right)$ $Q_1Q_2=M+AT$	$\Rightarrow$	Started at the bottom of the right column, multiplied both expressions by Q_1 and reduced on the right. Q_1Q_2 implies the product of two factors, that is, it's Q_1 times Q_2.
$Q_1Q_2-M=M+AT-M$ $Q_1Q_2-M=AT$	$\Rightarrow$	Moved up a row, subtracted M from both expressions and simplified.
$\dfrac{Q_1Q_2-M}{T}=\dfrac{A\cancel{T}}{\cancel{T}}$ $A=\dfrac{Q_1Q_2-M}{T}$	$\Rightarrow$	Moved up a row, divided by T and reduced on the right. The literal equation is now solved for A.

Homework 7.7 Use a two-column table to solve for the indicated variable.

9) Solve $P=s_1+s_2+s_3$ for s_2.

10) Solve $\dfrac{2A}{h}=b$ for A.

11) Solve $a^2+b^2=c^2$ for b.

12) Solve $r=\dfrac{C}{2\pi}$ for C.

13) Solve $F=\dfrac{-T}{\sqrt{n}}$ for T.

14) Solve $F=\dfrac{9}{5}C+32$ for C.

15) Solve $H=2\sqrt{t}+13$ for t.

16) Solve $V=\dfrac{\pi}{3}r^2h$ for r.

17) Solve $N=(2k-1)^2$ for k.

18) Solve $t=m\left(\dfrac{T-V}{6}\right)$ for T.

19) Solve $L_1q_1+L_2q_2=1$ for L_1.

20) Solve $x+2y+7=0$ for y.

21) Solve $yy_1+xx_1=0$ for y.

22) Solve $z=b\sqrt{\dfrac{a_1+a_2}{2}}$ for a_1.

23) Solve $W=\dfrac{P-2L}{2}$ for P.

24) Solve $ay^2+bx+c=0$ for y.

25) Solve $h=\sqrt{\dfrac{A(y+1)}{a}}$ for y.

26) Solve $D=\dfrac{(\sqrt{d}-1)^2}{9}$ for d.

27) Solve $Q_1=2\sqrt{x-100}-Q_2$ for x.

28) Solve $s=2\pi r^2+2\pi rh$ for h.

29) Solve $A=\dfrac{h(a+b)}{2}$ for a.

30) Solve $A=\left(\dfrac{mk_1+nk_2}{m}\right)^2$ for k_1.

31) Solve $E\sqrt{n}=2s$ for n.

32) Solve $\sqrt{(x-x_1)^2+y^2}=1$ for x.

Homework 7.7 Answers

1) $\dfrac{S-24}{3}=\dfrac{S}{3}-\dfrac{24}{3}=\dfrac{S}{3}-8$

2) $\dfrac{z-8x}{-4}=\dfrac{z}{-4}-\dfrac{8x}{-4}=-\dfrac{z}{4}+2x=2x-\dfrac{z}{4}=2x-\dfrac{1}{4}z$

3) $\dfrac{A}{\frac{b}{2}}=\dfrac{2}{2}\times\dfrac{A}{\frac{b}{2}}=\dfrac{2A}{\cancel{2}\left(\frac{b}{\cancel{2}}\right)}=\dfrac{2A}{b}$ or $\dfrac{A}{\frac{b}{2}}=\dfrac{\frac{A}{1}}{\frac{b}{2}}=\dfrac{2}{b}\times\dfrac{A}{1}=\dfrac{2A}{b}$

4) $\dfrac{A-\frac{b}{2}}{\frac{1}{2}}=\dfrac{2}{2}\times\dfrac{A-\frac{b}{2}}{\frac{1}{2}}=\dfrac{2\left(A-\frac{b}{2}\right)}{2\left(\frac{1}{2}\right)}=\dfrac{2A-2\left(\frac{b}{2}\right)}{1}=\dfrac{2A-\cancel{2}\left(\frac{b}{\cancel{2}}\right)}{1}=2A-b$ or

$\dfrac{A-\frac{b}{2}}{\frac{1}{2}}=\dfrac{\frac{\left(A-\frac{b}{2}\right)}{1}}{\frac{1}{2}}=\dfrac{2}{1}\times\dfrac{\left(A-\frac{b}{2}\right)}{1}=2A-2\left(\dfrac{b}{2}\right)=2A-b$

5) $\dfrac{mx-nx}{nx}=\dfrac{mx}{nx}-\dfrac{nx}{nx}=\dfrac{m\cancel{x}}{n\cancel{x}}-\dfrac{\cancel{nx}}{\cancel{nx}}=\dfrac{m}{n}-1$ or $\dfrac{mx-nx}{nx}=\dfrac{\cancel{x}(m-n)}{n\cancel{x}}=\dfrac{m-n}{n}=\dfrac{m}{n}-\dfrac{n}{n}=\dfrac{m}{n}-1$

6) $-\frac{x}{x-1}=\frac{-x}{x-1}=\frac{-1(-x)}{-1(x-1)}=\frac{x}{-x+1}=\frac{x}{1+-x}=\frac{x}{1-x}$

7) $\frac{\frac{yz}{1-y}}{z}=\frac{\frac{yz}{1-y}}{\frac{z}{1}}=\left(\frac{1}{\cancel{z}}\right)\left(\frac{y\cancel{z}}{1-y}\right)=\frac{y}{1-y}$

8) $\frac{\frac{2A-ah}{2}}{h}=\frac{\frac{2A-ah}{2}}{\frac{h}{1}}=\left(\frac{1}{h}\right)\left(\frac{2A-ah}{2}\right)=\frac{2A-ah}{2h}=\frac{2A}{2h}-\frac{ah}{2h}=\frac{\cancel{2}A}{\cancel{2}h}-\frac{a\cancel{h}}{2\cancel{h}}=\frac{A}{h}-\frac{a}{2}=\frac{A}{h}-\frac{1}{2}a$

9) $s_2=P-s_3-s_1$

Operations	Inverses
$+s_1$	$-s_1$
$+s_3$	$-s_3$

10) $A=\frac{1}{2}bh$ or $\frac{hb}{2}$

Operations	Inverses
$\times 2$	$\div 2$
$\div h$	$\times h$

11) $b=\pm\sqrt{c^2-a^2}$

Operations	Inverse
^2	$\pm\sqrt{\ }$
$+a^2$	$-a^2$

12) $C=2\pi r$

Operations	Inverses
$\div 2$	$\times 2$
$\div \pi$	$\times \pi$

13) $T=-F\sqrt{n}$

Operations	Inverses
$\times -1$	$\div -1$
$\div\sqrt{n}$	$\times\sqrt{n}$

14) $5/9(F-32)=C$

Operations	Inverses
$\times 9$	$\div 9$
$\div 5$	$\times 5$
$+32$	-32

15) $t=\left(\frac{H-13}{2}\right)^2$ or $t=\frac{(H-13)^2}{4}$

Operations	Inverses
$\sqrt{\ }$	^2
$\times 2$	$\div 2$
$+13$	-13

16) $r=\pm\sqrt{\frac{3V}{\pi h}}$

Operations	Inverses
^2	$\pm\sqrt{\ }$
$\times \pi/3$	$\times 3/\pi$
$\times h$	$\div h$

17) $k=\frac{1}{2}\pm\frac{\sqrt{N}}{2}$

Operations	Inverses
$\times 2$	$\div 2$
-1	$+1$
^2	$\pm\sqrt{\ }$

18) $T=V+\frac{6t}{m}$

Operations	Inverses
$-V$	$+V$
$\div 6$	$\times 6$
$\times m$	$\div m$

19) $L_1=\frac{1-L_2q_2}{q_1}$

Operations	Inverses
$\times q_1$	$\div q_1$
$+L_2q_2$	$-L_2q_2$

20) $y=\frac{-x-7}{2}$

Operations	Inverses
$\times 2$	$\div 2$
$+x$	$-x$
$+7$	-7

21) $y = \frac{-xx_1}{y_1}$

Operations	Inverses
$\times y_1$	$\div y_1$
$+xx_1$	$-xx_1$

22) $a_1 = 2\left(\frac{z}{b}\right) - a_2$

Operations	Inverses
$+a_2$	$-a_2$
$\div 2$	$\times 2$
$\sqrt{\ }$	^2
$\times b$	$\div b$

23) $P = 2W + 2L$

Operations	Inverses
$-2L$	$+2L$
$\div 2$	$\times 2$

24) $y = \pm\sqrt{\frac{-c - bx}{a}}$

Operations	Inverses
^2	$\pm\sqrt{\ }$
$\times a$	$\div a$
$+bx$	$-bx$
$+c$	$-c$

25) $y = \frac{ah^2}{A} - 1$

Operations	Inverses
$+1$	-1
$\times A$	$\div A$
$\div a$	$\times a$
$\sqrt{\ }$	^2

26) $d = (1 \pm 3\sqrt{D})^2$

Operations	Inverses
$\sqrt{\ }$	^2
-1	$+1$
^2	$\pm\sqrt{\ }$
$\div 9$	$\times 9$

27) $x = \left(\frac{Q_1 + Q_2}{2}\right)^2 + 100$

Operations	Inverses
-100	$+100$
$\sqrt{\ }$	^2
$\times 2$	$\div 2$
$-Q_2$	$+Q_2$

28) $h = \frac{s - 2\pi r^2}{2\pi r}$

Operations	Inverses
$\times 2\pi r$	$\div 2\pi r$
$+2\pi r^2$	$-2\pi r^2$

29) $a = \frac{2A}{h} - b$

Operations	Inverses
$+b$	$-b$
$\times h$	$\div h$
$\div 2$	$\times 2$

30) $k_1 = 1 \pm \frac{\sqrt{A}}{m} - \frac{nk_2}{m}$

Operations	Inverses
$\times m$	$\div m$
$+nk_2$	$-nk_2$
$\div m$	$\times m$
^2	$\pm\sqrt{\ }$

31) $n = \left(\frac{2s}{E}\right)^2$ or $n = \frac{4s^2}{E^2}$

Operations	Inverses
$\sqrt{\ }$	^2
$\times E$	$\div E$

32) $x = x_1 \pm \sqrt{1 - y^2}$

Operations	Inverses
$-x_1$	$+x_1$
^2	$\pm\sqrt{\ }$
$+y^2$	$-y^2$
$\sqrt{\ }$	^2